Jeremy Campbell is a graduate of Oxford University. He is the author of *Grammatical Man* and is the Washington correspondent of the London *Evening Standard*.

By the same author

Grammatical Man

JEREMY CAMPBELL

Winston Churchill's Afternoon Nap

PALADIN
GRAFTON BOOKS
A Division of the Collins Publishing Group

LONDON GLASGOW
TORONTO SYDNEY AUCKLAND

Paladin
Grafton Books
A Division of the Collins Publishing Group
8 Grafton Street, London W1X 3LA

Published in Paladin Books 1989

First published in Great Britain by
Aurum Press Limited 1988

A CIP catalogue record for this book is available
from the British Library

ISBN 0-586-08798-2

The quotation from Emily Dickinson is reprinted by permission of the
publishers and the Trustees of Amherst College from *The Poems of
Emily Dickinson*, edited by Thomas H. Johnson, Cambridge, Mass.:
The Belknap Press of Harvard University Press, Copyright 1951,
© 1955, 1979, 1983 by the President and Fellows of Harvard College.

Printed and bound in Great Britain by
Collins, Glasgow

Set in Caledonia

To my mother,
Alfreda Rose Campbell

Contents

INTRODUCTION   The Silent Orchestra   11

PART ONE
TIME DETHRONED   21

ONE   The Cloakroom Ticket and the Overcoat   23
TWO   The Unimportance of Time   42
THREE   The Importance of Time   63

PART TWO
RHYTHMS ANCIENT AND MODERN   81

FOUR   A Temporal Identity   83
FIVE   The Master Clock Revisited   94
SIX   A Biological Rainbow   115
SEVEN   Light and Time: Together Again   135
EIGHT   Timetable for Simple Minds   155
NINE   The Clocks of Sleep and Dreams   167
TEN   The Gates of Day and Night   186

PART THREE
TIME AND INFORMATION 201

ELEVEN	Time and the Biological Brain	203
TWELVE	Chemical Clocks and a Biological Miracle	216
THIRTEEN	The Conversational Waltz	229
FOURTEEN	Evolution and the Retreat into Time	247
FIFTEEN	The Cough That Floated	266
SIXTEEN	Temporal Intelligence	285
SEVENTEEN	Darwin and the Return of Martin Guerre	299

PART FOUR
ASSEMBLING THE SENSE OF TIME 319

EIGHTEEN	Taking Time Apart	321
NINETEEN	The Codes of Time	339
TWENTY	Putting Time Together	354
TWENTY-ONE	"There's an Idea Missing"	371

| | Notes | 392 |
| | Index | 420 |

Introduction

The Silent Orchestra

Ideas about the human experience of time, in flux for many decades, are starting to crystallize, to come together into a new synthesis—one could almost say a new science. As a result, inner time, like outer space, is less mysterious than it used to be. In fact, the study of biological and psychological time, a collaborative effort extending over several domains of knowledge, is in a state of creative ferment reminiscent of the heyday of research into the nature of physical time earlier in the century. It may lead to a similar leap of understanding.

Today's life sciences take time more seriously than ever before. By doing so, they have given us a new appreciation of the ingenious lengths to which nature herself has gone to take time seriously, almost from the simple beginnings of life on earth. Internal mechanisms evolved which made the body perform its operations on a schedule, synchronized with daily, monthly, seasonal schedules of the physical world outside, so that the body did the right thing at the right time and was prevented from doing the right thing at the wrong time, not as a luxury, an extra refinement, but as a basic necessity of existence. Timing was too important to be left to chance. Biological clocks, timekeeping devices built into living organisms as part of their anatomy, are extremely ancient, being found in the most primitive species. Evidently they had a high priority in evolution and were indispensable for survival.

It is only to be expected that where life of any sort is concerned, time should be taken seriously. Time is only a dimension, of course, a relation between events, and must not be thought of as a substance or a process. But in the world of life, events in time have real significance. Daybreak and nightfall, spring and summer, have a definite meaning for living creatures, and for some it may be more than their life is worth to confuse one with another. The environment has a temporal structure. It changes, and some of its changes are not random, but periodic. The rising and setting of the sun, the ebb and flow of tides, the exits and entrances of the seasons, are regular and predictable in time. And as the environment is periodic in this way, the creatures that inhabit it are periodic also. By means of inner clocks, the operations of the body and certain types of behavior as well are timed so as to be appropriate to changes in the surroundings. In general, animals have a daytime biology and a nighttime biology, a summer one and a winter one, and it is often of great importance to an animal or a plant that daytime biology does not coincide with nighttime in the outside world, and summer behavior is avoided during winter weather.

There is a temporal environment as well as a spatial one, and living things need to be masters of both, or run the risk of becoming extinct. The shaping force of natural selection has seen to it that they acquired their own internal clocks, and the clocks ensure that the body's internal timetables are relevant to the timetables of the outside world, because how the environment changes and when it changes have meaning for the organism. Correct timing of this sort is as automatic, as "natural," as breathing.

So living things adapt to the times of the external world as they adapt to its spaces, by means of built-in biological mechanisms. Such mechanisms are necessary, in part because the time of real life, unlike the time of mathematics, is one-way, irreversible. A day runs from dawn to dusk and never backward, from dusk to dawn, whereas a mathematician can reverse time easily, by the stroke of a pen. Such one-wayness means that unless an animal is securely located in time, the mistakes it makes may be irreparable, because there are not as many second chances in time as there are in space. The present is unique and does not repeat itself. A time cannot be revisited like a place, except in memory, and memory is fallible. Often there is a right moment to act or to change one's inner chemistry, and the moment must not be missed. Biological clocks make it possible to seize the moment by locating the animal securely in time. Being synchronized with the external "clocks" of daylight and darkness, tides or seasons, inner clocks are to the unconscious processes of the body what a wristwatch is to a businessman, who schedules his activities appropriately to his working day. Because the businessman's watch is synchronized with local time, he can be ready for a meeting before it begins, and is not caught unaware when it happens. He is prepared for important events in advance of those events. He can seize the moment because he can predict the moment, thanks to the fact that his watch keeps time with the watches of everyone else around him. In a different, but analogous, way the body "knows" what time of day or night it is, and so prepares itself for waking or sleeping, feeding and fasting. The changes going on inside the body keep a step ahead of the changes going on outside. This is a strategy for survival in evolution, as the use of a watch is a strategy for survival in business.

Inner clocks, organizing the body in time, are a constraint on freedom. They dictate the "when" of biological events and make the organism more predictable than it would be otherwise. That is what happens when nature takes time seriously. But now a complication enters. All species do not take time equally seriously, or they take it seriously in different ways. That is one of the fundamental distinctions between one species and another. The body may be organized tightly in time, or loosely. Some species have a narrow time window in which to perform certain important functions, such as mating or migrating, but other species are free to reproduce at any time of the year and migrate whenever the fancy takes them.

The tightness or looseness of the time constraints in various kinds of creatures is not arbitrary or capricious. It is logical, and the logic is that of evolution. As a species evolves, it acquires certain strategies for adapting to, for coming to terms with, the temporal environment. If it is important that some natural function, say mating, should be carefully scheduled so that offspring are born at a season of the year when the weather is mild and food plentiful, the time constraints will be strict. But the essence of evolution is change. A strategy of adaptation that is effective at one stage in the history of a species may become obsolete at a later stage, as biology or circumstances alter, and for the strategy that supersedes it, time may be less important. Since, in evolution, new structures of body and brain are often added on to existing ones, a function that once took time very seriously may be overlaid by a more recently evolved one that does not take time so seriously. A tension can exist between the old and the new. Such tensions are especially pronounced in the most highly evolved of all animals, humans. The human inclination to nap in the afternoon, which the conscious mind usually resists, is an example of an ancient timetabling mechanism, located in a primitive part of the brain, being overridden by higher, more modern centers of thought. Our ancestors may have napped in their tropical habitat to avoid the intense heat after midday, and a biological clock still programs an interlude of sleepability at that hour, but the present-day human, equipped with a powerful cortex, can say "no" to the program.

It is a major theme of this book that the ways in which time has been made important or unimportant in the evolutionary history of species, the extent to which time is emphasized or deemphasized, is an aspect of the "nature" of that species. Temporal structure, whether biological or psychological, helps to determine the built-in limits and the built-in freedoms that establish what a creature can do and be. We are no exception to that rule. Human temporal structure is a key to human nature.

Because much of our temporal structure is hidden or half-hidden from conscious awareness, it is difficult to learn about internal time constraints through introspection or common sense, which often misleads. One of the first intellectual platforms to collapse under the assault of new knowledge about biological time is the platform of common sense. Humans are highly rhythmic, but the rhythms are often far from obvious. Certain biological clock systems have been discovered only through the use of sophisticated computer programs, and when they are brought to light in this way, often surprise us. By showing us these invisible restrictions on our temporal freedom, scientists modify our knowledge of human nature, and they do not always do so in predictable ways. They are drawing a new map of the temporal anatomy of body and brain, and the map tells us truths we could not know otherwise.

It would be a great mistake, however, to assume that this time anatomy is simple, that the clocks of the body all tick to a single measure, like watches in a jewelry store. A better image is that of an orchestra, a silent orchestra made up of numerous players under more than one conductor, each contributing in special ways to the harmony and complexity of the whole.

In the body, time is measured by a multitude of biological oscillators which repeat themselves, cycle after cycle, in a steady rhythm. In an orchestra, waves of sound are produced by man-made oscillators, internal vibrations in various musical instruments, whether strings, woodwind, brass or tympani. The music an orchestra makes is full of interest for us because the sound waves are not simple, but richly organized in time. If all the waves were of one frequency, as in the sound made by a tuning fork, listeners would find the result intolerably boring. But in music there are waves of many different frequencies and many different amplitudes, or loudnesses. What is more, there is a definite, orderly relationship between frequencies, a harmonic relationship, which is present in musical sounds but is absent in mere noise.

Until quite recently, biologists devoted most of their attention to clocks in the body that oscillate at one particular frequency, namely the 24-hour rhythm which corresponds to the cosmic cycle of daylight and darkness. This emphasis on a single frequency simplified temporal anatomy, but also made it monotonous, as if a concert audience were to settle back in their seats as the lights go down only to see a man walk onto the stage and play a tuning fork.

Today, as a result of clever computer detective work, a spectrum of frequencies has been found in the time structure of the body, and even in individual body functions, such as the regular rise and fall of core temperature. Some are faster than one cycle every 24 hours, while some are slower. Some are "louder," more prominent than others. Their periodic ups and downs are more marked. And certain of these newly identified frequencies are related to others harmonically, in a manner exactly analogous to the spectrum of sound waves produced by a musical instrument. Just as we find an orchestra more interesting than a tuning fork, the new approach to biological time is vastly more exciting than the old, and leads to the possibility of more powerful and general theories. In the words of a leading biologist, we no longer have a single note, but a symphony, even though we cannot hear the music. The wellworn phrase "the harmony of the body" has acquired new meaning.

The silent orchestra helps us also to adapt to our uniquely human environment, which contains periodicities, temporal patterns, which are generated not by the rising and setting of the sun but by other people. These are the time patterns of speech and of music itself, the external orchestra we can actually hear. Here again, evolution has taken time seriously. We possess built-in psychological oscillators which make it possible to discriminate the time order of speech sounds automatically and without needing to think about what we are doing. Without these timekeeping mechanisms human language would be impossible, because the meaning of speech sounds is encoded in their order in time. Unconscious timekeeping of a high order of precision is essential, too, to the success of a conversation, where speakers and listeners interact with one another in a kind of concert, and where an entrance made slightly too late or too early can disrupt the whole ensemble. To be an effective partner in a conversation it is almost as important to "seize the moment" in the flow of talk as it is for an animal trying to survive in an environment where correct timing may mean the difference between eating and being eaten. The human brain also perceives the highly complex time structure of music in a natural, easy way, because of the orchestra within. Nature gives us a free ride, so to speak, in that respect. It seems that our minds are tuned to the temporal anatomy of music and speech in much the same way as our bodies are tuned to the rhythms of day and night by biological clocks. Rhythm is intimately linked to communication, and man is by nature a communicating—and a musical—animal.

Nature has let us down, however, when it comes to the conscious awareness of extended spans of time. The inner orchestra by itself is not enough. No built-in clock exists that can give us reliable knowledge of the passage of time, one that might tell us that exactly an hour or a week or a year has elapsed since some event occurred. That is why clocks were invented. The psychology of time as it is going by appears to be radically different from the psychology of time that has gone by. Perceiving the present and perceiving the past are separate mental processes. In special circumstances, one process may break down while the other remains intact. As far as the sense of time that has gone by is concerned, nature left us free to make our own mistakes.

Our subjective impressions of elapsed time are notoriously elastic and apt to be distorted by the accidents of mood, context and circumstance. Presumably evolution did not take that aspect of the relationship between inner and outer time as seriously as it took other aspects, and the brain was thrown back onto its own resources, left to do the best it can using any strategy that works. One such strategy is the mental construction of internal models of external time. Biological clocks are internal models, analogs of definite spans of world time, but they come ready-made, as a gift of nature. We do not need to acquire them. It is quite otherwise with the time models constructed by the brain, which require intellectual effort and tend to be idiosyncratic rather than uniform. One person might conjure up a mental image of a year as a straight line, while others picture it as a circle or an ellipse. But the mind is not content merely to make models. A whole arsenal of mental strategies is brought to bear on the task of being fully conscious of time. We think and reason about time, drawing on reserves of worldly knowledge and experience, using powers of logic to correct vague and variable "feelings" about the amount of time that has passed. No specific, innate mechanism guides us to a correct solution.

Our time, the time of personal experience, the philosopher Jean-Paul Sartre once said, is part imagination and part fact. It is like a mermaid, with a human head and the tail of a fish. The head symbolizes the power of the mind to construct a future which is "unreal," in the sense that it exists only as a possibility toward which we aim. Unreal time is not given to us as a fact. We do not swim in it, moving with its currents as our bodies move in synchrony with the rhythms of the external environment. The mermaid, we could say, extending Sartre's image, represents the dual character of human time, with its two domains, one biological, the other psychological. Biological time evolved as a way of adapting, unconsciously, to the temporal patterns of our surroundings, the cosmic cycles that are stable and endure. The mermaid's tail is a fitting symbol for biological time, which is immersed in nature and has a close, harmonious relationship with it. Psychological time is different. It disengages from the time of the physical world, rising above it, as the mermaid's head rises above the waves. The mind cannot measure objective time with the accuracy of a clock, but must construct its own subjective time, which is no more than a model of reality and is often a fiction, a metaphor, a work of the imagination.

It begins to be clear now why philosophers and others have talked so much about the complexity of time, its mysterious manysidedness. In fact, it is not time that is complex, but the heterogeneous array of strategies and mechanisms, biological clocks and ad hoc mental constructs, by means of which human beings take time seriously. What makes psychological time so complicated, so difficult to enclose in a neat definition, is the freedom of the human mind to manipulate the information it receives from the senses, or from memory, including information about time. Humans are the great generalists of the animal kingdom, and their approach to the task of being master of time is a general one. That is also the approach of this book.

Humans and time need to be studied together, not as two separate problems. Poets have long appreciated the force of that statement and biologists are beginning to, as they reveal the unexpected richness of the temporal structure of the body, and the intricate ways in which it is related to the temporal structure of the world. As for psychologists, they have come a long way since Albert Einstein, aware that his theory of relativity could answer questions about the time of physics, but not about the time of the mind, wondered aloud how a sense of the passage of time originates in a person, whether it is inborn or is a secondary phenomenon, derived from the experience of objects moving in space. Inner time is seen now as being not one thing but many things. It reflects the mixed nature of human beings, which is partly fixed and partly free. And that in turn is a result of our evolutionary history, of the particular strategies by which we know the world, adapt to it and feel at home in it. The hidden constraints, as well as the hidden freedoms, are emerging from their hiding places at last. Trying to understand them is part of the great and endless enterprise of trying to understand ourselves.

PART ONE

TIME DETHRONED

ONE

The Cloakroom Ticket
and the Overcoat

Albert Einstein, who did more than any single person to challenge and disturb our neat common-sense ideas about space and time, once summed up his opinions on the nature of scientific knowledge by talking about beef tea and overcoats. He remarked that the logical connection between the world as a scientist thinks about it and the world as ordinary people experience and observe it is a great deal more remote than we might suppose. The difference between what science says the world is and how the world appears to our senses, Einstein said, is not as straightforward as the difference between beef tea and beef. Beef tea is a drastically altered form of beef, but the relation of one to the other can still be detected by the senses; at least the tea smells of beef and tastes of beef. The connection between science and ordinary experience is much less direct than that. It is more like the connection between the number on a cloakroom ticket and the overcoat left with the attendant in the cloakroom. We can read the number on the ticket and know that it matches the number held by the cloakroom attendant, and can use it to retrieve the coat. But there is nothing whatever in the number that suggests in any way the idea of "coatness," let alone its actual texture, shape and color.

By temperament, Einstein was attracted to the sort of world knowledge represented by the cloakroom ticket number rather than to the beef tea type, which is knowledge close to ordinary common sense and observation. He refused to base his science on the evidence of the senses, and would have no traffic with any philosophy that did so. Reality, Einstein believed, is knowable only through the "free constructions of the human mind." It is essentially an intellectual puzzle. In his *Autobiographical Notes* he wrote: "Out yonder there was this huge world, which exists independently of us human beings and which stands before us like a great, eternal riddle, at least partially accessible to our inspection and thinking. The contemplation of this world beckoned like a liberation."

Time in particular, Einstein considered, has no reality on the level of being, of "life" and the senses, but only in the domain of thought. Its reality is the number on the ticket. The coat itself is an illusion.

Today, the gap between the two kinds of knowledge described by Einstein appears to be widening, especially with respect to time. A chorus of voices can be heard asserting that, in spite of what many intelligent people may suppose, there is nothing in modern physics that corresponds to such common-sense and generally held beliefs about time as its quality of onward passage, its continuous flow. Nor does physics admit the existence of a "now," a present moment which divides the past from the future, and which is the same for everyone. The Cornell University astrophysicist Thomas Gold states with complete finality that the concept of "the flow of time has been abolished." Sir Alan Cottrell, an English physicist, thinks the refusal to permit any place in physics for the passage of time or for the now has precipitated a crisis, because science has "brought us right up to the edge between the material and mental worlds, and it does look as if there might be an unbridgeable gap between them."

Yet while advances in theoretical physics have conspired to intensify the strangeness, the foreignness of time, compared with the time of human experience, a new science is establishing itself which offers the best hope of bridging Sir Alan's "unbridgeable" gap. This science approaches time from a very different perspective, that of the living organism. It might be called the science of life time.

It has been said of Einstein that a result of his theory of relativity was "to 'physicalise' space and time, to bring them, as it were, fully into the physical world, so that they no longer functioned as merely the *setting* for the world, but became actual *parts* of the world interacting with matter and motion. Space and time were 'dethroned' and endowed with physical, changeable properties. They lost their immemorial 'untouchability' and began to enter the domain of the comprehensible."

The dethronement of time has now entered a new stage. Time is more than physicalized; it has also been biologized and psychologized. Time interacts, not merely with matter and motion, but also with the fluid life of the body and the information circuits of the brain. If, in the theory of relativity, time, metaphorically speaking, got its hands dirty, in the new science of life time one might say that it gets its hands wet. Einstein stated that "time and space are modes by which we think and not conditions by which we live," yet biologists are showing that the body contains a great collection of clocks of various kinds which regulate changing states of the body and the brain in time, and these clocks are set and corrected by what are *exactly* the "conditions by which we live"—namely, the 24-hour alternation of daylight and darkness and the cycle of the seasons.

This descent of the concept of time from its "untouchable" status as a semidivine cosmic principle, down into the hurly-burly of the physical universe and finally into the cells, tissues and organs of the living system itself, has taken millennia to accomplish. The history of ideas about the nature of time is long, contentious, untidy and often wearisomely repetitive. But one persistent theme, appearing and reappearing across the centuries, masquerading in a number of different disguises, is the notion of a special, privileged time, "out there," enthroned, a universal master clock which never needs correcting, never stops, and is unaffected by anything that happens in the physical universe, in the world of life, or in the minds of men. This sort of time is aloof, perfect, uniform and self-sufficient. It is one thing, not many things.

For Greek thinkers, the master clock was given a unique place in the imagined universe. It was located at the outermost boundary of the cosmos, where a great, translucent sphere was supposed to rotate, carrying round on its inner surface the fixed stars. Plato, an aristocrat who hated democracy, especially in his younger years, took an aristocratic view of time, and this had a tremendous influence on later thought. In his *Timaeus*, the first Greek attempt to explain the universe as the deliberate creation of a rational, divine intelligence, Plato gave time an exalted position by proposing that it is one with the heavens, that time and the heavens were created together as a unity. One cannot exist without the other, and if one perishes, both perish.

In the fourth century B.C., when Plato lived, the universe, earlier thought to be a great dome arching over the earth, was conceived as a hollow sphere, made of some special substance, such as fire or air, or later crystal, through which light could pass. Such a sphere, whose surface was at the farthest limit of the cosmos, rotated once in nearly 24 hours, and with it the fixed stars also rotated. The earth was a smaller sphere suspended in the exact center of the larger one. Thus each star, and each point on the surface of the outermost sphere, was an equal distance from the earth. The orderliness and regularity of Plato's universe reflected his belief that it was fashioned by a reasonable mind, a supernatural craftsman or demiurge who aimed at perfection even if perfection was not quite within his grasp, and took pleasure in what he had done. This account differed from earlier theories of the universe, which viewed it as being biological and organic in nature, and explained it by using such images as the labors of childbirth and the hatching of an egg.

Plato, like Einstein some 2,500 years later, saw a wide gap yawning between the abstract truths of mathematics and logic, on the one hand, and the knowledge human beings gain directly from the experience of living in the world, on the other. The first kind of truth could be proved, and therefore had permanent value, while the second kind was uncertain and true only for the time being, a "likely story," which might be contradicted by new evidence at some time in the future. We know that the universe itself is not just a likely story, but was planned along rational lines, Plato argued, because it contains enduring, unchanging features, lawful regularities. The most obvious of these regularities are the motions of stars and planets in the heavens; they are proof that a mind ordered the cosmos, and it was exactly in this supremely rational domain of the heavens that Plato placed the master clock.

Ideally, Plato's demiurge would have preferred to make the whole universe eternal and perfect, like himself, but that was not possible, given the imperfect materials at his disposal. Of all things in the universe, the creator chose to make the heavens as nearly eternal as he could manage. When the divine craftsman ordered the celestial region, Plato says, he "made in that which we call time an eternal moving image of the eternity which remains for ever at one. . . . So time came into being with the heavens in order that, having come in being together, they should also be dissolved together if ever they are dissolved; and it was made as like as possible to eternity, which was its model."

The demiurge did not start from scratch in his task of fashioning the universe, but went to work on an already existing state of chaos. In Plato's account of creation, a peculiar feature of this primordial chaos is that, while it is timeless, it is not spaceless. Space existed before time began, because space is a necessary framework even for the formless muddle on which the divine artificer imposes order and system, a system that includes time. Space is given in advance. Time is different. It arises as a result of the organizing activity of the artificer and is a product of reason, rather than a necessary condition that must be satisfied before the work of building can begin. Unlike space, time is not independent of the mind that fashioned the universe. In other words, time is inherent in the order that is brought out of the mindlessness of chaos. It is associated with meaning, harmony, intelligence and design. Time cannot exist without the master clock. That is Plato's scheme.

The sphere of the fixed stars represented the simplest and clearest standard of time, because it rotated always in the same place. The stars never wander about in the sky, unlike the planets, whose irregular motions can only be understood by relating them to the regular movement of the outermost celestial sphere. This idea, that circular movement is the most natural, fundamental and best mode of motion, had an extraordinarily long run in scientific thought, and was not completely overthrown until the age of Sir Isaac Newton, in the seventeenth century. Newton's debunking of the notion that circular motion is primary and all other kinds are inferior was an act of intellectual bravery which won Einstein's unstinted admiration.

Aristotle, Plato's pupil, took a more democratic and pluralistic approach to time. His world was a world of process and becoming, where genuinely new things came into being. These things develop, grow and change in such a way as to realize a plan that is implicit in and special to the thing in question, determining its final form. Thus the concept of a chicken is contained in the egg, and an oak tree can be said to be present in an acorn. The world is, and process is, the fulfilling of a plan, the actualizing of a potential development toward an end already specified in the beginning. Time plays an important role in nature, because realizing a plan occupies a certain amount of time. A plan has a definite end state, when, for example, a human being is mature, or an oak tree has done all the growing it ever will do, and has no more potential left to actualize. At this stage of completion, humans and trees reproduce themselves rather than go on to become something different and new. What is more, Aristotle made clear, the time of process has a structure. It is not just a succession of separate instants following one another, from present to future. If the final form of an object is implied by the plan, which is there at the start, then the future must be part and parcel of the present.

Aristotle recognized that time is closely connected with motion, and motion in his philosophy includes not just movement through space, but the growth of plants and animals, the process of things realizing their plans, ceasing to exist, or changing their quality or size. Time is not the same as motion, however. There are many different kinds of process, some faster than others, and time is common to all of them. Yet we cannot say that time has different speeds. To Aristotle, motion is more fundamental than time, which is a secondary effect, the measure or "number" of motion, that aspect of it which allows us to count successive stages.

Though he was not as forthright as Plato in stating that time depends on clocklike processes, Aristotle did subscribe to the authority of the celestial master clock, because the number, or countable property, of the sphere of the fixed stars was the best known. Aristotle accepted the doctrine that circular motion, with its simplicity and completeness, is the most perfect kind, unlike the rectilinear motion of bodies on the earth. Circular motion continues indefinitely. Earthly objects travel in a straight line because their matter is less uniform than the matter of the celestial sphere and is open to a wider range of possibilities. Straight-line motion would continue without stopping if space were infinite in extent, but Aristotle did not believe in the endlessness of space, and therefore assumed that circular motion was the only kind which can proceed forever without changing direction. So the master clock of the outermost sphere was prior to and more basic than the human sense and experience of time. It was also divine, since Aristotle maintained that the rotation of the sphere was due to the action of God. God, as the first mover, was changeless and eternal, and the sphere of the fixed stars was the first moved. Since the first mover is self-same, identical with himself, the movements of the heavens must also be uniform and universal. The alternation of day and night is the result of the entire universe completing a full revolution every 24 hours, around an axis that passes through the center of the earth. To account for the motions of the planets, the sun and the moon, Aristotle proposed the existence of 54 lesser rotating spheres, a great, interlocking assembly of heavenly clockwork, driven from the sphere of the stars.

The concept of the master clock as time enthroned did not run into very serious trouble until the end of the Middle Ages. As the medieval epoch came to a close, however, the ideas of Copernicus and the Italian Renaissance philosopher Giordano Bruno began to spell disaster for the whole picture of the earth as the hub of the universe and man as the point of the whole enterprise of creation. As long as the earth was seen as a sphere hanging in the exact center of a cosmos bounded by a larger sphere which carried the stars round and round, the master clock was safe. The theory of time as a privileged, aristocratic unity "out yonder" was intact.

The game was up, though, when Copernicus and Bruno showed that the earth was nothing special and nothing central. Copernicus proposed a sun-centered universe in which the earth revolved, which meant that the sphere of the stars did not go round and round, but was at rest, and so could not drive the clockwork of the planetary spheres. He did not accept the idea of an endless universe, but he prepared the way for the new view of the earth as one of an infinite number of worlds in a universe that went on forever. The stars were scattered across boundless space. At one stroke, this dagger at the heart of established ideas about the cosmos put in question not only the status of man but also the nature of time. The fixed stars, it became clear, are not all an equal distance from the earth, but are located at vast and varying distances from it. So there could be no one, perfect, circular motion of the heavens to serve as an ideal standard of reference for time, as Plato's demiurge had carefully arranged. There must be many different kinds of motion. Giordano

Bruno, who was finally burned at the stake for heresy in the Piazza del Fiori in Rome, his tongue "imprisoned on account of his wicked words," made the unsettling proposal there are as many times as there are worlds.

The only escape from such a ticklish predicament—one that had been encroaching gradually, even before the dawn of the Renaissance, as the theory of celestial spheres became increasingly infirm—the only way to preserve the unity of time, was to detach time completely from its dependence on the motion of material bodies. This was the course that began to be adopted in the early years of modern science. It was the culmination of a trend of thinking that had emerged during the two hundred years before Copernicus and Bruno, and held that we can speak of true "mathematical" time without tying it to an actual celestial clock. The result of all this was, in the words of the modern historian of science Milič Čapek, the "separation of time from its physical content."

Time became more fully mathematicized through the work of another heretical genius, Galileo Galilei, who was born in 1564, the same year as William Shakespeare, and died in 1642, the year of Sir Isaac Newton's birth. Inspired by the success achieved by astronomers in applying mathematics to the behavior of the heavenly bodies, Galileo decided to analyze the movements of less grandiose, earthly bodies and created a new science called terrestrial dynamics or "local motion." Since the speed of motion is arrived at by dividing the distance traveled by the time elapsed, time was of great importance in Galileo's new science, perhaps more important than he fully realized at the outset of his researches. Galileo believed that time was uniform and continuous, and as a champion of geometry, he naturally represented it as a straight line. He was not the first person to symbolize it in this way, but he was the first scientist of such high celebrity and stature to do so, and his fame did much to make linear, geometrical time widely accepted. Time became a continuum that could be measured. In Galileo's analysis of motion, time was an axis, a geometrical coordinate on which intervals were marked off, starting from zero.

Galileo espoused the cloakroom ticket theory of knowledge. He believed that "this grand book, the universe," was
written in the language, not of sense experience, but of mathematics, especially geometry. He delighted in the way that
the ingenuity, or "sprightliness" of judgment, of theoretical
scientists produced results that shocked and outraged common sense. The idea of moments in time forming a continuous sequence, like points on a line, was of great importance
both in Galileo's work and for subsequent thinking about
time itself. As concepts go, it was highly abstract, and it emphasized the spatial aspects of time. Galileo's taste for geometry meant that his natural instinct was to think of motion
in terms of space rather than in terms of time, so much so that
he may have been led to make the initial mistake of assuming
that when a body is in free fall its speed increases with the
distance of fall, the amount of space it has traversed, when in
fact its speed is proportional to the time in seconds that
elapses during the fall. Accelerted motion must be based on
time, not on space. Alexander Koyré thinks that Galileo was
forced to revise his approach when he realized not only that
the idea of time is contained in the idea of motion but also
that time is paramount in any theory of what *causes* accelerated fall. If the cause is a succession of tiny spurts, which
accumulate and build up, making an object fall faster, then
the spurts follow one another in time and only incidentally
follow one another in space.

In his *Dialogue Concerning the Two Chief World Systems,* Galileo shows that he remained a believer in the superiority of circular motion. He suggested that if a body in
motion is left to its own devices, its path will be a circular
orbit, not a straight line, and it will move along this orbital
path forever. Arthur Koestler chided Galileo for being unable
to "rid himself of the old circular obsession." Nevertheless,
the fact that Galileo did study straight-line motion by means
of geometry spelled the downfall of the view that circular
motion is the fundamental form of movement in nature.

Turning time into geometry was an apparently simple
innovation which led to momentous discoveries in mathe-

matics and shaped the thinking of generations of scientists. It sired such distinguished intellectual offspring as logarithms and the calculus. In the hands of Newton, geometrical time set the stage for the appearance of a new and highly questionable candidate to replace the wrecked concept of the universal master clock which had wheeled in impeccably uniform circles around the demiurge's almost perfect heaven. This candidate went by the name of absolute time.

Absolute time was even more pristine and aristocratic than the celestial clock time of Plato, since it was not required to have anything whatever to do with matter. It was a reality in its own right, and had its own nature. Events in the world did not affect absolute time, and it, in turn, did not affect events. Each was indifferent to the other. The concept arose logically out of the idea of geometrical time, which was fully discussed in England in the *Lectiones Geometricae* of Isaac Barrow, published in 1669. Barrow was the first occupant of the Lucasian chair of mathematics at Cambridge University, a post now held by Stephen Hawking, one of the most original of post-Einsteinian theorists of physical space and time. Barrow, who would be much more famous than he is today if Newton had not appeared on the scene to overshadow him, took time more seriously than some of his contemporaries, and certainly much more seriously than the Greeks had. Mathematicians who were content to have a hazy notion of time he called, derisively, "quacks." Barrow came to appreciate the importance of mathematical time, especially in relation to motion, after perusing the work of Galileo's pupil Torricelli. An enthusiastic cheerleader for the superior power of geometry over arithmetic, Barrow, like Einstein, preferred the continuous to the discrete. He decided that time was an essentially mathematical concept, best represented by a line, "for time has length alone, is similar in all its parts and can be looked upon as constituted from a simple addition of successive instants, either a straight or a circular line."

The most characteristic feature of time, in Isaac Barrow's view, was its unstoppable, even flow. Time proceeds in this way no matter what else may be going on and regardless of our conscious awareness of it. He declared that "whether things move or are still, whether we sleep or wake, time pursues the even tenour of its way."

Barrow vacated the Lucasian chair in order to devote himself to theology, which at that time was the ruling science. He was succeeded in the professorship by Newton, who had attended Barrow's lectures on geometry and helped to prepare them for publication. Newton had been rescued from a life of farming in the wilds of Lincolnshire by the intervention of a perceptive uncle, and was sent to study at Cambridge. Here he came under the influence of Barrow, whose researches into tangents to curves pointed ahead to the discovery of the calculus, and Newton adopted the "even tenour" hypothesis of a time independent of events. Like Barrow, Newton conceived the moments of time as forming a continuous sequence, like points on a line. "Absolute, true and mathematical time," Newton said in a celebrated passage not always quoted in its entirety, "of itself, and by its own nature, flows uniformly on, without regard to anything external, and by another name is called duration: relative, apparent and common time is some sensible and external (whether accurate or inequable) measure of duration by means of motion, which is commonly used instead of true time; such as an hour, a day, a month or a year."

Opinions among historians of science are divided as to whether Newton needed to introduce the principle of absolute time. Some say it was required as a universal standard against which to measure relative and common time. Others insist that the concept was completely superfluous and unnecessary. Einstein believed that Newton took the idea less seriously than many of the scientists who scolded him for espousing it. Newton, Einstein said, was "better aware of the weaknesses inherent in his intellectual edifice than the generations which followed him. This fact has always roused my admiration." Whatever may be the case, absolute time was an important feature of the Newtonian universe, because it enabled all events, wherever they may occur, to be assigned definite moments in the evenly flowing temporal stream. It affirmed the existence of a universal "now," the same for all observers.

Absolute time did not meet its scientific Waterloo for some two centuries after Newton, though it became involved in some skirmishes. The Wellington who brought about its defeat was Einstein, who reverenced Newton greatly for his intellectual courage. In his special theory of relativity, published in 1905, Einstein showed that the independent status of time is a fiction.

Once, when he was asked how he had made the leap from classical doctrines of space and time to his revolutionary new ones, Einstein replied, "By refusing to accept an axiom." The axiom he discarded was exactly that of the universal now. Einstein rejected the assumption, so unshakably self-evident to common sense, that when we describe two events as being simultaneous, they really do happen at the same time, and that is all there is to be said on the matter. This axiom follows inescapably from Newton's description of "true, mathematical" time, which is the same everywhere and for everyone in the cosmos. Einstein had the independence and daring to assert that when Newton defined absolute time he was talking about the overcoat, the time of common-sense experience. The number on the cloakroom ticket tells us that time is relative, that there is no universal now.

Einstein toppled time from its high perch as a primary, irreducible category in the description of the world. That is one of his great contributions to philosophy, quite apart from his accomplishments in physics. In fact, in the theory of relativity, neither space nor time is the most fundamental concept, although each looms so large in interpretations of the theory's consequences. More basic than either space or time is a new absolute, the speed of light, which has an upper limit of 186,000 miles an hour. If there is a master clock in the Einsteinian universe, it is the unvarying maximum speed of light, a uniquely "privileged" clock which is the successor to Plato's celestial timepiece and the indifferent flow of Newton's unruffled timestream. Nature imposes a constraint on the speed at which light can travel, so that even if the source of light is moving in the same direction as the light, the speed of light through space does not increase. What dooms the principle of absolute time is the absolute value of the speed of light. The reason is as follows.

Suppose a man called Box is sitting in a moving train carrying a clock which works by means of light. The clock consists of a source sending pulses of light vertically up to a mirror on the ceiling, and the mirror reflects the light back down again to a photoelectric cell in the clock. The passage of the light from source to ceiling and back down again to the photoelectric cell is treated as a single unit of time, like the tick of a mechanical clock. A second man, called Cox, is standing at the side of the railroad tracks. Cox is carrying an identical light clock, which is designed to complete a cycle in exactly the same amount of time as Box's clock on the train. As the train rushes past, Cox notices a curious difference in the behavior of the light clock on the train. Since Box's clock is in motion with respect to Cox, its mirror appears to move forward a little between the moment when the light pulse leaves the source and the moment when it strikes the mirror. The light on the train, therefore, as observed by Cox, who is standing by the tracks, has a greater distance to travel than it has in his own clock, which is at rest with respect to him. If one cycle of Box's clock on the train is to be completed in the same unit of time as one cycle of Cox's clock, we would expect that its light beam travels faster than the light beam in Cox's clock. Yet this is forbidden by the principle of the absolute speed of light. The only conclusion to draw is that if each cycle of Box's clock, as observed by Cox, takes longer to complete than each cycle of Cox's clock, then as far as Cox is concerned, time must slow down, because the light cannot increase its speed beyond a universally fixed limit. For Box, however, Cox's clock is in motion with respect to him, and his own clock is at rest, so that he observes time by the side of the tracks slowing down and his own time keeping its normal rate. Accepting the constant speed of light means throwing out the hypothesis of a uniform flow of time, the same for all observers, and that is exactly what Einstein did.

Einstein's time is almost the opposite of Newton's time. It is not logically prior to the world of physical events, and has no objective meaning when considered apart from those events. Unlike the speed of light, the rate of measured time and the extent of measured space may vary according to the state of motion of the observer. The distance between two events is not the same for everyone, nor is the time interval between the events the same. In a sense, each is "private" and local. A state of total anarchy, in which there would be as many different times as there are observers, is avoided by joining the single coordinate of time to the three coordinates of space to form a four-dimensional space-time interval, and it is this interval that is the same for all observers. The space-time interval is "public" and general, and represents a new kind of absolute. It is a continuum, but of a different sort. So time and space must mix together if either is to have objective meaning.

In the General Theory of Relativity, published in 1915, gravity, since it affects light, also affects time. Gravity is no longer a mysterious force, making apples fall to the ground and keeping the planets in their orbits, but is simply the curvature of four-dimensional space-time. Massive bodies, such as the earth and the stars, distort the space-time continuum. They stretch and warp it, so that other objects, and light itself, follow the undulating, irregular contour created by the mere presence of matter. These objects take the simplest and easiest route, bending in toward the massive body, "downhill" along the space-time continuum, thus making it appear that they are drawn to it by an invisible force of attraction. In fact, time is affected in the neighborhood of a massive body for two reasons. One reason is that light has farther to travel at the same speed, because space is stretched and distances are increased. The second reason is that gravity makes time pass more slowly. Atoms on the sun, for example, where gravity is very strong, vibrate more slowly than on the earth. Their "clocks" tell a different time. The wavelengths of the light they emit are different. On the edge of a black hole, a region of space where the collapse of a star into a small, extremely dense mass creates a gravitational field of tremendous intensity, time stands still. Objects that fall to the center of the black hole are squeezed out of existence, and it is here, at what is called a singularity, that time and space themselves cease to exist in the sense that they cannot be defined and thus have no meaning.

So, in Einstein's theory, time must be associated with something else; with space, with the way matter is distributed in a given region of space, with motion and the speed of light, to be worth talking about at all. In extreme circumstances, time makes an exit from the stage of nature altogether. In the singularity at the heart of a black hole, at the first stage and the final stage of the universe, in the microcosm of matter at the subatomic scale, time has no role to play and cannot be discussed in any meaningful way. John Wheeler, an American relativist, a friend of Einstein's and a protege of Niels Bohr, argues that time is an approximate category in physics, and applies only within a limited range. It is a derived concept, not a basic one. When we talk about events that occur at very small distances, or at the genesis and the denouement of the cosmos, the word "time" makes no sense.

Even in everyday circumstances, between those two exotic extremes, the meaning of "time" is the product of a line of reasoning that does not begin by referring to time. A deep description of nature, Wheeler says, must transcend the category of time. He adds: "Time does not today stand in splendid isolation, a concept with an independent existence of its own, free of entangling alliances with the rest of physics. General relativity has subdued the concept time to membership in a larger kingdom: space-time."

Wheeler has noted the strong influence on Einstein of the seventeenth-century philosopher Benedict Spinoza, who was excommunicated in 1656 from the Amsterdam synagogue for denying the biblical story of an original divine creation of the world. Spinoza, a "hero and role creator" for

Einstein, rejected the Genesis account of the making of the world, because it raised an insoluble question: "In all the nothingness before creation, where could any clock sit that should tell the universe when to come into being?" For that reason, Spinoza concluded that the universe has no beginning and no end, but must go on forever. Such an argument, Wheeler believes, was in large part responsible for Einstein's deep reluctance to accept the idea of a universe that expanded from a singular beginning, and would contract into a singular end, an idea that forces us to conclude that time, too, has a finite existence. Einstein changed his theory of general relativity by adding a term to his equations which had the effect of holding the universe static, without a start or a finish. Ten years later, when Edwin Hubble proved that the universe is expanding, Einstein expressed bitter regret for his action, calling it "the biggest blunder of my life." A universe with a beginning and end would certainly imply that time has a beginning and an end, if, as Einstein believed, time cannot exist without a universe, or in a universe that is completely empty. Asked to condense into a few words the core of his theory of relativity, Einstein said: "If you don't take my words too seriously, I would say this: If we assume that all matter would disappear from the world, then before relativity, one believed that space and time would continue existing in an empty world. But, according to the theory of relativity, if matter and its motion disappeared, there would no longer be any space or time."

The new science of life time makes an even more extreme departure from the ideas of Plato and Newton than the thinking of Einstein and his successors did. If time physicalized is not an absolute, neither, and most emphatically, are time biologized and time psychologized. Life time is a democracy, not an aristocracy. Until very recently many scientists believed that body time is a Platonic universe, with a supreme chronometer governing the whole system. That faith fell with a crash in the 1970s. There is no one master clock which regulates the timetables of the human body; at least two biological clocks have been identified which can be regarded as having a privileged status. Presumably there are more still to be discovered. A host of lesser clocks exist in the body, some more powerful than others, some more independent than others. Nor is there a psychological master clock in the head, ticking away to give the mind an always accurate sense of the passage of time in the external, physical world. In the bodily and mental life of human beings, as in the theory of relativity, time has no independent status.

Psychological time is certainly not absolute or "evenly flowing." It is variable, an effect of multiple and various operations of the mind, in which beliefs, emotions, logic and memory all play a role. The present and the past are dealt with by means of different mental strategies. Certainly the passage of time and the "now," both unaccounted for in physics, are very much part of the psychological experience of time. But the brain has one kind of system to deal with time as it is going by and other kinds for time that has gone by. Copernican diversity rather than Platonic simplicity is very much the ruling principle of psychological time. Even when the brain deals with rapid temporal patterns such as those of music, presumably by means of some mechanism in the brain which operates automatically and with high precision, it is not good at perceiving pure time, unrelated to other properties of musical structure. If, as John Wheeler argues, the time of physics is a secondary concept derived from physical happenings, psychological time is derived from mental happenings, which means that we are unlikely to have a satisfactory theory of subjective time until we have a more complete theory of the human mind.

The collapse of the doctrine of a biological or psychological master clock robs life time of its simplicity; still, it opens up exciting possibilities. The temporal architecture of body and brain is heterogeneous, made up of many different components and structures, some of which have not even been identified, let alone described in detail. Making sense of all the parts, explaining them in relation to the workings of a harmonious whole, may seem a task as daunting as that which faced post-Copernican thinkers, when the architecture of the universe was revealed as heterogeneous, and the suspicion arose that there might be as many different times as there are worlds.

Yet biologists and psychologists have made remarkable progress in dissecting the parts of temporal anatomy, and even in the more difficult task of explaining the whole, so that they are taking over where physicists left off, making us think in new ways not only about time but about life itself.

TWO

The Unimportance
of Time

During the long saga of evolution, living organisms adopted two seemingly contradictory strategies with respect to time. One strategy was to reduce the importance of time in the life of a species, so that when a biological or behavioral event happens has little or nothing to do with why that event happens. A second strategy was to increase the importance of time, to such an extent that anyone trying to account for the behavior of an individual member of the species could simply say: "The animal did this or that because it was time to do this or that." Under the first strategy, time is of minor relevance to behavior. Under the second, time is a reason for behavior; it has explanatory power. Time may not be the only or the most profound reason, but it is certainly one that must be taken into consideration.

Paradoxically, whereas the strategy for making time important arose earlier in evolution than the strategy for making time unimportant, modern science arrived at an understanding of the time-unimportant strategy about a century before it established firmly the existence of the time-important strategy. This accident of history has helped to delay a full appreciation of the role played by time in the world of life, and to encourage a prejudice against time as an explanation for biological events.

In the simplest sense, time is important for most living species. As time goes on, the environment changes in a periodic fashion, and the changes have meaning for the organism. An animal's whole way of life, indeed its life history as an individual, may be designed to take advantage of changes that are beneficial and to avoid changes that are harmful. Day alternates with night, and the environment of day is very different from the environment of night. In temperate zones the seasons follow their annual cycle, from warm to cold and back to warm again, as predictably as a clock. A simple type of animal, lacking any biological means of opposing these changes, would have to adjust its way of life accordingly. Time would be a dictator, a tyrant. The animal would be cold when the weather is cold, and warm at an hour of the day and a season of the year when the weather is warm. An earthworm's temperature is the same as that of the soil in which it moves, and a fish is neither warmer nor cooler than the sea in which it swims. Such "cold-blooded" animals, which are not necessarily cold but take on the temperature of their surroundings, are at the mercy of the clock and the calendar. Only if the physical environment were to remain the same 24 hours a day, 12 months a year, would such an animal's behavior be unaffected by time.

There is a way out of this dilemma, however. If a means could be found to ensure that the internal conditions of the body do not change, even while the surroundings continue to change, the behavior of the animal would be independent of calendar time, almost as if it lived in a constant environment. It is sameness, either external or internal, that emancipates life from the tyranny of time, and mechanisms that produce internal sameness have appeared during the course of evolution, just as mechanisms for breathing, sleeping and eating have arisen, as part of the basic biological equipment of a species.

Nature's device for making time unimportant for the body is called homeostasis. The word is derived from the Greek *homeos*, meaning similar, and *stasis*, meaning state or condition. Homeostasis is an evolutionary strategy for preserving internal sameness by resisting and smoothing out changes. If the weather outside should suddenly turn cold, homeostasis compensates for the change by raising the internal temperature of the body just enough to keep it close to a set value, as a thermostat maintains roughly uniform warmth in an apartment, regardless of conditions outside.

The evolution of sameness in living organisms, in the form of mechanisms that detect change and take steps to counter it, was first recognized as a definite principle of medical science in the middle of the nineteenth century. The discovery was made that the human body possesses an internal environment which is more constant than the external environment, and it was given the name *milieu intérieur*, a term first used in a lecture at the University of Paris in 1857, just two years before the publication of Charles Darwin's *Origin of Species*. The author of the phrase was a French scientist named Claude Bernard, one of the most original thinkers in the whole of nineteenth-century medicine.

Bernard brought a unique blend of daring imagination and brutal realism to the study of the living system. He had a sort of reverence for the blood-spattered actuality of the dissecting room, which at that time was not a suitable place for sensitive souls. Two of Bernard's biographers describe it as a "confusion of dispersed limbs and grimacing heads, the blood underfoot, the revolting odors." Another young medical student of the period was so appalled by his experiences that he jumped out of a window and ran away, abandoning medicine forever and turning to a career in music. The student's name was Hector Berlioz. Bernard suffered no such revulsion. He was perfectly at home stooping over cadavers. In later life he would stand for hours in a cramped laboratory, his hands drenched with blood, a top hat on his head, holding an umbrella up against the rain which dripped through cracks in the ceiling. Claude Bernard was one of the most perceptive observers of what the body actually is, how its various parts are put together; one of his students said of him that he "seemed to have eyes all around his head." His analytical approach gained him the reputation, in certain quarters, of being a materialist who made biology into a branch of engineering and reduced the mystery of life to the banality of a machine. In Dostoevsky's novel, *The Brothers Karamazov*, a character shouts the name "Bernard!" as a shorthand expression of contempt for contemporary science and its mechanistic spirit. Bernard's own wife and daughters organized an antivivisection league as a protest against the animal experiments he considered essential to an understanding of the human body.

Yet Bernard was not simply an observer, a mechanic. The philosopher Henri Bergson, one of the major thinkers about time in that or any other age, said of him that his constant thought was to show how fact and idea go hand in hand in scientific research. Bernard reveled in the intricacy of living matter, but one of his great gifts was the ability to simplify a mass of complex information, using it to arrive at overarching concepts, and this came naturally to him because, though a detail man, he believed in the universal harmony of things, and this belief strengthened as his analytical work advanced. He could use facts won in the squalor and discomfort of the dissecting room to generalize, not just about the body itself, but also about its relationship to the rest of the world. This breadth of comprehension makes Bernard an exciting thinker even in today's very different atmosphere, because while he cut and separated the parts of the body with his knives, he also had a clear vision of the whole. He appreciated that a living creature is organized with respect to its own inner workings as well as to its exterior surroundings. What is now called "adaptation" is this second kind of outward-looking organization, which enables a species to make a close fit between its own biological structure and behavior and the patterns, including the time patterns, of the particular worldly niche the species inhabits. It is interesting to note that the first example of the use of the word "organize" in the *Oxford English Dictionary* refers to the human body. The fifteenth-century author quoted there assumed that the body was organized, not for its own sake, but for the specific function of receiving the soul.

Bernard realized that in the relationship between life and the physical world, the body has an autonomy that a lifeless piece of matter, a stone or a lump of clay, does not possess, a theme emphasized later in philosophy by Bernard's fellow countrymen, the French existentialists. One cannot hope to understand life processes from the outside, comparing what goes into the body with what comes out. That, Bernard said, is like trying to tell what happens inside a house by watching what goes in through the door and comes out by the chimney. Bernard was impressed especially by the way many of the properties of the living system stay the same over time even though conditions in the outside world may fluctuate. He proposed that the body has not one but two environments, an external one in which it moves and acts, and an internal one which surrounds and bathes the tissues of the body, as the sea bathes the fish that swim in it. At first, Bernard defined this internal milieu as the blood plasma, but later he extended the term to include all the circulating fluids of the body. Living things do not respond to external change in the same way as lifeless matter does, Bernard declared. They respond rather to changes in the internal environment. In this respect life is an entity profoundly different from nonlife, which is modified by outside forces only. Bernard talked about the "organizing and creative idea," a hallmark of the living organism, which is subject to special laws and obeys rules that do not apply elsewhere. Henri Bergson maintained that Bernard's unique contribution to science and thought was to show that life is not governed by a collection of laws that fit together according to human logic, but by laws that need to be looked at in the light of nature's logic, which is quite a different thing. What seems absurd to us may not seem absurd to nature.

The concept of the internal milieu was central to Bernard's whole theory of life. The milieu is a source of sameness and stability, whereas the external environment is full of differences and change; that is part of its "logic." Measuring the temperature of the blood, he found that it rises slightly when traversing the liver and falls a little when passing through the lungs, but on the average, blood temperature remains surprisingly constant. It was as if the organism had enclosed itself, Bernard said, "in a kind of hothouse, where the perpetual changes in external conditions cannot reach it." Since many of the cells of the body are highly specialized and very sensitive to change, they need to be cocooned and coddled by the relatively changeless, continually circulating, internal fluid which transports food and oxygen around the body, removes waste products and keeps temperature on an even keel.

Early in his career, Bernard treated the internal milieu as something separate and distinct from the world outside, but later he came to believe that inner and outer environments are related in various intricate and ingenious ways. Life itself, Bernard surmised, results from a kind of tension between the internal structure of organisms and the external, "cosmic" conditions. The organism does not war against those conditions, however. It maintains itself by an adaptation, an accord with them. It "participates in the universal concert of things." Bernard divided living species into three distinct categories, based on the nature of their relationship with the cosmic conditions.

In the first category of life are those species that have broken off completely their connection with the external environment. This is the extreme case of a type of organism that, to all intents and purposes, is outside time. A plant seed, for example, remains inert for centuries unless it is placed in the soil and supplied with air and water.

The second category Bernard called "oscillating" life. Oscillating forms shift from one state to another as and when the external environment dictates, because they are linked so closely to it, and must alter their ways of life in step with the world as it varies. Such species are captives of world time. They live according to a timetable that is determined entirely by external patterns of change. Plants become dormant when the summer season gives way to winter. The dormouse and the marmot oscillate between hibernation during cold weather and activity when the season is warm. An American mammal, the tenrec, does almost the opposite, entering into a state of lethargy at the height of summer. In dormancy, bodily processes function at a low level. The heart beats more slowly and motion comes to a stop. The animal has no choice in the matter, no way of resisting torpor, because time-dependent changes in the external environment rule its existence.

Bernard's third category of life, into which fall the most highly organized animals, including human beings, is "free life." Creatures in this category are able to act spontaneously, because they are not chained to the general cosmic forces. Their life is never suspended under any circumstances. They may seem indifferent to the external environment, but they are certainly not disconnected from it. Quite the reverse. The constancy of the internal milieu is maintained by forces, always active, which detect and cancel out external variations. As a result, far from being cut off from the world outside, "the higher animal is on the contrary in a close and wise relation with it, so that its equilibrium results from a continuous and delicate compensation established as if by the most sensitive balances."

Inner sameness, Bernard stressed, is the primary condition for a free existence, one that is not at the mercy of every chance shift of a variable in a world that is never the same from one moment to the next. There is a price to be paid for this freedom, however. Higher forms of life are relatively independent of the external environment, but the greater the degree of this independence, the more dependent a species becomes on the stabilizing forces of the internal milieu. This spectrum of dependency-independency, extending throughout the entire domain of life, is a unifying principle in nature.

In the twentieth century, the theory of internal sameness

was refined and elaborated by an American physiologist named Walter Cannon, of Harvard University. Cannon was the son of a Wisconsin railroad engineer; he was a man filled with a sense of obligation to leave human existence better and more enlightened than he found it. When Cannon was ten years of age, his mother, on her deathbed, said to him: "Walter, be good to the world." He did not let her down. Like Bernard, Cannon was an experimentalist of dexterity and skill. One of his most notable accomplishments was to detach and remove the entire sympathetic nervous system, and this led to new ideas about the means by which constancy is maintained in the internal milieu, or, as Cannon called it, the fluid matrix, the "watery environment" of the body. In 1926 Cannon coined the word *homeostasis* to refer to the "continuous and delicate compensation" that Bernard had described. The work of homeostasis, Cannon realized, was a cooperative one, in which brain and nerves as well as bodily organs all played a role.

In Cannon's view, the most important effect of homeostasis is the freedom it gives to the organism, which is able to exist and go about its affairs in widely varying conditions without needing to vary its behavior drastically or worry about compensating for every external change; that task is taken care of automatically and unconsciously by built-in mechanisms of the body. The routine functions of the body display a uniform stability, day in, day out and from year to year, in spite of the fact that the body is made of highly unstable material and is sensitive to the slightest changes in its surroundings. Cannon was struck by the lengths to which an organism will go to preserve its own internal constancy, using up reserves of energy in an apparently spendthrift fashion so as to keep temperature, say, or the composition of the blood from changing by more than a specified amount, throwing away valuable substances such as water, salt and sugar in this enterprise. "In critical times," Cannon said, "economy is secondary to stability."

The conditions that permit a healthy life are quite narrowly circumscribed. In humans, the normal temperature of the body is 98.6 degrees Fahrenheit. It decreases slightly at night and increases a little during the day, but the variation is only about one or two degrees in 24 hours, even though the temperature of the external environment may move up or down dramatically according to hour and season. If body heat rises just a few degrees above normal, it is a sign of fever. Similar limits are imposed on variations in the composition of the fluid matrix. Glucose concentration usually does not fall below 0.6 percent or rise above .18 percent. The heart will stop if the amount of potassium in the blood increases even slightly, while a small rise in the normal quantity of magnesium blocks nerve function, which explains why magnesium salts can be used as an anesthetic. And unless losses of water are balanced by water taken in, the consequences for life may be serious.

There is a good reason why the fluid matrix is watery. In the body, water is the most abundant substance, the medium in which all the essential chemical processes take place. When early creatures first emerged from rivers, lakes and oceans, they were able to adapt to life on dry land only because they internalized their former watery habitat, incorporating it into the body itself, so that cells were immersed in the fluid matrix as fish are immersed in the sea. The first one-celled organisms evolved in the sea, from which they drew nourishment and into which they deposited wastes, so that they were entirely dependent on it for existence. When many-celled life forms arose, cells were not in direct contact with the sea, with the result that it became necessary to create an inner, biological sea as an analog of the ocean outside. The organism's chemical apparatus produced a fluid with properties similar to those of seawater. Today, in many species, the resemblance of the fluid matrix to the ancestral seawater environment is plain to see, especially in its high concentration of sodium. Human beings are about two-thirds water molecules, and most of the other molecules that compose the body's chemistry are dissolved in water. The amount of a given substance dissolved in a specific volume of water is known as its concentration, and concentration is a measure of how densely packed the molecules of the substance are in the solution.

One of the most important properties of water, as far as homeostasis is concerned, is that it does not change its temperature very easily; quite a large amount of heat is needed to warm it appreciably, so that the ups and downs of temperature in the outside world do not produce immediate and equivalent changes in the internal milieu.

The fluid matrix provides a context of sameness in which the cells of the body are free to be different. Each individual cell in a complex organism such as a human being is not versatile, but highly specialized, like a typecast movie actor who can play only one kind of role. Such a cell by itself is unable to survive in widely varying conditions. Paradoxically, by being so limited in the range of its functions, the cell helps to make the organism as a whole sophisticated and resourceful, and therefore more independent of external circumstances. In order for an organism to become versatile, cells must not all perform the same services. There must be specific kinds of cells for muscles, other kinds for nerves, still others for glands. What enables cells to specialize and be different, and so makes possible the emergence of complex life forms, is the sameness of the fluid matrix, providing uniform conditions which free the various types of cells to give full time to their particular role. The fluid supplies food and oxygen to the body cells, just as water in a stream transports the staples of life to simple, one-celled organisms anchored on the bed of a river. These stationary marine forms, stuck fast, perhaps in some obscure crevice, are unable to move around in search of food, but instead must have it delivered to them by the swirling waters surrounding them. In a similar way, the cells of the body are fed by the circulating streams of the blood and the lymph. Into these streams the cells discharge their waste products, which are carried away. Oxygen is brought to cells by the red corpuscles of the blood, and nourishment by the plasma, which forms about half the total content of the blood. The plasma delivers food from the intestines to the cells and carries waste away to the kidneys.

The lymph is a sort of middleman or broker, mediating exchanges between the blood vessels and the cells; it contains white corpuscles, and these protect the body against harmful invaders and alien substances.

The temperature and pressure of the fluid must be held within prescribed limits, and its ingredients (sugars, fatty acids, vitamins and electrically charged substances such as calcium, sodium and potassium) must be present in the proper concentrations; otherwise the workings of cells may be impaired. If cells do not function as they are supposed to do, neither will the body and brain. Homeostatic control over these and other variables is exercised largely by nerves and hormones, acting together, sensing departures from the prescribed values and correcting them by sending messages to various body systems. If the level of glucose falls, for example, perhaps as a result of eating too little, certain cells in the pancreas release glucagon, a hormone that signals to the cells of the liver, "telling" them to release glucose; if the level rises too high, usually after a meal, other pancreatic cells release the hormone insulin, which results in glucose being taken out of the blood and stored in liver and other cells. Homeostatic devices in the brain, known as "glucostats," also keep track of glucose levels, and set in motion a train of body processes, adjusting appetite up or down as needed.

Creatures that lack an efficient biological means of keeping their internal temperature constant must exist on schedules imposed by the timetable of the external environment, unless that environment is constant. They are restricted to a narrow range of temperatures within which they can compete successfully with other species. Many of the simpler forms of life, such as bacteria, fungi and yeast, suspend metabolism and essentially cease to live when the temperature does not suit them. Numerous plants and insects live only during the most favorable season, and when the season comes to an end, they die off, leaving eggs or seeds behind. Deciduous trees reduce their exposure to extremes of hot and cold by shedding all their leaves. Snakes and lizards bask in the sun in the morning so as to warm up their bodies, and in this way become active, but in the more intense afternoon heat they must retreat to the shelter of rocks and burrows and lie still. That routine is forced upon them. A desert lizard will perish if kept for only about fifteen minutes in full sun.

Homeostasis is an aspect of the relationship between time and freedom, and that relationship is as important in biology as it is in philosophy and metaphysics. Walter Cannon believed that homeostasis, by removing the need for the organism to spend all its time consciously and deliberately reacting to change, leaves it free to seek "unessentials," things not directly related to moment-by-moment survival. It is from such unessentials that civilized life is made. Sameness of internal temperature broadens the range of an animal's experience. An inner milieu that does not change enables the body to function as usual even in external conditions that are unstable and very unlike those of the inner milieu, as a spacesuit makes it possible for an astronaut to walk on the surface of the moon. A creature that possesses such a milieu can tolerate wide differences outside because of the sameness inside, and for this reason its options are more plentiful. It is able to go almost anywhere, almost anytime. By making time matter less in the life of a species, homeostasis is a source of freedom, increasing the possibilities of doing something new and being something new. A distinguishing mark of those vertebrate species whose body temperatures are homeostatically maintained, usually at a level higher than that of the surroundings, is that they tend to lead active and adventurous lives. One reason for the disappearance of the dinosaurs, which lacked sophisticated systems of internal temperature control, may be that they lost their competitive edge during cold weather, when food was scarce and body temperature fell. This need not have mattered when all other animals were equally lethargic in winter, but it might have been a liability when mammals appeared on the scene, their biological thermostats reducing their dependence on time of day or season of the year. The earliest mammals, precursors of man, probably escaped the predations of the dinosaurs by becoming nocturnal, foraging for food at night and sleeping during the day, an option, an exercise of freedom in time, not open to cold-blooded creatures.

Homeostasis is especially necessary for the proper functioning of the central nervous system of animals on the higher rungs of the evolutionary ladder. Before intelligent life could appear, and well before the culminating event of human consciousness, the mechanism to ensure the sameness of the internal milieu had to be in place. Severe disruptions of homeostasis will kill the body, but before that terminal stage is reached, smaller fluctuations impair the nervous system and brain. In fact, the internal milieu seems to be more constant for the cells of the brain than for other parts of the body. Possibly, the evolution of better and better biological thermostats played a part in the rise of human culture. The earliest civilizations, with few exceptions, arose near a line, or isotherm, on the map of the earth's surface that connects all points having a mean temperature of 70 degrees Fahrenheit. Later, the center of population shifted to regions where the temperature was nearly nine degrees less. A cooler temperature may have sharpened and stimulated the wits of prehistoric man without causing him serious discomfort, and this was probably due more to homeostasis than to the discovery of fire. The historian Arnold Toynbee theorized that man needs a stimulus from the environment in order to rise to his full potential, and of the many forms such a stimulus may take, one is a bracing climate.

It may happen in the course of evolution that a particular biological function, which was once strongly time-dependent, breaks free of the time constraint, due to the pressure of natural selection, so that what was once periodic becomes constant. Such a deemphasis of time, increasing freedom of choice, is an event that can lead to interesting and unpredictable consequences for the future of the species. When temporal restrictions on behavior are removed, new constraints may need to be acquired—otherwise chaos might ensue—

and these constraints in turn can become new pressures leading to evolutionary change.

A dramatic example of this process, not involving homeostasis, is the loosening of the calendar constraint on mating behavior. In vertebrate animals other than primates, reproductive activity occurs only during certain definite and widely spaced periods of the year and not during others. Such activity is date-determined according to what is called a diphasic sexual cycle. Primates are not bound by this rule, however, and are in a reproductive state almost without interruption throughout the year. Changing from time-bound to time-free sexual behavior seems to have been an event of great importance, beause it led to the existence of groups of primates, young, middle-aged and old, all thrown together, in circumstances where strong erotic attraction was present more or less the whole time. As the anthropologist Ernest Becker has noted, when "mateability" ceased to be seasonal and intermittent, vertebrate behavior piled up on itself. A jumble of statuses was created, to which members of the group were forced to adjust. "This provides a welter of confusion and stimulation, a new environment that must be like Times Square to someone raised on a farm," Becker says. "At each point in the growing animal's life, he must find a new adjustment to make to those around him: young to young, male to female, male to young, young to female and so on. This need for continuing adjustment provides part of the stimulus for the emergence of a larger-brained animal. Nothing is so unpredictable as are other living organisms."

The preservation of sameness in the routine operations of the body, reducing the importance of time, was an essential step if complex creatures were to emerge, adventurous and active, free to expand the range of their experience and to lead more varied and more interesting lives. But the evolution of mechanisms for producing sameness was not confined to bodily processes alone. Strategies were acquired by the brain in its fundamental operations of knowing, learning and remembering, which mediate the relationship between the internal environment of mind and the external environment of the world. They supply psychological samenesses, as homeostasis provides biological samenesses.

Impressions reaching the senses, for example, are different from moment to moment, because the world is never still. Objects move, and we move with respect to them. Light is brighter in the afternoon than in the evening, and shapes expand and contract as they approach us or recede into the distance. Yet the brain seeks out constancies amid this flux, and imposes sameness on what it sees and hears, in order to stabilize experience and create a consistent world. Colors may shimmer and shift from moment to moment as the light changes, but we experience them as being constant over time. A yellow dress still looks yellow, even when it is drenched by sunlight streaming through the gorgeously tinted panes of a stained glass window in a cathedral. Again, while the brightness of objects changes with the amount of light they reflect, they seem always to have a constant brightness. Snow is still seen as white, even at night, because the brain is so organized that it perceives an unchanging property of the surface of the snow, the property of reflecting most of the light that falls on it, even though the intensity of the light may alter from moment to moment and from hour to hour. An entire theory of color vision has been derived from the basic fact of these constancies of human perception.

The physicist P. W. Bridgman once remarked that however much the world changes, human beings will always find, or try very hard to find, some elements of sameness in it, and will do so in fairly uniform ways. That is one of the characteristics that define thinking creatures. If the world were to turn strange on us all of a sudden, Bridgman surmised, if stones started to fall upward and the moon spun backward, we would try to make sense of the new state of affairs in exactly the same way as before. We would go on counting the objects around us, measuring their lengths by fixed standards and observing the time of events by means of clocks, in the hope of discovering certain unchanged relations amid the seeming chaos. The conscious mind finds, or imposes, invariants in its traffic with a world that varies over time. What we do in attempting to make the world intelligible, in Bridgman's view, has a greater permanence and stability than the world itself displays.

Sameness also seems to lie at the heart of that most abstract of all systems of human knowledge, logic. Modern logic is interested in relations between symbols, and chiefly in those relations that do not change. For example, the logical propositions

> *This thing is bigger than both of us.*
> *Breakfast is eaten before lunch.*

are very different in content, but their logical form is exactly the same. Each is making the statement, "x has some relation to y." Bertrand Russell, writing in prison, where he had been sent for six months in 1918 as a result of an article published in a pacifist journal, suggested that this element of sameness in our thinking is perhaps more profound than we realize. Russell proposed that there exists a proposition so fundamental that the whole of logic and mathematics can be derived from it. He called such a proposition "primitive." In order to qualify as primitive, a proposition must consist of constants as well as variables. Constants represent sameness, the unchanging laws that govern the ways in which variables vary.

Yet constants and variables are not enough by themselves to make a proposition primitive. In addition, it must possess a further property. The proposition must be tautological. To the uninitiated, a tautology may seem like needless repetition, containing too much sameness and not enough difference. A tautology may say something like "It is either raining or it is not raining." Or, more simply, "x is x." But while such statements appear superfluous and not worth saying, they are nevertheless extremely useful and important to logicians, because whether a particular array of propositions taken together does or does not result in a tautology gives valuable information about the structure of the propositions in question. One striking feature of the tautology is that it is independent of time. Time is supremely unimportant in the statement "It is either raining or it is not raining," since the statement will always be true no matter the hour of the day or the season of the year, and it will be as true a thousand years hence as it was a thousand years ago, whether we are awake or asleep, looking out of a window or sitting in a darkened movie theater. Bertrand Russell conjectured that if we were to make a mental voyage out to the farthermost frontier of existing knowledge about the logical foundations of mathematics, we would come face to face there with this apparently repetitious, timeless epitome of nonchange, the tautology.

In physics sameness, or invariance, was a concept that came to be of high importance as science became increasingly mathematical and concerned with abstract entities which have no counterpart in ordinary experience; with cloakroom tickets rather than with overcoats. Scientists studied the behavior of objects, often fictitious and impossible to visualize, as they underwent various kinds of change, and looked for properties or quantities that did not change under such transformations. Such highly sophisticated invariances are the descendants of those discovered during the Renaissance, when painters needed to have a theory of how objects in the real world can be represented convincingly in a pictorial work of art. Projective geometry had been developed in the seventeenth century as part of the movement in Renaissance painting to escape from the flat, stylized pictures of the medieval period and represent scenes on canvas in a more realistic and lifelike way. In order to accomplish this, painters needed to know what does not change, as well as what does change, when a three-dimensional scene in the actual world is transformed into a two-dimensional image on a flat square of canvas. A surface that is "really" square, say a tabletop, may be distorted by the artist into a four-sided figure whose edges are not parallel, and a circle may be represented as an ellipse. Yet an observer looking at the painting knows at once that what he sees is a square and a circle, respectively, because the invariants have been preserved in such a way that figures which are in some respects quite unlike each other are still recognizably the same in art as they are in life. The invariances found in many of the basic mathematical theorems of projective geometry, inspired by painters who were also engineers and mathematicians, true Renaissance men, are the ancestors of modern relativity theory, whose new invariants, or absolutes, such as the speed of light, replaced the false absolutes of classical mechanics.

In the past few years, new insights into the nature of human memory have been made possible through the recognition that the brain has more than one strategy for remembering. And the most important difference between these memory strategies is the emphasis or lack of emphasis on time. In one type of memory, apparently the more ancient one, which has the greatest practical usefulness and was probably selected in evolution because it helps species to survive and master the external environment, time is of little significance. This memory system is a storehouse of knowledge about facts, formulas, names, words and their meanings, symbols, rule for manipulating symbols, concepts and relations. Since information of this kind is usually good for all occasions and all places, and remains the same even though the world may change in other ways, it does not need to have a time context. We do not usually remember the day, month or year in which we acquired such knowledge, or the time order in which various items of knowledge were learned. The brain may organize these items into a sequence that has nothing to do with time, say, by listing the days of the week in alphabetical rather than in chronological order. This is the normal way in which a modern digital computer operates. Another type of memory, which seems to have appeared later in evolution and is more of a luxury, a frill, suitable for a life in which leisure and culture, the arts and intellectual pursuits are valued, is highly time-dependent, storing information in a unique context of time and place. This memory emphasizes time and uniqueness, rather than timelessness and universals.

Inner sameness, whether biological or psychological (the two cannot be separated in any clear-cut way), is an evolutionary invention peculiar to advanced forms of life and necessary if living creatures are to avoid being the slave of time. As a result, there seems to be more sameness inside the higher types of organisms than there is outside, in the world. An extra ration of sameness means an extra ration of freedom.

William James, brother of the novelist Henry James, in his extraordinary book *The Principles of Psychology,* which though published in 1890 is so crammed with insights that reading it can be an exciting experience even in the 1980s, coined the metaphor of a "stream of consciousness." The metaphor expressed James's conviction that thinking is continuous, that to say of the mind that it thinks is like saying of the wind that it blows, or of the rain that it rains. The stream provides the basis for what James called the principle of constancy in our meanings, the ability to recognize an event or idea or object when it appears more than once, and thus also the capacity to recognize constrast and differences. He considered the sense of sameness to be "the very keel and backbone of our thinking," and a prerequisite for the very existence of a sense of self. The mind owes its special architecture, its design, to the fact that it makes continual use of this notion of sameness. As it is, James's principle decrees that constancy is so important to the mind that the mind constructs the experience of sameness even though it may not be "really" present in the world. James stressed that he was speaking as a psychologist, not as a philosopher, and used the term "sense of sameness" only in connection with the structure of the mind, not the structure of the universe. "We do not care," he said, "whether there be any *real* sameness in *things* or not, or whether the mind be true or false in its assumptions of it. Our principle only lays it down that the mind makes continual use of the *notion* of sameness, and if deprived of it, would have a different structure from what it has."

Because it is an aspect of the structure of the mind, the principle of constancy may not apply to all living species universally. James goes on to say: "Not all psychic life need be assumed to have the sense of sameness developed in this way. In the consciousness of worms and polyps, though the same realities may frequently impress it, the feeling of sameness may seldom emerge. We, however, running back and forth, like spiders on the web they weave, feel ourselves to be working over identical materials and thinking them in different ways. And the man who identifies the materials most is held to have the most philosophic human mind."

It is almost a cliche to say that time is an effect of change, and that without change, the idea of time would cease to have any meaning. Modern physics, after Einstein, supports the view that time depends on events, on happenings in the material world, and that to argue otherwise is to flirt with the discredited Newtonian concept of absolute time. A universe empty of events would be a universe without time. Yet there is sameness as well as change in our world, and the chief supplier of sameness is the living system, with its mechanisms, both biological and psychological, for seeking out and creating constancy. Such sameness does not make time stop dead in its tracks, but it alters the meaning that time has for the body and for the mind. The mental experience of time when sameness predominates differs significantly from the experience of time when there is a great deal of change. And the body's homeostatic controls, some of them highly complex, smooth out differences, so that an "event" in the external world as drastic as a plunge in temperature of 30 degrees Fahrenheit does not result in a comparable change in the internal milieu. Thanks to mechanisms of compensation, the milieu remains almost constant and therefore nearly eventless, and so, in a sense, almost "timeless," according to the modern definition of physical time.

Living organisms have evolved, too, the means to emphasize time, to increase its importance in their lives, but in a highly selective fashion. In the body, that is the role played by biological clocks, inborn devices for measuring time and bringing temporal order to the body's vast array of functions. Such timekeeping mechanisms appeared almost at the dawn of life on earth, and are present in extremely primitive organisms. Thus time was emphasized in evolution long before it was deemphasized. Biological clocks preceded homeostasis. Over millions of years, more highly organized species emerged which were able to manipulate the meaning of time in their own existence by making time important in some repects and unimportant in others, so that "time biologized" is an advantageous and successful balance of the two.

THREE

The Importance of Time

The story of modern biology is essentially the story of the discovery of more and more structure in the living organism. Until fairly recent times, the new vistas of structure revealed by science were mostly spatial. Through more detailed knowledge of how the body is organized in space, at smaller and smaller scales, the realization dawned that structure, the way in which the parts of a system are arranged, is closely related to function, the way in which that system behaves.

To the naked eye, the tissues of the body appear to be smooth and continuous masses of biological material. On closer inspection, however, they turn out to be highly differentiated, being composed of tiny, individual building blocks which themselves are complex and highly organized. What seems to be sameness actually conceals a wealth of difference. Many centuries ago, Aristotle said that all animals and plants are made up of a limited number of parts which are repeated in each organism, but he was talking about the macroscopic anatomy of living things, gross components such as leaves and flowers, legs and arms, which the ordinary observer can easily see. It was not until the invention of increasingly powerful optical instruments, starting in the seventeenth century, that the true extent of biological structure on the small scale began to be apparent. In 1665, the Englishman Robert Hooke peered into a primitive microscope at a sliver of cork, whose surface had looked smooth and uniform to the unaided vision, and saw that it was actually full of tiny holes. Hooke named the holes "cells," after the little bare rooms in a nunnery; the holes were in fact the remains of dead cells. Later, he extended the term to include the microscopic parts of living tissue. As time went by, Hooke's nun's room metaphor was replaced by a model of the cell as a closed sack holding a liquid in which objects floated. Then, in 1839, the German microscopists Matthias Schleiden and Theodor Schwann presented the first definitive version of the "cell theory," which states that all plants and animals are constructed from cells, and these are the basic units of life. The cell theory is one of those overarching principles of biology, like Darwin's doctrine of natural selection and the discovery of the double helix of DNA, that make it possible to progress beyond mere description and frame powerful, general theories of the nature of life.

The realization that organisms contain much more built-in structure than was previously suspected was a result of increasing technical ingenuity on the part of microscope designers. The cell is the most elementary entity that can sustain life, and yet this basic component is itself full of structure, on a tiny scale. Just as astronomers had to magnify celestial objects that are very far away in order to have more knowledge of them, so also biologists needed to resolve, or discriminate between, a number of minutely small objects within the cell. The naked human eye has a resolving power of about a tenth of a millimeter. Objects any smaller than that either are invisible or can be seen only as a blur. The modern electron microscope, which uses high-speed electrons instead of light waves, can magnify an image 500,000 times, or even 2,000,000 times by photographic processing, while enlargements of up to 2,000 times are possible with the light microscope. It was the invention of the electron microscope that confirmed what before had been a mere inference—namely, the existence of a plasma membrane bounding each cell, separating it from the internal environment. Most components of the living cell are so small that they can be observed only through the resolving power of the electron microscope, which has made it possible to visualize the individual DNA molecule.

The search for spatial structure has been highly successful, though it is far from finished. As improvements are made in the design of microscopes, it is certain that more structure will be found. Only in the past few years, however, has a serious search been undertaken for the temporal structure of life forms, the way they are organized, not in space, but in time. There are powerful reasons for undertaking such a search, because today most of the big, unanswered questions about the living organism are temporal ones. In this endeavor, as in the search for spatial structure, progress has been made, not by means of the naked eye, but through instruments that increase the eye's resolving power. In the study of biological time structure, the chief instrument is not the microscope but the modern high-speed computer, with its ability to analyze and perform mathematical operations on millions of observations of bodily changes over time, conjuring significant temporal patterns out of a mass of data which might otherwise appear to have no pattern at all. A given function of the body may undergo regular, rhythmic change, but the extent of that change may be too small to be detected merely by looking at a chart of sample readings, or it may be hidden by other variations. By means of computer analysis, however, the rhythmic ups and downs can be made to stand out clearly, as spatial patterns, otherwise too small to see, are resolved by the lenses of optical instruments. Such computer programs have unmasked a time structure in the living system that is richer by far than could have been imagined only three decades ago, and the work of investigation is only in the early stages. There are rhythms whose existence is only just coming to light, and doubtless others still in hiding, which will be revealed at some future date.

States of the body, from the beating of the pulse to the ebb and flow of chemicals in the internal milieu, and even something as basic to the maintenance of the body as the cycle of growth and division of cells, are not "smooth," or homeostatically constant in time, as Robert Hooke's slice of cork, unmagnified, appeared smooth in space. Such states vary in a regular, periodic, wavelike fashion, each complete cycle of change occupying a definite span of time and repeating itself cycle after cycle.

Even in Claude Bernard's day it was clear that physiological states were not uniform throughout each 24-hour period. Temperature rises and falls during the course of a day and a night, though by so small an amount that Bernard dismissed the variation as "rather feeble fluctuations," not realizing that a slight warming up or cooling down of the body can have momentous consequences for health and even for survival. Walter Cannon pointed out that homeostasis provides, not strict sameness, but variability within limits. Today, homeostatic constancy is seen to be an appearance, concealing beneath the surface of sameness many kinds of change, and the changes are not disorderly or random, but regular in time. Scientists recognize that rhythmic variation is the rule rather than the exception in all living things, even the simplest and most primitive ones. Some fluctuations are not feeble at all, as Bernard supposed, but show a robust, clearly marked pattern of increase and decrease, which is regular and predictable over time. These "large" rhythms, of high amplitude, which need little magnification by the computer, include changes in the levels of hormones such as cortisol and insulin, in blood pressure, in the rate at which the kidneys produce urine, and in the alertness of the brain. Blood pressure readings may rise and fall during a 24-hour period by as much as 20 percent, and the number of infection-fighting white cells in the bloodstream increases and decreases by 50 percent.

Body functions that might be expected to be the most uniform and constant turn out to be inconstant, though in a rhythmic fashion. The human pulse, for instance, does not remain the same at 70 beats a minute throughout the 24-hour span, but slows down at night and speeds up during the day; the difference between the top and bottom of the cycle may be as great as ten beats a minute. Blood pressure is lowest first thing in the morning, and rises as the day proceeds. Virtually everything in the body fluctuates rhythmically, including the operation of organs such as the kidney and the liver, the immune system (which defends the body against harmful invaders), the manufacture of proteins (building blocks of organic material which play an essential role in just about every aspect of anatomy and physiology), secretions of the endocrine glands, and the presence of neurotransmitters (chemical substances that mediate communication between nerve cells in the brain, and therefore influence states of consciousness and states of sleep). There are many, many more examples.

The internal environment, the fluid that bathes the cells, and whose "fixity," Claude Bernard said, provides the body with its independence from changes in the outside world, is not fixed at all; its composition alters cyclically over time. The ingredients of the fluid are strictly controlled within specific limits, but the limits vary rhythmically as time passes.

Bernard held that the internal milieu is different from the external milieu because it is constant, while the world outside is full of differences. That is part of the body's peculiar logic. The more recent science of chronobiology finds that there exist also mechanisms which, like those of homeostasis, are innate. These mechanisms organize the body in time, so that it is one thing at one hour of the day and quite a different biochemical entity at another hour, just as the dynamics of spatial structure see to it that the organism is different from place to place in its anatomy. There is a Dr. Jekyll and Mr. Hyde aspect to the temporal life of any organism. Homeostasis tends to make a creature indifferent to external time. Biological clocks elevate time to a definite and inescapable importance. Yet the logic of biological time, like that of homeostasis, is special to the world of life. It is not the same as the logic of physical time, and is different from one species to another. In that sense, Bernard's general thesis holds.

We have seen that creatures which crawled out of seas, rivers and lakes to make a life for themselves on dry land managed to survive in their new habitat by internalizing the water that was once their external spatial environment, incorporating a chemically similar version of it into the body in the form of the fluid matrix. Starting very early in evolution, primitive species also internalized certain aspects of their external temporal environment, including such basic geophysical periodicities as the cycle of daylight and darkness, caused by the rotation of the earth on its axis, the succession of the seasons, caused by the orbit of the earth around the sun, and the regular changes in moon and tides. Perhaps, when life was just beginning about five billion years ago, the thinness of the earth's atmosphere did not contain, as it does today, a protective upper layer of ozone which screens out harmful ultraviolet radiation from the sun. In order to survive, primordial life forms had to restrict certain essential metabolic operations to the safe temporal haven of the hours of darkness, so that ultraviolet light would not disrupt them. The fittest organisms, therefore, would be the ones that evolved ways of imitating the 24-hour cycle internally, and dividing up the cycle in such a manner that fragile biochemical activities were scheduled in the nighttime hours. Likewise, the fit marine creatures that first ventured onto land were those whose internal milieu imitated the sea.

Perhaps the earliest species merely responded passively to the "clock" of the earth's rotation, being driven directly by it rather than internalizing its time patterns. But even if that is the case, biological timekeeping mechanisms were built into the organism while life was still at a very simple stage, when the only species around consisted of single cells. There are inborn daily rhythms of mineral metabolism in corals and in marine and freshwater molluscs such as cockles and scallops.

A body rhythm, say in temperature, with a period of about 24 hours, is an internal model of the cosmic cycle of day and night. It mimics geophysical time, so that the wisdom of the body, which Walter Cannon celebrated in his writings on homeostasis, must be extended now to include intrinsic biological sources of information—one might almost say "knowledge"—about the time patterns of the external world. Possessing such information as a birthright, the body can act on it and arrange its own affairs accordingly, without needing continual input from the environment.

One very important feature of biological clocks is that the world outside may be used as a source of information about time, but such information need not be the immediate cause of biological or behavioral events. If it were a direct cause, the body would simply dance to the world's tune, like Claude Bernard's oscillating life forms, reacting to external changes as they happen, but not anticipating them, not keeping one step ahead of them. Clocks enable the organism to prepare, to put itself into a physiological state that is appropriate not just to the present moment but to the future.

Thus, an animal sleeping in a dark burrow is woken up not by a shaft of daylight but by an innate, self-winding biological clock which is corrected and reset every day by light signals from the environment, so that it is in step with, or "entrained" by, the local 24-hour cycle of day and night. The body clock itself triggers physiological processes which rouse the animal, not exactly at dawn, but a little earlier, so that it is ready for the day. The clock, not the light, is the creature's wake-up call. Chanticleer, the conceited cock of fable, thought that by crowing he caused the sun to rise each morning. In one version of the story, he nursed the illusion that the more glorious his song, the more splendid would be the dawn. A flock of conspiratorial night birds, congregating in a nearby wood, plotted to eliminate Chanticleer, thinking that by doing so they would prevent the sun from rising ever again, plunging the world into endless night, their element. We know now that Chanticleer's body clocks timed his crowing just before the sun came up, because they supplied information to the body about the entire day-night cycle, making it possible to set in motion a train of biological events which culminated in crowing ahead of sunrise. Without such built-in, self-winding clocks, Chanticleer could have crowed only in response to sunrise, after the fact, which would not have inflated his vanity to nearly the same proportions. Chaucer came close to the truth in the *Canterbury Tales* when he said that Chanticleer was an expert in the movements of cosmic cycles and in his crowing was more accurate than a clock. If homeostasis turns the body away from unpredictable changes in the external environment, smoothing them out, biological clocks turn it toward predictable changes, of which the most predictable is the alternation of day and night, not in order to respond to such changes, but rather to be ready for them when they happen.

Forces in the environment do not generate body rhythms. The cosmic cycle of daylight and darkness acts rather as a synchronizer, a time-giver or *zeitgeber*, to use the technical term, which entrains body clocks so that their period, the time taken to complete a single cycle, matches the 24-hour period of actual day and night, the period of the earth revolving on its axis. Like a quartz watch, which is powered by an internal battery, body clocks are self-sustaining, "self-winding." There is no need for an external source of energy to keep them going. But they do need to be reset, because left to themselves they have a period that is nearly but not quite 24 hours. That is why they are called "circadian" clocks, a term coined by Franz Halberg (a highly respected chronobiologist at the University of Minnesota), from the Latin *circa dies*, meaning "approximately a day."

Cut off from the time-givers in the environment, circadian clocks revert to their own intrinsic, approximate period. That means body functions are scheduled not at the same time each day, but later and later or earlier and earlier, depending on whether the intrinsic period is longer or shorter than 24 hours. Disentrained, the human system, which has a natural period of about 25 hours, schedules its daily routines progressively later. On weekends, for example, many people manage to ignore the resetting effects of time-givers, sleep late and let their body clocks free-run. The timetable preparing the body for waking runs an hour later each day of the weekend, so that by Monday morning, biological time is so out of step with cosmic time that when the alarm says 7 A.M. it is only about 4 A.M. by the circadian clocks. The Monday morning blahs occur because inner clocks, uncorrected, disentrained and allowed to go their own sweet way, have made inner time irrelevant to outer time. If, in the real dark night of the soul, it is always three o'clock in the morning, as the saying has it, breakfast time on Monday for those who sleep late during the weekend is only about an hour ahead of that spiritual nadir, biologically speaking.

When entrainment is broken, body clocks do not stop, but the timetables of body functions under the control of the clocks no longer have any logical connection with what is happening outside. Their temporal organization becomes meaningless. Entrainment not only adjusts the period of circadian pacemakers, but also manipulates the timetables, so that biological events occur at moments or "phases" in the circadian cycle that coincide with specific moments or phases of the day-night cycle itself. When the circadian system is fully entrained, any given biological event has a definite temporal or phase relation to events in the world, as well as to other events within the body. The body is organized in time both with respect to the world and with respect to itself, and entrainment is essential to the success of both kinds of organization.

Many functions in the body cease to make sense if their periodic rises and falls do not coincide with specific hours of the day and night. Core temperature, for example, in the entrained state, peaks late in the afternoon, then declines, bottoming out in the early hours of the morning, about that time when the soul is in the depths of its famous dark night. It starts to climb shortly before waking, so as to prepare the body for a day of activity. We wake up when the temperature cycle has already started its climb. Disentrained, the temperature cycle could not play its important, practical, outward-looking role of anticipating the waking day, because its clock would be wrong with respect to external time. It might crest at three o'clock in the morning instead of sinking to its lowest level, and bottom out the next day, just when we need to be most alert. If Chanticleer's circadian clocks had free-run, the fabulous bird might have saluted the dawn in the afternoon, which would have made him look ridiculous.

There are many examples in nature of a cause that is internal and temporal rather than external and physical. Insects go through a long, quiescent period called diapause, enabling them to survive adverse conditions such as cold weather and drought. Yet the event that starts diapause is not the actual onset of cool or dry spells of weather, because an insect must anticipate the change of season in order to build a shelter and conserve food supplies, not just react to such a change when it happens. A biological clock sets off this chain of activity; the insect's biology uses information about the changing length of daylight as season follows season to keep in step with environmental time. Body clocks can use even the tides as information about time. A certain type of diatom, a tiny, single-celled plant which inhabits the edges of oceans, rises to the surface of the sand during ebb tide, and then descends just moments before the incoming tide reaches it. This rhythm is internally generated. It is not "caused" by the tides. Placed in a vessel of sand in a laboratory, the diatom's periodic journeys up and down in the sand are so precisely synchronized with the tides that scientists planning field trips to the shore at the time of low water sometimes watch the movements of the diatom instead of looking up the tide tables.

It is here that the dividing line between biology and psychology begins to wobble. If body rhythms internalize specific, invariant features of the world, so also, though in a vastly more sophisticated way, do certain mental processes, in human beings and perhaps in animals as well. In both cases, the underlying mechanisms appear to be inborn. For example, Roger Shepard, a psychologist at Stanford University, believes that the mind internalizes the laws of motion which govern objects in external space. The brain can manipulate imagined objects according to rules of geometry which are exactly the same as the rules that apply to objects in the real world. Such rules have been incorporated into the structure of the mind in the course of evolution, and are encoded in the genes. We can move mental models of things around in our head, whether or not the actual objects are present before our eyes or, indeed, whether they exist at all in the real world. The image of a unicorn is as easy to manipulate mentally as the image of a famous racehorse or the family cat. In the fictional space of the mind, these images are rotated, tilted and transformed in lawful fashion, so that they behave just as they would if they were material objects in the actual space of the world.

Scientists and artists have arrived at important creative ideas through this apparently innate ability to perform realistic operations on imaginary things. The German chemist Friedrich Kekulé, for example, is often thought of as having made new discoveries about the nature of chemical bonds, most notably those of the benzene ring, during vivid night dreams. In fact, however, Kekulé began by visualizing atoms dancing about in space, hooking up to form chain molecules, while he was fully awake but daydreaming by his fireside. The aftermath of his conscious imaginings was a full-blown nighttime dream in which Kekulé saw a coiling, serpentine chain suddenly twist into a closed loop, like a snake biting its own tail. The experience led the dreaming chemist to the answer, elusive for so long, to the puzzle of the structure of benzene.

It was Kekulé's ability to manipulate, bend and fold the writhing chains of molecules lawfully in his mind's eye, and finally in sleep, that provided him with new knowledge, largely independent of direct sense experience, and the ability itself came out of an inborn power of the mind to internalize certain constraints that apply in the physical world. Among these constraints are the three dimensions of space and the single dimension of time, the upright direction in space due to the force of gravity, and the rule that there is a simplest way in which an object may move from one position to another, being rotated about a unique axis in space. Without these rules, the objects we visualize and move around in mental space might behave fantastically, lawlessly, and be quite unreliable as a guide to how the same objects would behave in the real world. Their legitimate motions in the human imagination enable us to make plans and anticipate the results of actions before those actions are actually taken. A person can learn how to maneuver a bed up a narrow flight of stairs, by performing the task first in the rehearsal rooms of the mind, confident that the mental constraints that apply to an imagined bed will also be valid when manhandling an actual bed up a real staircase.

Many biological rhythms internalize time rules of the physical world, as the mind's eye internalizes space rules. Rules are constraints on freedom, limiting possibilities, and it is those particular constraints that have been the most stable and enduring over millions of years of evolution, such as the regular schedules of day and night, spring, summer, autumn and winter, the phases of the moon and the predictable ebb and flow of the tides, which are now internalized most deeply in the biology of living things. Such built-in rules are a guide to the patterns of the world's temporal order, as the constraints on mental imagery are a guide to what is possible or impossible in the space of the real world. They have the power to generate information about external time internally, on the basis of very little in the way of information from outside. A single pulse of light may be enough to synchronize biological time with local time, yet the system of biological clocks provides the body with a guide to the entire 24-hour cycle, so that the organism can behave appropriately in time throughout that cycle, without further feedback from the outside world. Homeostasis operates by means of feedback, sensing departures from the norm in this or that bodily variable, using such information to undertake corrective action. It works from moment to moment, acting on data that must be updated continually. Biological clocks are not like that at all. They do not depend on constant feedback. They are autonomous, in the sense that while information about external time, often in the form of daylight and darkness, is needed to keep the clocks in step with local time, the amount of such information is small compared with the wealth of temporal data the clocks supply to the organism internally. Thus the organism achieves a measure of independence from the environment, because the body "knows," metaphorically speaking, what the environment is going to do next and when it will do it, and acts accordingly. Irving Zucker , a Berkeley psychologist, has commented on the "seemingly mad" independence from feedback which is such a striking property of biological clocks, but such independence is not really mad at all. Body clocks just attach paramount importance to time, and other considerations have a much weaker influence on their operations.

New insight into the nature of biological clocks has been provided by the rather recent discovery that there are rhythms in the body with periods for which no counterpart can be found in the environment, and which do not seem to be models of cosmic rhythms. Like the mental image of a unicorn, they are rule-bound, and do have an artificial connection with reality (we can look at pictures of unicorns in books), but they are a pure biological invention. They have to do with the internal logic of the body, not with the external logic of the world.

Perhaps the most intriguing of these "unicorn" body rhythms are those that have a period of about seven days. These circaseptan, or about weekly, rhythms are one of the major surprises turned up by modern chronobiology. Fifteen years ago, few scientists would have expected that seven-day biological cycles would prove to be so widespread and so long established in the living world. They are of very ancient origin, appearing in primitive one-celled organisms, and are thought to be present even in bacteria, the simplest form of life now existing. An innate rhythm of about seven days occurs in a giant alga some five million years old, *Acetabularia mediterranea,* popularly known as mermaid's wineglass, because of its graceful stem and cupped crown which looks as if champagne could be drunk out of it. If this primitive alga is subjected to an artificial schedule of alternating light and darkness, and this schedule is changed every so often, the jet-lag effect reduces the plant's rate of growth if the change is made every seven days, but increases growth if it is made at other intervals, say every five days. Somehow, nobody knows just how yet, mermaid's wineglass is able to translate the manipulation of a 24-hour cycle of light and darkness into the measurement of a seven-day week.

In humans, about weekly rhythms have been found in the fluctuations of several body chemicals, in blood pressure and heartbeat, and in the response to various external challenges such as infection, chemotherapy and organ transplants. When a human patient receives a kidney transplant, there is a rhythm of about seven days, a predictable rise and fall in the probability that the body's immune system will reject the new kidney. A major peak of rejection occurs seven days after the operation, and when a serum is given to suppress the immune reaction, a series of peaks occurs, with increasing risk of rejection, at one week, two weeks, three weeks and at four weeks, the time of the highest risk of all. Seven-day rhythms of response to malaria infection and to certain bacteria have also been discovered. In the days before wonder drugs such as penicillin came on the scene, doctors were acutely aware that a crisis in pneumonia occurred on the seventh or ninth day of the illness. There are seven-day rhythms in the acid content of human blood, in oral temperature, in the amount of calcium in the urine, in the number of red blood cells and in the quantity of cortisol, the "coping" hormone in humans.

Computer analysis is usually needed to detect rhythms of about seven days, which tend to be small in amplitude, not rising and falling as much as many other biological rhythms, especially those with a period of about 24 hours, and sometimes they are obscured by other, irrelevant variations. They need the resolving power, the "magnification," of computer analysis to make them apparent, just as the fine cellular structure of living tissue could not be seen without the resolving power of a microscope. One result of using computers, therefore, has been to find components of biological time structure that seem not to be models or internalizations of external time patterns, but rather constructs in their own right, providing a common seven-day schedule for various processes in the body, coordinating them in time. Franz Halberg is a figure of great importance in this enterprise.

A striking feature of scholarly investigations into the origins of our modern calendar week is the failure to trace it to a cosmic cycle. Independence from geophysical time patterns is one of the week's most curious features, which researchers have often remarked upon. The month, as measured by the phases of the moon, was a basic unit of social time for many primitive peoples, and it was subdivided in different ways, often in the form of intervals between market days, which varied greatly in their length. Market days occurred every four weeks among some West African tribes, every five days in Central Asia, every six days for the Assyrians and every ten in Egypt. Italy had an eight-day market "week" until the fourth century A.D. In all the ancient world, so far as is known, there was no seven-day calendar cycle except for the Jewish week, which existed at the very beginning of the monarchical period in Israel and perhaps even earlier than that. A seven-day week was unknown among the ancient Greeks, whose holidays were held at very irregular intervals, since they fell on the days of religious feasts in different cities up and down the country. Certain months, such as the Greek version of our February, were so crammed with holidays as to be a schoolchild's delight.

At the beginning of the Christian era, there was a seven-day week associated with the seven "planets" of the ancient world, which at that time included the sun. The planets were reverenced as divine, being called by the names of gods: Saturn, Jupiter, Mars, Sun, Venus, Mercury, Moon. Each planet was given control over a specific day of the week. The planetary week, however, was a relative newcomer compared with the Jewish week. Its origins appear to be Egyptian, and it did not come into general use until early in the third century A.D. The planetary week, far from being an invention in its own right based on ancient star worship and thus closely linked to cosmic matters, may have evolved from the Jewish week, and was undoubtedly influenced by it. Presumably the seven-day structure of the Jewish week came first, and later people began to call the days of the week after the names of the planets. Our modern week is a blend of both traditions.

In a new study, Eviatar Zerubavel concludes: "One of the most distinctive features of the week is the fact that it is entirely dissociated from the lunar cycle. It is essentially defined as a precise multiple of the day, quite independently of the lunar month. . . . Those who believe that our seven-day week has derived from the lunar cycle seem to forget that the latter is not really a twenty-eight-day cycle. In fact, approximately twenty-nine days, twelve hours, forty-four minutes and three seconds—that is, about 29.5306 days—between any two successive new moons." Zerubavel goes on to say that "the dissociation of the seven-day week from nature has been one of the most significant contributions of Judaism to civilisation. Like the invention of the mechanical clock some 1,500 years later, it facilitated the establishment of what Lewis Mumford identified as 'mechanical periodicity,' thus essentially increasing the distance between human beings and nature."

Willy Rordorf, in his excellent book *Sunday*, echoes that view when he says: "The fact that the Israelites in the early monarchical period already knew the seven-day week is all the more astonishing, as this institution runs contrary to every natural arrangement of time. Neither in the natural processes on earth, nor in the revolutions of the stars, is a periodicity of seven days to be found."

The discovery of seven-day rhythms of the body offers a solution to the puzzle of the origins of the calendar week. A case could be made for saying that the week is not dissociated from nature at all, as these authors suppose, but only from *cosmic* periodicities. The week is related to natural rhythms, but such rhythms are internal and biological, not external and physical. They are the unicorns of the body's time structure. Franz Halberg proposes that body rhythms of about seven days, far from being passively driven by the social cycle of the calendar week, are innate, autonomous, and perhaps the reason why the calendar week arose in the first place. In fact, the social week may entrain the biological week, adjusting it to precisely seven days, as the day-night cycle entrains body rhythms of about 24 hours, making them 24 hours exactly. But the social week does not explain those rhythms; rather, the rhythms explain the social week. The calendar week synchronizes and coordinates the work and leisure activities of human beings in a civilized society in much the same way as the biological week coordinates functions of the human body.

The living system organizes time on its own terms and to its own advantage. What time means in the biology of a species depends on the particular way in which that biology is organized in time. Homeostasis alters the meaning of environmental time by making it less intrusive, while biological clocks ensure that certain selected elements of world time structure, some more "worldly" than others, acquire special significance for the organism. The two regulatory systems, one imposing sameness in time, the other providing orderly change, are complementary rather than being in conflict. A body function alters in a rhythmic fashion, and homeostasis stabilizes the altered state of that function. The clocks are able to generate regular periodic variations because homeostasis resists random, irrelevant variations. Both systems collaborate in maintaining the special time structure of the body rather than simply surrendering to the time structure of the environment. The meaning of biological time, therefore, is in part a construct, an evolutionary strategy, and since strategies vary from species to species according to circumstances and the logic of a particular way of life, so too does the biological meaning of time.

PART TWO

RHYTHMS ANCIENT AND MODERN

FOUR

A Temporal Identity

We have come a long way from the sort of thinking which supposes that time is measured by oscillations of heavenly bodies, while life simply abides by the tick of those cosmic clocks. Inner time structure, in certain of its manifestations, seems to determine outer time structure, rather than the other way round. Rhythms of about seven days arose in living creatures millions of years before the calendar week was invented, and may conceivably be the reason why it was invented. The existence of such elaborate and highly differentiated temporal structure, with more structure still that escapes detection until researchers are sufficiently ingenious to coax it from its hiding places, means that each living species, and perhaps each individual within a species, has its own unique and complex temporal relationship with the environment. There are many different rhythms in the body, at every level of organization, and they are highly interrelated, affecting and being affected, modulating one another in a dynamic web of dependencies and independencies. So many different components, so intricately connected, compose the network of rhythms that it is sometimes difficult to know whether a biological rhythm is influenced by another rhythm in the body or by a cycle in the external environment. Thus it is an extremely risky undertaking to say whether the explanation of what an organism does at a certain time is internal or external; it may be a subtle mixture of the two. What goes into a structure from the outside in the form of information from the environment affects what comes out of the structure in the form of behavior, but the link between input and output may be highly indirect, "nonlinear," in the language of mathematics.

The systems scientist Gerald Weinberg has said that innate biological structure "stands between input and output, so that the system does not simply follow the environment's behavior. . . . A system's variables may well be one or two steps removed from the direct influence of the environment, protecting the system's identity." This concept of an "identity," intervening between what goes into an organism in the form of information, and what comes out in the form of adaptation, either to the outside world or to the organism itself, is an important one.

To say a living species has a temporal identity means that its body functions are organized in time in such a way as to serve the needs of the organism. Biological events happen according to a timetable that suits the biochemistry and the way of life of a particular plant, animal or person, and do not just "follow the environment's behavior." In fact, it is often more satisfactory to explain the "when" of behavior in terms of temporal identity than to explain it in terms of the environment, because the identity includes an individual's relationship with the environment. Structure, Gerald Weinberg points out, determines the meaning that the outside world has for the organism; the same external event can affect different individuals in different ways and affect the same individual differently at different times.

Common sense might suggest, for example, that the reason human temperature is high in the evening and low in the early morning is that a day of activity makes the body warm, and a night of lying relatively still in a dark room cools it down. Such an explanation is highly plausible, but it happens to be wrong. The apparently direct connection between activity as cause and body heat as effect is a bogus one. Waking does not warm the body up, and going to sleep does not cool it down. Temperature is time-dependent, under the control of biological clocks, and it rises and falls as these inner clocks dictate. Normally, the clocks are synchronized with the day-night cycle in such a way that a person is warmer during the day than at night, so as to be energized to meet the demands of the day. One is then in an appropriate relation with the world, and this, too, is part of temporal identity. But even when such synchrony between inner and outer is lost, the body's clocks tick on, though at their own "natural" rate, instead of at the rate of the cosmic day-night clock.

What is more, the exact timing of the rise and fall of temperature may vary from person to person. "Larks," people who feel most lively in the morning, have body temperature cycles that crest early in the day and show a large difference between the top and bottom of the cycle, whereas "owls," or evening people, peak later in the day. Larks tend to suffer more than owls from disturbed sleep if they work at night, and when they fly across several time zones, from east to west. The difference between the peaks of temperature in owls as compared with larks is only slightly more than an hour, but the contrast between the two types is made more marked by the fact that larks secrete a certain adrenal hormone called epinephrine in significantly greater quantities in the morning than owls do. Epinephrine is a potent stimulant. It increases the flow of nutrients and oxygen in the body and steps up production of glucose, mobilizing resources so as to function well under stress and in emergencies. Temporal identity also seems to be linked to personality. Extroverts are usually owls while introverts tend to be larks. It may be relevant here to note that star golfers turn out to be more "larkish" than star water polo players.

Another way in which individual time structure may differ from one person to another is the stiffness or looseness of daily rhythms, their changeability or constancy over an extended period of time. In Russia, researchers are trying to establish how variable human body rhythms are from day to day. People whose rhythms are relatively flexible have less trouble adjusting to unusual shift-work schedules, and are good candidates for space missions. Also, temporal identity changes with age. As a child grows up, the peaks of certain rhythms, especially that of a brain chemical called serotonin, shift drastically in relation to the peaks of other rhythms in the body. Middle age is a time of life when it becomes less easy to adjust to changes in the temporal patterns of the environment, the disruptions of jet travel and shift work; and adjustment is more and more difficult after the late forties. In old age, temporal identity alters again. The amplitudes of some rhythms shrink; they do not rise or fall as much during a single cycle, and sometimes rhythms disappear altogether. In humans this flattening out of cyclical ups and downs can be seen in potassium excretion, growth hormone and sex chemicals such as testosterone and luteinizing hormone. An exception to this trend is the rhythm of cortisol in the bloodstream, which tends to retain its amplitude throughout life.

The nature of temporal identity does not always allow us to say, even of an organism fully entrained to the 24-hour day-night cycle, that a particular body function will peak at a specific time of day or night. So interrelated are the parts of biological time structure that a given rhythm may have no absolute time of rise and fall, but only a time relative to one or more of the other rhythms in the body. This relativity principle of biological time was highlighted when Dr. Hugh Simpson, of Glasgow University in Scotland, compared the daily variations in the levels of hormones of the adrenal cortex in two groups of people: some Indians living in an equatorial rain forest and several medical students at Dr. Simpson's own university, in Glasgow. The puzzling fact emerged that the peak of hormone excretion took place between 9 A.M. and noon in the forest people and between noon and 3 P.M. in the medical students. How could this large difference be explained? Biological time seemed to bear no absolute relation to local time in these two cases. The mystery was solved when it was found that the forest people and the Glasgow students did not have the same sleeping and waking habits. Since the sleep-wake rhythm appears to be a fairly independent one in the human body's network of biological clocks, and the hormone changes a fairly dependent rhythm, the timing of hormone changes was synchronized with the sleep-wake cycle. The students and the Indians went to sleep and woke up at different local times, so the peaks of their hormone rhythms would also occur at different times. The absolute time of a hormone peak may vary according to the individual and his habits, but its relationship to the time of the sleep-wake cycle in humans appears to be universally the same, since the meaning for the organism of the increases in hormone levels is to prepare the body for waking activity, and would be meaningless if it were not linked to the sleep cycle.

The principle of the relativity of body time is not just a philosophical abstraction. It has proved of great practical use in the treatment of illness. Today, implantable devices which time doses of drugs so that they enter a patient's system at those hours of the day or night when they are most effective against a given disease are synchronized with hospital clocks; but a third generation of such devices will eventually time drug doses with reference to the "clock" of one of the patient's more independent rhythms, perhaps his temperature cycle, thus respecting his temporal identity. The marker rhythm may vary according to the nature of the disease.

The importance of biological time structure, modifying as it does many previous assumptions about the "why" and the "when" of events in the body, has profound consequences for medicine. In the days when homeostasis was sovereign, and rhythmic change was ignored, physicians assumed that it makes no difference what time of day a drug is given to a patient. The chief consideration was the convenience of the hospital staff. The effect of the drug was thought to be the same whether given at morning, noon or night. Chronobiology operates on quite another set of assumptions. It considers the timing of a drug dose as being of great significance, as important as the amount of the dose, because of the fundamentally different view it takes of biological cause and effect.

A drug is an external event, the "cause" of some change, one hopes a change for the better, in diseased tissues. But we know now that the same drug may have different effects on the body depending on what time it is by the body's clocks. The biological "meaning" of a drug, therefore, may alter as body time alters. Biological time structure, "temporal identity," stands between input, which is the drug, and output, the body's response. In some cases, time of day can mean the difference between life and death. Another complication is that a drug itself can act as an external synchronizer, a time cue, affecting the timing and amplitude of body rhythms.

At the Masonic Cancer Center in Minneapolis, scientists have learned that giving the widely prescribed anticancer drug adriamycin at six o'clock in the morning results in fewer harmful side-effects than when the same drug is taken in the evening. Almost the reverse is true of Cisplatin, another popular cancer-fighting agent, which should be given in the late afternoon, when the body's potassium excretion is at its highest level. Cisplatin is a powerful and effective platinum substance which kills cancer cells, but may also destroy or severely injure cells of healthy tissues. If the drug is given on a schedule that ignores biological time, the patient may suffer irreversible loss of as much as a third of his kidney as the price of the remission of the disease. By means of suitable timing, however, the toxic effects of Cisplatin have been avoided almost entirely, and adriamycin patients with bone marrow cancer recover as much as a week earlier than they would if the drug were given at arbitrary times.

During the 1960s, when chemotherapy for cancers such as those of the gut and bone marrow was being developed, researchers found that there are clocked rhythms of cell division, with a period of 24 hours, among others. In a large number of tissues in the human body, cells divide on roughly the same 24-hour schedule. This has been a discovery of great importance for chemotherapy.

The cycle of the cell, the basic unit of organization of the body, is one of the core processes of the living system. In fact, the cell cycle is itself a sort of clock, a programmed sequence of events in which time is of the essence. In order for the body to grow and maintain itself, new cells must be produced, and this is done by existing cells dividing into two. Some specialized cells never divide, but a tremendous number do, and they go through a cycle in which there are intervals of apparent rest and also a period of division, called mitosis. But cells do not divide at the same rate at all times of day or at all months of the year. Human skin cells, for example, divide mostly between midnight and 4 A.M., during sleep.

Within each cell there is a store of information, a message, in the form of DNA, which contains coded instructions for making the key molecules of life. Before the cell divides, forming two daughter cells, it must make a copy of the DNA message, so that each daughter cell is provided with a complete text. There are four main phases in the cell cycle. First is a quiescent stage, in which no DNA is produced. During this time, enzymes needed for copying the DNA message are formed. Next comes the DNA synthesis stage, in which an exact new copy of the text is produced. The cell now contains precisely twice the original amount of DNA. In another apparently quiet period, synthesis of DNA stops, while new proteins are made which provide the apparatus needed for cell division. Finally the cell splits into two, producing a generation of daughter cells.

All cells must go through the stage of DNA synthesis, healthy and deranged ones alike. That is why anticancer agents that act by preventing DNA synthesis are so toxic. They kill almost as many normal cells as cancer cells. Leukemia, for instance, is a cancer of the bone marrow. Bone marrow produces blood, and it is also part of the immune system. If bone marrow cells are killed, the patient may suffer serious illness from infections which otherwise would be rendered harmless, because his immunity is impaired. Cancer patients can die even while the drug itself is working.

In plants, animals and people, cell division is rhythmic. Most cells are in the division stage at particular times of the day, in a sort of group synchrony, and since the cell cycle is an orderly sequence of events, there are also specific times of day when the same cells are at the stage of DNA synthesis. There are waves of division and waves of DNA synthesis, and the crests of the waves are usually separated by a few hours. The peak of synthesis in animals occurs somewhere between the middle and end of the night. An anticancer agent such as ara-C works by blocking DNA synthesis, and is therefore most effective in destroying cells when a large number of the cells are at the peak of the synthesis stage. Administration of ara-C can be timed to coincide with the crest of the synthesis wave, the trough, or anywhere in between the two.

Luckily for the progress of cancer treatment, the temporal behavior of cells in most cancer tumors is often highly erratic and unpredictable. The cell cycle tends to escape from the control of the biological clock which drives the cycle in healthy tissues, at least when it is in the stage of exponential growth. Group synchrony is lost. Time structure is lost. Even if a rhythm does persist in cancerous tissue, its peak may come at a different time from the peak of the rhythm in normal cells. Such a separation, or asymmetry, of the two types of cell cycle behavior is a boon to medicine, because it enables physicians to protect normal cells from the toxic effects of an anticancer agent by giving it at times when large numbers of such cells are at a stage of the cycle when the drug has little effect. If there is no group rhythm of synthesis and division among diseased cells, there is no best time to administer the drug. One hour of the day or night is just as good as any other hour. The number of sick cells that are at a particular state of the cycle is random and hence unpredictable, so that the time of administration might as well be random. Ara-C, therefore, can be taken when the wave of synthesis in healthy tissue is not at its crest, and still kill roughly the same amount of cancerous tissue as it would if given earlier or later in the 24-hour cycle. Time is unimportant in the work of killing sick cells, but of the utmost importance in sparing healthy cells. This procedure is known as shielding the patient in time.

If sick cells do show a daily rhythm of synthesis, but one that is desynchronized from the rhythm of healthy cells, a drug can be given at the peak of the sick rhythm, secure in the knowledge that the peak of the healthy rhythm comes at a different time of day. In the future, scientists may be able to manipulate the synthesis rhythm of sick cells with artificial synchronizers, so that it peaks at a time when the normal cells are at their least susceptible to the toxic effects of drugs or radiation therapy.

The toxic effects of at least eleven widely used anticancer drugs have been found to depend on the hour at which they are given, and for each particular drug there may be a best time for administration, probably because different drugs are effective at different stages of the cell cycle. Timing may also be the answer to better results in transplant surgery, since cells of the immune system responsible for rejecting foreign materials in the body appear to behave in a rhythmic fashion across the 24-hour span. The transplant is best carried out when the rhythm is at its nadir, with a diminished ability to start the rejection process.

Since a disease is often rhythmic, drugs can be taken in a rhythmic manner, to synchronize with the illness. People suffering from allergies are most sensitive to substances such as pollen and house dust at night and in the early morning. These are also the times when asthma is apt to strike, a fact physicians have known for more than a thousand years, although in the seventeenth century, when the idea of time as cause in the body was unknown, it was thought that the reason for nocturnal asthma attacks was that the body overheated at night. We know today that asthma attacks are due in part to the narrowing of the tubes leading air in and out of the lungs, and that there is a 24-hour rhythm in the width of the tubes, so that they are at their widest between 4 P.M. and 6 P.M., and narrowest at the time of waking up in the morning.

Just as there are rhythms with daily or longer periods in the response of the body to powerful drugs and radiation therapy, so, too, food is handled well at certain hours of the day and not so well at others. When we eat, as well as what we eat, has important consequences for health. A single meal of 2,000 calories, taken every morning for a week, will result in loss of weight. But exactly the same meal, eaten every evening for a week, is likely to cause a gain in weight, because carbohydrates are burned more rapidly in the morning than they are in the evening. A calorie seems to be used in a different way by the body at different times of day. At a certain hour a calorie may be used to maintain body weight, but not at other times. "A calorie is not the same calorie at breakfast as it is at dinner," Dr. Franz Halberg finds.

The body is able to "predict" certain events in the outside world because the environment is partly rhythmic and contains regular, repeated cycles that are stable in time. That confers an evolutionary advantage, since to be one jump ahead of anything that presents opportunities as well as dangers is the key to survival and an effective way of life. The eighteenth-century philosopher David Hume declared that we cannot know whether the sun will rise tomorrow morning, no matter how many times it has risen on previous mornings. The body, however, may be wiser than Hume. Biological clock systems in living organisms treat tomorrow's sunrise as something to rely upon, and arrange their inner timetables accordingly. To oversimplify, clocklike mechanisms that prepare the body for activity are set to reach the peak of their cycle in time for waking, and rhythms that shut the body down for inactivity are coordinated to peak in time with sleep. Similarly, physicians can predict responses within the organism, because the organism is rhythmic, and while it may change from one hour to the next, it is possible to know in advance when and how it will change. Thus the strategy of modern medicine is analogous to the strategy of the living system, which is to act at the most opportune moment, relative to the rhythms of nature, in this case human nature. Physicians adapt to the temporal structure of the body as the body adapts to the temporal structure of the world. That is possible because time has descended from Plato's heaven and Einstein's universe into life itself, and must therefore be part of any attempt to explain why life is the way it is and why living creatures behave in the way they do.

FIVE

The Master Clock Revisited

The notion of a single, sovereign master clock had considerable appeal. It survived in various forms for centuries. Plato's celestial timepiece, the sphere of the fixed stars, aloof and aristocratic, representing time and the heavens as a unity, influenced the ways in which people thought about time until Copernicus dealt the concept a knockout blow with his suggestion of a universe of which the earth is not the center.

Until quite recently, biologists entertained the idea that the timing of many functions in the body was under the control of one master clock in the brain, which drove the lesser clocks, making them run to its own measure. For many years the location of this supreme chronometer remained a mystery, and when in 1972 an important pacemaker, which seemed to drive several biological rhythms, was found in a definite region of the brain, many scientists exclaimed, in effect: "Aha, the master clock!"

The new clock, discovered independently by two teams of researchers, was in the hypothalamus, an important control center which excels at coordinating bodily processes that have some independence from the outside world. The hypothalamus is in the middle of the brain, just above the roof of the mouth in humans, and has a dominant, central status as a regulator of the body. It occupies less than 1 percent of the total volume of the brain, but the importance of the hypothalamus far exceeds its size; it has star quality, and is often called the brain center. It is the headquarters of homeostasis, adjusting body temperature, blood pressure, pulse rate and other elements of the internal milieu. The hypothalamus was known to play a leading role in maintaining the sameness of the milieu. Now, all at once, it was shown to be involved in regulating, not just the constancy of bodily states, but their rhythmic, predictable time-dependent differences. The hypothalamus seemed to contain the engine of temporal identity.

If this was the master clock so keenly awaited, it was not much to look at. The clock was located in the front part of the hypothalamus, where specialized groups of nerve cells, called collectively the suprachiasmatic nuclei, or SCN for short, show sustained oscillations. The SCN, a pair of small, egg-shaped clusters of nerve cell bodies, are independent of each other at birth, but become coupled during the period of growing up. They are linked by a tract to the retina of the eye, which picks up light and dark signals from the environment, signals that for the SCN are really messages about local time.

To be a true master clock, the SCN would have to satisfy certain strict criteria. A master clock must be self-sustaining, like a watch that winds itself, ticking on whether or not it receives regular time cues from the outside world. Also, it must be a generator of all the rhythms in the body, which would cease to be rhythmic if the master clock were to stop. At first, it did seem as if the clock in the SCN met those requirements. The two newsmaking experiments in 1972 found that rhythms of hormone secretion from the adrenal cortex and rhythms of drinking and moving around in rats simply vanished when the SCN were removed.

It was not very long, however, before doubts began to arise. For the very reason that certain eminences in the community of biologists were so sure they were right, and so harsh in dealing with doubters, some scientists began to wonder if anything in biology could be as simple or as certain as all that. One of the doubters was Lawrence Scheving, a professor of anatomy at the University of Arkansas Medical School in Little Rock. "All of a sudden, it became dogma overnight that this was the master clock," Scheving remembers. "I started looking at rhythms of body temperature, and I wasn't a bit convinced that they were eliminated if the brain clock is put out of action. To say so at that time, however, was just anathema to the biological establishment. It was assumed that if the rhythms didn't disappear, you hadn't done a proper job of removing the SCN."

Scheving went to see a colleague at Arkansas named Ernest Powell, who was a specialist in the structure of the central nervous system. Powell, too, was dubious about the new doctrine of the supremacy of the master clock. It flew in the face of his whole way of thinking about the brain. In the 1960s, Powell had investigated the nervous system of cats, and he came to the conclusion that the brain is not something in which each part has a particular function separate from other parts, or that a circuit is switched on only when needed, being switched off when its work is done. The focus of the nervous system, Powell thought, is more like a Mexican jumping bean, dynamic and always on the go. What happens in one part of the brain has an effect on the rest of the system. His ideas fit well with the metaphor of the brain as a hologram, put forward by the California neurophysiologist Karl Pribram. Pribram's metaphor is strongly holistic, because a hologram, a sort of photograph which stores information in the form of wave patterns, rather than images, possesses a curious property: if a piece of the hologram is torn off, the piece remaining can still be made to generate a complete, three-dimensional picture of the object photographed, albeit a rather fuzzy one. Pribram saw an analogy between holograms and human memory, where the retrieval of part of a past experience may summon up the entire experience. To Ernest Powell, it made no sense to suppose that removing the SCN, two little clusters of nuclei in a small cranny of the brain, would cause all rhythmicity in the body to disappear. He thought it unlikely that one part of the central nervous system could have so much power and be so indispensable. Eliminating the SCN ought to be something like breaking off a piece of a hologram; the part that was left could still produce roughly the same result as the whole, even if the quality was slightly impaired. "I could not accept that one single group of nuclei was the whole story," Powell said. "My feeling is that the SCN are important in mediating the messages about time that arrive at the eyes from the day-night cycle. They seem to be a bridge between outer time and inner time. But other structures in the central nervous system are at least as important as the SCN. I see the nervous system as a conglomerate of clocks."

Scheving asked Powell if he would take part in an experiment to test the master clock hypothesis. He knew that the man who had the best system in the world for monitoring temperature rhythms was Franz Halberg in Minnesota, and he arranged for Halberg to monitor the rhythms, while Powell did the work of removing the SCN.

What the team found was a severe blow to the ruling wisdom. Computer analysis at Minnesota revealed that rhythms were not obliterated. Putting the SCN out of action did not dissolve time structure. When observations were made over a sufficient span of time, they showed that the rhythms remained, though in an altered form. The rhythms did not rise and fall as much as when the SCN were left intact, and the peaks occurred at a different time. In challenging the master clock dogma and asserting the existence of independent rhythms, Dr. Halberg quoted the words used by Galileo in 1633 after he was ordered by the Roman Inquisition to abandon his heretical belief that the earth moves around the sun: stamping his foot on the ground, Galileo

looked up at the heavens and declared, *"Eppur si muove—*
And yet it moves."* Without the SCN, rhythms of the body
continue to move. "The whole world thinks the SCN are a
generator," Halberg says, putting the nuclei firmly in their
place. "They are not. They are a fine-tuning device."

Later, Scheving carried out various studies of rhythms in
the gut in the absence of the brain's so-called master clock.
"They were altered there too," he said. "But they did not go
away." A number of other researchers examined various
body systems and obtained similar results. The rhythms
might disappear for a while, perhaps for as long as several
weeks, but they came back. In certain cases, removing the
SCN caused rhythms to separate into multiple components,
so that unity of biological time was lost. It was clear that the
SCN played an important role as an organizer and modifier
of body rhythms, perhaps synchronizing many of them with
the light-dark cycle, but they could not be said to be solely
responsible for rhythmicity itself. That took some of the
glamour away from the SCN. Martin Moore-Ede of Harvard,
a more conservative thinker on the subject of this brain clock
than Franz Halberg, believes now that it is an open question
whether the SCN act as a true pacemaker or merely as a sort
of way station, a mediator between the light-dark signals
from the outside world and various internal biological clocks,
perhaps synchronizing them, creating temporal unity in the
body.

In the 1980s, the trend of new research has been to de-
mocratize the time structure of the body, to show it as a collaborative enterprise, in which there are many different
players interacting in complex ways, rather than being under
the command of a single, supreme controller. Plato's master
clock ensured simplicity, unity and permanence, reflecting
the rational intelligence of a creator. By contrast, biological
time is not simple or immutable, nor is it always and in every
circumstance a unity, and it reflects the opportunism and
adaptability of the evolutionary process. The SCN are not *the*
master clock; they may be *a* master clock, one of several, but
the altered view of their role means that "master" must be

defined in a different way. The Copernican dilemma makes its entrance once again.

As a way of making sense of the diversity now known to exist in biological time structure, it has been suggested that there are two major pacemaker clocks in the brain, each in charge of different sets of biological rhythms. The two clocks are called "X" and "Y." The Y pacemaker is in the SCN and is associated with cycles of rest and activity, changes in skin temperature (as opposed to those of deep body temperature), rhythms of calcium excretion, and the pituitary gland's secretion of growth hormone, which stimulates metabolism and growth, acting directly on the cells of the body. Growth hormone surges strongly about an hour after sleep begins, and there are other peaks during the day that appear to be random. The X clock, whose whereabouts has not been established yet, is supposed to control the daily rise and fall of deep body temperature, the timing of that stage of sleep in which the most vivid dreams occur, the fluctuating amounts of cortisol in the bloodstream and the excretion of potassium. These two proposed clocks normally tick together in synchrony, being connected by some kind of coupling device which enables information about time to pass between them.

A mathematical model which is able to account for many oddities and anomalies in body time structure finds that pacemaker X is not in touch directly with external time signals of daylight and darkness. These signals reset biological clocks every day, synchronizing them with local time, so that rhythms peak at appropriate hours of the day or night. Instead, X is entrained by pacemaker Y, which receives light-dark signals at first hand. Of the two pacemakers, if they can be called that, X is thought to be the stronger and more stable. In humans, though apparently not in other species, the asymmetry between the strong and weak pacemakers is considerable. The strong X exerts about four times as much influence on the weak Y as Y does on X. When the body is cut off from all external signals about time, so that the system is out of step with the 24-hour day-night cycle, both the strong and the weak pacemakers free-run at first, with periods of about

25 hours. If temporal isolation continues for several weeks, the two pacemakers uncouple, and are no longer in communication. Each goes its own way. What happens then is that the weak pacemaker gradually lengthens its period, but the strong one remains close to 25 hours even after many months of isolation.

This model explains why the body rhythms of an airline passenger who crosses several time zones adjust to the new local time of his destination at different rates. Clock Y, being less "stiff" and more plastic than the stronger clock X, recovers more rapidly from the disruption of air travel, usually adapting to the altered day-night cycle within two days, whereas the more stubborn X clock takes several days to catch up. The delay between the two creates the effect we all know as jet-lag. When body time is disrupted by a change in shift-work schedules, rhythms of digestion realign themselves with the new sleep-wake cycle relatively quickly, while the temperature rhythm falls into place more slowly, and rhythms of cell division even more slowly than temperature. The rhythms that adjust late are less easy to manipulate.

When the coupling between the two clocks is in the process of breaking down, the clocks affect each other in a curious fashion. The weak clock tends to slow the strong one down because of its longer period, and the strong one tends to speed the weak one up, because of its shorter period. The result is that for a few cycles their periods match, then diverge, then match again.

The breaking away of X from Y in conditions of isolation was vividly illustrated in the case of Michel Siffre, a French geologist who stayed 100 feet down in the aptly named Midnight Cave, near Del Rio, Texas, all by himself for six months in 1972, the year the SCN clock was identified. Siffre was cut off completely from all sources of information about local time. He slept in a nylon tent and wore electrodes which monitored his heart, blood pressure, brain and muscle activity, as well as the extent and structure of his sleep, data that were transmitted to scientists in tents aboveground. Siffre was linked to the surface by telephone, but his conversations were kept brief and curt, to avoid any remark that might give a clue as to time. He carefully preserved the whiskers he shaved from his face, so that they could be weighed, since beard growth is supposed to reflect the fluctuating activity of certain hormones in the bloodstream. The cave was stocked with frozen food, the same as that given to the Apollo 16 astronauts. To entertain himself, Siffre had a record player which broke down and books which soon became mildewy and unpleasant to handle.

Siffre went down into Midnight Cave on February 14. By the 21th day, unknown to him, he was living on a 26-hour biological schedule instead of a 24-hour day. His 63rd biological cycle, on April 30, actually coincided with the 77th day of his ordeal. When Siffre finally climbed out of the cave to freedom on September 5, the record of his Y pacemaker, controlling the cycles of activity and rest, showed that it free-ran at periods of between 25 and 32 hours, interspersed on two occasions with doubled cycles 45 to 50 hours in length. The weaker Y clock had sometimes roughly matched the period of the stronger X, which remained close to 25 hours, but also broke away, compensating by going through much longer cycles. The average Y cycle was 28 hours. Yet in spite of the changes in the length of Siffre's rest-activity rhythm, each "day" had seemed like a normal one to him. Only when he came to the surface did he learn how his days had varied.

The experience took its toll on Siffre. On April 30, the 77th day of his stay, he noticed a certain fragility of memory, but was otherwise in excellent form. On the 79th day, he began to crack up. Overcome by despair and self-pity he shouted into his telephone: *"J'en ai marre!*—I've had enough!" The experiment proceeded just the same. On the 80th day, he planned to commit suicide, making it look like an accident, enabling his wife to collect the insurance. The plan was abandoned. On the 94th day he disconnected the electrodes and the deep body temperature probe. "I am living through the nadir of my life," he wrote in his journal. Then guilt assailed him and he reconnected the monitors.

Long after he emerged from the cave and returned to a normal way of life, he said: "I still suffer severe lapses of memory. My eyesight has weakened, and I am the victim of a chronic squint. I suffer psychological wounds that I do not understand."

Certainly boredom, loneliness, discomfort and general misery contributed to Siffre's lasting hurts, but probably desynchronization of his body rhythms also played a part.

The X and Y model, while it has considerable explanatory power, does not satisfy everyone. Some practical biologists, those who "get their hands wet," in the jargon of the profession, find it too abstract and too limited. They would rather talk about actual biological entities, those they can see and touch. Franz Halberg refers to the X and Y model as "a scaffolding, not a building. And in Minnesota in the winter you can get very cold working in a scaffolding." There may be more than two pacemaker clocks in the brain, perhaps many of them. The total number is quite unknown today. The SCN might be just one element in the network of elements in the hypothalamus, a small cog in a large machine. Knocking out one cog does not necessarily stop the entire machine from running. Chronobiologist John Pauly says: "I think our bodies operate with a host of clocks. There may be hundreds or millions of them. Some rhythms are totally autonomous, and others are more dependent."

Another blow to the concept of the master clock is the fact that certain body systems, when removed and kept in a dish in the laboratory, still show daily rhythms. They generate their own periodicity, which is one characteristic of a master clock. One such system is the adrenal glands. The adrenals consist of two glands, one located over each kidney. Each gland secretes two different types of hormones, one type from the medulla, or core of the glands, and another type from the cortex, or outer layer. One of the most important hormones produced by the human adrenal cortex is cortisol, a potent chemical with a wide range of effects in the body. Cortisol helps a person cope with stress, especially stress that goes on for an extended period of time. Mental conflict and frustration, as well as annoyances that are hard to ignore, such as noise, cause levels of cortisol in the blood to rise. Air traffic controllers tend to secrete more cortisol than workers in other occupations. A study of people living near the Three Mile Island nuclear power plant in Pennsylvania found that their cortisol was higher than normal, even though the accident at the reactor had taken place 17 months before the survey was made. An increase in the amount of hormones secreted by the adrenal cortex is not a sign that one specific emotion or mood predominates, but reflects a general state of arousal in readiness for coming to grips with challenge, harm or threat.

Quito aside from the vagaries of external events, however, the secretion of cortisol varies according to an internal logic of the body, in a regular, clocked fashion, displaying a powerful rhythm. Cortisol starts to rise about four or five hours before the end of sleep, cresting near the time of awakening, and sinking during the day to reach low levels at bedtime. Under normal circumstances, the rhythm is tightly synchronized with the day-night and sleep-wake cycles. The cortisol rhythm is a robust and stable one. In fact, the adrenal hormone system is so important to the fitness of animals, in the Darwinian sense, that it remains strong even under conditions so adverse that the estrous cycle, the female rhythm of sexual receptiveness to the male, shuts down. This suggests that the adrenals, which are critical for the survival of the individual, are given a higher priority in evolution than sexual functions, which are basic to the propagation of the species, a new insight into the concept of "fitness." So potent is the cortisol rhythm that it appears to drive another body rhythm, that of the changing numbers of eosinophils, a type of infection-fighting white cell in the blood, which rise at the time of day when cortisol falls and fall at the time when cortisol rises. In this local context, the adrenals could be described as a sort of "master clock" for the eosinophil cycle. They also have a marked effect on the group rhythm of cell division and DNA synthesis, on fluctuations in enzyme levels and on the quantity of charged metal particles in the urine.

Sections of human adrenal glands have been placed in a

nutrient fluid in the laboratory, where they still secrete cortisol in a periodic fashion, though with a less stable and weaker rhythm. The clock that generates the rhythm appears to reside in the gland itself. This discovery places the adrenal system in a wholly new light, because it has long been thought of as a simple feedback arrangement, in which the hypothalamus sends a chemical message to the pituitary gland, and the pituitary tells the adrenal cortex when and when not to secrete hormones; the pituitary orders the adrenals to slow down when hormone levels in the blood are too high, and to speed up when they are too low. But as research into biological clocks advances, time enters the system with a vengeance. Each step in the "feedback" chain is rhythmic. The adrenals on their own are periodic and secrete cortisol. What is more, the chemical substance made by the hypothalamus, the first link in the chain, also has a daily rhythm, and this is under control of a clock in the nervous system. The hormone signal put out by the pituitary is likewise periodic, and both of these rhythms are superimposed on that of the adrenals. All three vary in their response to chemical messages, depending on the time of day. Meanwhile, the pineal gland, whose importance in biological time structure is only just beginning to be appreciated, influences the amplitude of the cortisol rhythm and may shift the time of the rhythm's peak. The pineal, which has its own self-winding clock, appears to increase the strength of the cortisol rhythm at a time of day when the adrenals are especially responsive to the signal from the pituitary, and to reduce the strength at a time when the adrenals are relatively unresponsive. Franz Halberg calls this a "feed-sidewards," as opposed to a feedback, effect. It is a mechanism that alters, in a periodic fashion, the nature and extent of an interaction between the adrenals and the pituitary, which are themselves periodic. Of such sophisticated networks of mobile relationships is the fabric of biological time woven.

In the body there are many different clocks, but not one that is all-powerful, and not one that by itself can be used to explain the behavior of all the rest.

Certainly, the clock in the SCN is a very distinguished one. It is the first to receive information about daylight and darkness from the environment. Light signals arrive at the retina of the eye, and are encoded and sent along a specific tract to the SCN, which passes the message on to other clock systems in the body, synchronizing them with each other and with the day-night cycle. There are connections that link the SCN to the pituitary, the pineal and a number of parts of the hypothalamus and the central nervous system. Another feature of the SCN clock is that it has a more stable period from peak to peak of the cycle than that of the adrenal clock, even when it has no access to external time cues, and is more consistent in the timing of its peaks and troughs. Secondary clocks usually have a period which differs from that of a primary pacemaker. For that reason, the adrenals may not qualify as the site of a major pacemaker clock.

This intrinsic, natural period is symbolized by the Greek letter tau. A tau is innate, "hard-wired," in the language of computers. It is not imprinted from outside, nor can it be acquired by learning or experience. A tau is specified by instructions encoded in the genes, and it can be altered by changing a single gene. Each tau is peculiar to its owner, like a fingerprint. It is similar among members of the same species, but there are slight differences from one individual to another. The tau fluctuates around the species average.

Entrainment by a *zeitgeber*, or "time-giver," in the environment, such as the cycle of daylight and darkness, shortens or lengthens a tau, so that it becomes exactly the same as the period of the external cycle. In animals active during the day, the tau is usually longer than 24 hours and needs to be shortened, generally in the morning, whereas in nocturnal species it is typically less than 24 hours and needs to be extended, usually in the evening. The human tau is roughly 25 hours, as we have noted, so that our pacemaker clocks need to be advanced by one hour every day, but the precise length of the tau may vary from one person to another. "Owls," who tend to be extroverts, have longer taus than "larks," who are often introverted. Even when the body's clocks are entrained to an external cycle, the match between inner time and outer time may not be exactly the same for each individual. In the case of a very short tau, the *zeitgeber* may set it back too much, upsetting the circadian system. All humans whose taus are longer than 24 hours, however, adapt more readily to circumstances that lengthen the day-night cycle than to circumstances that shorten it. Flying from New York to London is more difficult to adjust to than flying from Los Angeles to Tokyo, because one shortens the day-night cycle for the traveler, who must set his watch back, while the other lengthens the cycle, so that we are flying toward our natural rhythm, with its tau of longer than 24 hours, making adjustment easier. The human system is already sliding in the direction of a longer day, as part of its temporal identity, and when a day really is longer than usual, the stress on the body is not as great as when it is shorter. In theory, it may be more comfortable for the circadian system when we set our watches back in the fall, giving ourselves an extra hour in bed, than when we set them forward in spring, eliminating an hour. The change to daylight savings time is like jet-lag in miniature. A survey taken in California showed that a small number of people, between four and five percent of the population, experience high levels of stress after the change to daylight savings time. Owls, with their longer taus, are especially suited to a change that delays the circadian clocks. On shift work, extroverts adjust to a new shift schedule that lengthens the day about 50 percent faster than to one that shortens the day.

The hallmark of the SCN clock is its reliability. Its role in the circadian system, therefore, may not be to generate rhythms, but to coordinate and synchronize them, keeping less stable clocks from drifting out of step. It is more like the conductor of an orchestra than the orchestra itself. In its absence, rhythms do not cease, but they are not held together in a single framework. Temporal relationships tend to become chaotic. An orchestra can still make music without a conductor, if need be, but the performance is likely to be ragged and less tightly organized.

It would seem, however, that the orchestra of biological time, which is made up of a plurality of players, has more than one conductor. The clock in the SCN may not even be unique as a receiver of time cues from the outside world. Daylight and darkness are not the only source of information about local time for living organisms. Other time cues, some strong, some weak, are capable of entraining body clocks, and they need not involve the SCN. Nor do they all have the same effect on the circadian system. No one clock in the body is able to respond to every sort of time cue the environment may send. One type of clock may be receptive to a particular type of time cue, which it is specially tuned in to, and be indifferent to other cues.

Sometimes *zeitgebers* change, suddenly and drastically, and they may not all change at once or to the same extent. That spells trouble for the circadian system, because of the varying responses of its various parts to specific *zeitgebers*. In the case of jet travel, all the *zeitgebers* do change simultaneously. Cycles of daylight and darkness, noise and quiet, work and rest, mealtimes and social routines, are shifted with respect to the traveler's own body clocks, so that he feels sleepy when everyone else is lively, and up when others are down. But these cycles are all shifted by the same amount, making it relatively easy to adjust to the new schedule. That is not true for shift workers, however. When a factory worker rotates to a different shift, his sleep-wake cycle changes, but the other *zeitgebers* in his environment, the cycles of light and dark, noise and quiet, social rhythms, remain the same. That may be one reason why shift work is so stressful, and the quality of sleep so poor. The change to daylight saving time is a milder example of this phenomenon. Here, the day-night cycle remains as it was, but the others are advanced by one hour. Probably body clocks respond differently to each of the *zeitgebers,* and the effect of one *zeitgeber* even on a single clock is not the same as the effect of all of them together.

The ability to entrain to more than one time cue is not confined to human beings. Even such lowly species as cockroaches and crickets can be entrained by cycles of external temperature as well as by those of daylight and darkness. Day is usually warmer than night, so that the rise and fall of temperature acts as a sort of environmental clock, albeit a noisy and unreliable one. It sometimes happens, however, that night is warmer than the day, in which case the temperature cycle gives misleading information about local time. The cockroach, a resourceful creature, follows the temperature cycle as long as it is reliable as a clock, but when it ceases to be reliable the animal switches and follows the light-dark cycle instead. When the situation returns to normal, and nights are cooler than days, the cockroach's clocks go back to following the temperature cycle. If the peak time of an environmental cycle conflicts too much with the peak time of a circadian rhythm in the body, a different cycle may begin to exert a stronger effect on the animal.

Meal schedules are an important time cue for several animal species, and can act as an environmental synchronizer for circadian rhythms. There may be some ingredient in the diet that entrains a specific pacemaker in the brain, other than the SCN. Animals stay entrained to cycles of food and no food even when the SCN is out of action. They show a self-sustained, circadian rhythm of feeding when food is made available only at certain times of the day. Even if an animal has not eaten anything for several days, it can predict the hour when food is normally obtained, by means of its inner timekeeping system. The brain clock that tells the animal when to expect food is not entrained by the light-dark cycle, but by the meal schedule itself. When that schedule is changed, either by shortening its period or by altering the intervals between meals but keeping the period the same, some of the body's circadian rhythms synchronize with the new food cycle, while others synchronize with the light-dark cycle. Presumably there is a separate pacemaker, organizing rhythms of metabolism in the body, which is receptive to food time cues but insensitive to light cues. Exactly where in the brain this clock resides is not known, but there are some hints that it may be in the ventromedial nuclei of the hypothalamus, the VMH for short, which communicate with the SCN via several nerve pathways.

Lesser clocks, of a lower order than these brain pacemakers, may also be able to entrain to the meal cycle. Cells in the liver and in the intestine, where food is processed, contain self-winding clocks which generate circadian rhythms even when removed and placed in fluid in the laboratory. As Martin Moore-Ede points out, the times of day and night when food is available in the wild may vary greatly, so it is entirely plausible that a separate food pacemaker or pacemakers would arise in evolution, and be reset by a change of the meal schedule, leaving the day-night pacemaker unaffected. If there is a regular, predictable cycle of food availability, it will entrain the food clocks, and if the cycle changes, but settles down to a new, regular rhythm, the clocks synchronize with the altered regimen. It seems that these clocks are definitely specialized for food cues, because they do not respond to drinking cycles, no matter how predictable the cycles may be.

Human circadian rhythms can be manipulated by changing the meal schedule. Different hormone rhythms react to such a change in different ways, so that the peak time of one rhythm is shifted more than the peak time of another, causing the rhythms to separate. If a person is given only one meal a day and that meal is breakfast, the circadian rhythm of insulin peaks at about 10 A.M., whereas that of glucagon, a hormone that increases the amount of glucose in the blood, peaks at about 11 A.M. If the one meal is dinner, insulin crests at about 6 P.M. and glucagon at about 11 P.M. The shift in the peak of blood iron is even more dramatic, from just before breakfast on the first schedule to just before dinner in the second. By contrast, the rhythms of body temperature and cortisol are affected only slightly.

In the future it might be possible to manipulate rhythms in cancer patients by altering mealtimes. The daily rhythm in the body's resistance to toxic drugs and the rhythm of sensitivity of cancerous tissues to a specific anticancer agent could be shifted and separated in this way, so that the drug can be given at a time of day when both rhythms are at or near their peaks.

An important time cue for humans is the rhythm of social contact, which can entrain body clocks. People living together synchronize not only the rhythms of sleep and waking, which are under the control of the weak Y pacemaker, but also other rhythms associated with the strong X pacemaker, so that everyone's body time is in step with everyone else's. Oddly, there are small but significant differences between the period of rhythms in people living alone and the period of rhythms in people who maintain social contact with others. In one study, the average circadian period of solitaries was 24.87 hours, while that of people living in pairs was 25.21 hours. Also, there are huge differences in the amount of internal synchrony between singles and groups. Whereas the strong and weak pacemakers split apart in only 10 percent of solitaries, they separate in 40 percent of people living as couples. Young women who live together for some time develop synchrony in their menstrual cycles.

Light-dark cycles and the periodic rise and fall of surrounding temperature are the most powerful time cues entraining the biological clocks of insects, but for mammals, especially primates, regular social encounters may be more effective. The symptoms seen in people artificially isolated from *zeitgebers*, such as Michel Siffre, who became a temporal hermit to increase our knowledge of biological time structure, are similar to the symptoms of bereaved people, who have lost a husband or wife, or close friend. Myron Hofer thinks it possible that among the important regulators withdrawn by bereavement are the daily rituals of personal contact that served to entrain and synchronize the biological clocks of the survivor. The bereft person may be suffering from internal desynchronization of biological clocks. Some of the major, chronic effects of bereavement, such as poor appetite interrupted by sudden cravings for food, sleep disturbance, malaise, depression, impaired attention span, periodic fatigue, are also symptoms that appear after jet travel across time zones and as a result of a change in shift-work schedules, both of which disrupt the internal time relations of the body. Travelers who stay alone in their hotel rooms after arriving at their destination take longer to recover from jet-lag than people who go out and meet people in the streets, restaurants and offices. It may be true, also, that the elderly suffer disturbances of their body rhythms because they tend to be more isolated than younger, more socially active people.

Conscious awareness of time has some power to entrain human pacemakers, but this is not as simple as it might seem. One heroic volunteer who lived all by himself in a cave brought a wristwatch with him, intending to consult it so as to keep himself on a strict 24-hour schedule. As the days went by, however, the watch had less and less influence on the man's behavior. When it was time to get out of bed, he felt increasingly disinclined to rise. At bedtime by the watch, he experienced less and less sleepiness evening after evening. In the end he gave up the struggle and ignored the watch, letting his circadian clocks free-run with a period of about 24.7 hours. It seems that just knowing the time by itself is not enough to entrain a circadian pacemaker. There must be some sort of relationship with the outside world which makes it worthwhile and meaningful to entrain to the time of the outside world, and thus to counteract the built-in tendency of the circadian clock to run slow.

Within the body, messages about time travel from clock system to clock system, synchronizing them with one another and with the day-night or other cycles in the environment. The messages are carried by hormones or along pathways of the nervous system. Biologists talk about certain hormones as being "temporally active," which means that they communicate information about time from place to place in the body, so that each circadian rhythm "knows" what time it is by the world's clocks outside the body. Our old friend cortisol is a temporally active hormone. This internal network of circulating time messages, encoded and reencoded in more than one form, gives the circadian system its flexibility, and the system seems to be more flexible in humans than it is in other species, although humans are extremely rhythmic in certain respects. The "wiring" that connects clock with clock is less reliable in its performance, and less uniform from one person to another, than the clocks themselves, whose properties are remarkably similar in all living species. The special features, the surprises, the evolutionary innovations that appear in biological time structure tend to occur in the mediating pathways, the temporal communications network inside the body, which provides internal synchronization for the system as a whole. It may be that humans are uniquely prone to illnesses of mood and emotional state because such illnesses result from breakdown or disturbances in the wiring rather than in the clocks. The relationship between one rhythm and another may go askew, while the clocks themselves function as they are supposed to do.

The strength of the links that connect body clocks depends in part on the way of life of particular species. In some animals, two pacemakers are coupled tightly, almost as if two pendulums in a grandfather clock were tied with a rubber band. In other species, the pacemakers are independent, with no direct communications link between them. Instead, the organism relies on the synchronizing power of external time cues to keep its clocks running in step. Both pacemakers are entrained to the day-night cycle, but are not coupled internally. It seems that evolution makes a tight link between the two pacemakers if the animal is active at night, and is therefore unlikely to see a time signal from the sun. By contrast, pacemakers in animals who go about their business in the daylight are not linked. In the intertidal marine snail *Bulla*, which is nocturnal and spends the day buried under gravel, where it may not see a day-night cycle for many weeks at a time, the connection is tight. In the mollusc *Aplysia*, a snail very closely related to *Bulla*, however, the two pacemakers are independent of one another. The difference is that *Aplysia* does not bury itself and is active during the day.

In the same way, the cockroach, a night-crawling creature, has two primary pacemakers which control its activity at different times during the 24-hour cycle, one in each optic lobe. The clocks are firmly linked, and the output from each clock is integrated in the brain. Always the clocks remain in synchrony, even in the absence of day-night cues which would provide a common time reference. While it is risky to generalize about the evolutionary logic behind various types of circadian couplings, it does seem that in nocturnal animals internal connections between important pacemakers are strong, while in day-active species they are weak or nonexistent. In six species studied, pacemakers are coupled in nocturnal animals, and either are not coupled at all or are only very loosely so in day-active ones.

Body time is a multicomponent, multimedia affair. It is put together from a conglomerate of clocks, some clearly important, others not so important and still others occupying an ambiguous rung on the ladder of temporal status. But the clocks are not isolated units, each one responsible for driving a particular rhythm independently of all the rest; clocks interact, influence and are influenced, modulate, entrain, couple and uncouple, in a highly mobile set of relationships, which are mediated by messages encoded as chemistry and as electrical impulses in the nervous system. Some clocks are strong, some are weak, some are more independent than others, some are tuned to environmental time cues to which other clocks turn a blind eye. We cannot really talk about the "regulation" of biological time, although people often do, because that is the language of the master clock hypothesis. We must speak rather about the coordination of a multiplicity which under the right conditions can become a unity.

Life time is malleable. It takes many different forms in the animal kingdom because it is the result of a process of evolution, just like any other biological system, and the form it takes in a particular species is a result of strategies of survival developed over millions of years to suit the peculiar needs and way of life of that species. How the many clock systems in the brain and body interact with one another, and how pacemaker clocks relate to information about time coming from the outside world, say something important about the species itself, what its priorities are, what sort of being it is. Humans, the most versatile of species, are highly rhythmic, but their rhythms are held together in a unity that seems to be more fragile than that of other animals, more easily ruptured, with consequences for health both mental and physical, so that the delicacy of the structure of man's biological time structure, like the complexity of the structure of his brain, has not been bought without a price.

SIX

A Biological Rainbow

Searching for biological rhythms hidden amid the thickets of random or irrelevant variation in the body is a task not altogether unlike that of probing the sky for messages from intelligent beings elsewhere in the universe. That is more than just a fanciful analogy. A body rhythm, say, core temperature or the white blood cell count, is a mixture of sameness and change. It rises and falls in the course of a day, a week, a month, but the fluctuations are not haphazard. They display a recurring pattern, and the change repeats itself predictably, cycle after cycle, not exactly, but approximately. Some rhythms, notably that of the hormone cortisol, are strikingly similar from one day to the next. But change that is regular and predictable is often masked by change that is random and unpredictable, just as electrical noise in a radio receiver perturbs and even drowns out the message that is being processed in the receiver's circuits.

Biologists are not alone in trying to disentangle certain

kinds of periodicity from noisy sources. Other scientists and mathematicians have been devising ways of doing so for years. For instance, Dr. Kent Cullers, of SETI, the U.S. government's Search for Extraterrestrial Intelligence project, is perfectly at home talking about his mathematical techniques with a colleague like Jim Stevenson, a statistician who studies biological rhythms at the U.S. Space Agency's Ames Research Center in California, where SETI is also based. Dr. Cullers scans the cosmos for messages from other civilizations in our galaxy, using huge, dish-shaped antennas. He has been blind from birth, and spent much of his childhood listening to signals from around the world, and even from the sun, on his shortwave radio set instead of watching television like most of his peers. Today, Dr. Cullers creates sophisticated computer programs which can distinguish a genuine signal from a plausible impostor or even a hoax. The search is strewn with red herrings, like most detective work. One may be fooled into thinking that a periodic transmission is a message from a faraway planet, when in reality it may emanate from an earthly source or from lifeless objects in space. In 1965, Russian scientists announced the detection of signals from an advanced civilization, but the signals turned out to be coming from a quasar, a type of star that emits powerful radio waves. (One version of the story says it was a satellite.)

Cullers, who listens for word from aliens, and Stevenson, who dissects biological time structure, are both detectives of a sort, and they use similar mathematical devices in their sleuthing. Each hunts for oscillations with a specific frequency and amplitude, and they need to single these out from what Cullers calls "junk," irrelevant distractions, whether or not the distractions are periodic. Statistical methods are used in both cases, and there is a heavy reliance on computers. One of the basic mathematical procedures is Fourier analysis, by means of which a complex wave is broken down into a set of very simple and regular variations in time called sine waves. This type of analysis is named after Joseph Fourier, the French mathematician and physicist who went with Napoleon on his expedition to Egypt in 1798. Fourier himself was interested in the behavior of heat flow, but he was able to show that any periodic change in a variable can be represented as the sum of several simple sine waves of different amplitudes, phases and frequencies. The frequency of a wave is the number of cycles it completes in a given period of time and is a measure of how fast or slow the wave is. Amplitude is how much the wave goes up and down, and phase refers to the time at which the peak occurs relative to zero on the clock. The variable in question might be the ebb and flow of tides, the phases of the moon, a vibrating violin string, the cycle of the seasons, the temperature of an electric oven or of the human body, the current in a telephone wire, yet its intricate risings and fallings over time can be unraveled into simple components by Fourier analysis.

Kent Cullers, like Jim Stevenson, is highly selective in his listening. He knows that nature in general produces signals that contain many different frequencies, and what he is looking for is an *unnatural* signal, consisting of a few frequencies or only one. A signal displaying the entire range of frequencies is called "white noise." The sun and stars emit white noise, which can also be produced artificially simply by blowing into a telephone. By contrast, the type of signal Dr. Cullers is trying to detect would oscillate with a single frequency, pulsing like a lighthouse beacon or be continuously on like the carrier wave of a radio or television station. It would be the radio analogy of a whistle or a tuning fork.

A full understanding of the nature of a fluctuating process in the body cannot be reached simply by looking at a chart made up of readings taken at intervals over a span of hours, days or months. Of course, some rhythms, the dramatic and obvious ones, like the circadian cycle of sleeping and waking, will stand out vividly for all to see. Other rhythms, however, especially those of low amplitude, will hide themselves in the data. That is why it seemed at first that rhythms disappeared when the supposed "master clock" in the SCN was removed; the rhythms were still there, but they had a smaller amplitude and so tended to be obscured. The contour of a body rhythm on a time chart, its "wave," is likely to meander all over the place, jumping up and down, irregular and jagged, in fact thoroughly junky. Such a complicated curve might contain many different, simple sine waves, each with its own particular frequency and amplitude, but there would be no way of knowing, just with the naked eye, by means of what scientists call a "look-see" approach, how much of the total amplitude of the complex wave is due to this or that sine wave of a given frequency. By Fourier analysis, however, using computers, an enigmatic time plot of a rhythm can be transformed into a spectrum of the various frequencies of the simple sine waves that made up its complex curve. The frequencies are spread out, each one distinct and separate, instead of being superimposed on one another. The strength, or amplitude, of the complex wave is the sum of the amplitudes of the sine waves of different frequencies that together compose the wave. In the 1950s, the best that biologists could do was to look-see at the complex wave on a time chart, and it was not good enough.

A spectrum of frequencies, when read correctly, can reveal hidden aspects of nature. In chemistry, a spectrum is an encyclopedia of information about the atomic structure of matter. A chemical element, if it is heated to incandescence and the light passed through a sort of prism, emits its own characteristic spectrum of frequencies, which is unique to that element, as a face is unique to a person. It was by reading such spectra that scientists were able to identify and classify the elements, and detect their presence even in the sun and on distant stars. Today over a hundred chemical elements have been found, and nearly all of them are classified according to their own distinct spectra. As Buckminster Fuller once said, "The physical universe is an aggregate of frequencies."

A frequency spectrum also provides a window of sorts into what is going on in the human brain. Electrical impulses in the brain, when detected and amplified by a device called an electroencephalograph, or EEG, and traced out by pens on a moving sheet of paper, appear as oscillations, waves, which vary in their frequencies and amplitudes. A wave track on the EEG chart may be a complex rising and falling curve, made up of different simple waves of different frequencies, all superimposed on one another, needing to be teased apart into component sine waves by Fourier analysis. In the brain, there are four main frequency bands, and they are associated with certain types of mental activity. The slowest of the brain waves, those with the longest interval of time between peaks, are delta waves, with a frequency of between one and three cycles a second. A little faster are theta waves, at between four and seven cycles, then alpha waves at eight to thirteen cycles, and beta waves which are faster than thirteen cycles a second.

The amount of simple wave activity of a certain type, its share of the total complex EEG wave, changes as time goes on, and these changes reflect alterations in internal states of consciousness or in the structure of sleep. Beta and alpha waves predominate in the wakeful brain of a healthy adult, while slower rhythms come to the fore during sleep, in infancy and at times of serious illness. The EEG also gives information about the effects of brain damage and drugs that can impair alertness. Computers count the number of waves in a particular frequency band and determine their amplitude, so that the percentage of alpha, beta, delta and theta in any given time span can be known. It might be, for example, that in a five-minute period, 75 percent of the EEG wave is alpha, 20 percent is theta, and the rest is distributed among the other frequencies.

Perhaps the most important development in chronobiology since the 1950s has been the discovery, made possible by high-speed computers and ingenious programmers, of a spectrum of frequencies in the rhythmic activity of most functions of the body, with different frequencies being prominent in different functions. Often, but not always, it is the circadian frequency of one cycle in about 24 hours that is the most conspicuous. Before the arrival of computer methods, people assumed that the circadian frequency was the basic unit of biological time, and that the rest of the variation seen in the raw data was simply noise, junk. They did not talk about a spectrum of frequencies.

Franz Halberg, a leader and pioneer of high stature in the computer analysis of biological rhythms, compares the new techniques for detecting rhythms in noisy data with the revolution in optics that took place in the seventeenth century. Before that time, it was thought that white light was the simplest and purest form of light, while color, far from being intrinsic to white light, was something added to it, contaminating the purity. In 1666, however, Newton, then aged twenty-one, placed a prism in the path of a beam of sunlight as it shone into a darkened room, breaking it up into a rainbow band of different colors—red, orange, yellow, green, blue and violet. This band was named the spectrum, the Latin word for image, and it showed that white light was less pure and less simple than it had seemed. Newton realized that light, like sound, is periodic; the different colors of the spectrum of light correspond to vibrating waves of various frequencies. Sunlight is a mixture of a large number of component waves, differing slightly in frequency. A prism unmixes that mixture, and it does so because, while all components are bent as they pass through the prism, each one is bent by a different amount, and the higher frequencies are bent more than the lower ones.

Until recently, the prevailing view of biological rhythms was rather like that of white light before the arrival of Newtonian optics. The frequency of one cycle about every 24 hours reigned supreme. That cycle was to the time structure of the body what the cell is to its spatial structure, a fundamental unit of organization. Today the modern version of Newton's prism is the computer, with its power to handle vast amounts of data quickly, and the notion of one basic frequency has given way to that of a rainbow, a spectrum. Among the components of this biological rainbow are these:

FREQUENCY	TYPE OF RHYTHM
Less than 20 hours	Ultradian
Between 20 and 28 hours	Circadian
	Infradian
3½ days	Circasemiseptan
7 plus or minus 3 days	Circaseptan
14 plus or minus 3 days	Circadiseptan
21 plus or minus 3 days	Circavigintan
1 year plus or minus 2 months	Circannual

Dr. Halberg coined the terms ultradian and infradian as a deliberate allusion to ultraviolet and infrared light, whose frequencies are too high and too low respectively in the electromagnetic spectrum to be seen by the unaided human eye, and also as an allusion to ultrasound and infrasound, which are beyond the range of the human ear. The metaphor refers to the fact that certain rhythms in the body cannot be detected by a look-see approach using a simple time plot. Often ultradian rhythms are not as prominent as circadian ones, and they manifest themselves in subtle ways. They may be small in amplitude or overlaid with noise, in the form of unidentified or irrelevant biological variation, or both, like a faint sound of special meaning which we strain to hear amid the hubbub of a crowded room. A biological rhythm can be regarded as a "signal" in this sense. But just because certain rhythms are difficult to detect, they are not necessarily trivial or uninteresting.

It is now a basic tenet of chronobiology that much of what used to be considered random, meaningless variation in body functions actually consists of meaningful rhythms of various frequencies, which may interact with one another. Sameness has been conjured out of change. Not all the noise of thirty years ago has been resolved into rhythmicity. Some noise still remains and always will remain. Yet a remarkable amount of pattern and structure has emerged, much of it unexpected and of a kind that raises new philosophical questions about the nature of biological time.

Modern high-speed computers can analyze all the variation over time in oral temperature, say, or in the rise and fall of cortisol, or in the amount of iron in blood serum, and show how much of the variation is regular and how much of it is noise. They can also estimate the extent to which rhythms of a given frequency—circadian, ultradian or infradian—contribute to the total amount of change, just as electrical activity in the brain is analyzed in terms of the percentage of alpha, beta, delta and theta waves present in the EEG. By such procedures, it is possible to explain a sizable amount of variation in body functions as being orderly and predictable, traceable to components of the biological spectrum, and also to say how much of the variation cannot be accounted for in this way.

"It is superfluous to keep on discussing whether noise is in the nature of things, or whether it arises from our imperfect ways of measuring," Dr. Halberg maintains. "What I think is important is that there is a beautiful spectrum of biologic rhythms unfolding, which complements the electromagnetic spectrum, and in which we find certain prominent frequencies. There is a spectrum of rhythms in each physiologic function. We have not just a single note in the time structure of the body now, but a symphony."

Statistics plays an important part in the detection of these concealed temporal patterns in the body. In biology, nothing is certain, nothing is absolute and can rest on a single measurement. Even a prominent circadian rhythm with a large amplitude, entrained by the cycle of daylight and darkness, is not 24 hours to the minute, but usually a little longer or a little shorter. Also, more than one entraining agent may act on the mammalian system at one time, exerting different influences on the body's clocks. Dr. Halberg stresses that a biological rhythm is a statistical entity, in which there will always be an element of randomness. He refers all his students in Minnesota to Einstein's famous reply to those physicists who proposed the existence of an irreducible element of chance in the universe: "God does not throw dice." In the living organism, Halberg insists, God does throw dice. There will always be some uncertainty in our knowledge of the body's operations. Each cycle of a rhythm may not be an exact repeat of the previous cycle, but in a true rhythm there must be a pattern of repetition that is fairly regular, and mathematical analysis should show that the odds against such a pattern appearing merely by chance are sufficiently high.

A branch of mathematics called the statistics of inference is used by chronobiologists to analyze the spectrum of body rhythms. This type of statistics is also used by physicists to separate signals from noise in communications channels. It should not be confused with descriptive statistics, the kind that takes a complete corpus of information and summarizes it in a convenient form, as when a baseball player's batting average is calculated for an entire season. There are no missing data in the baseball case, because every detail of the batter's play, in every game, is known. The average is an exact description of his performance, with no margin of error.

The statistics of inference goes farther, making statements that are inexact on the basis of information that is incomplete. What makes this second kind of reasoning so powerful is its ability to measure the uncertainty content of a statement, to compute the amount of error that lurks in it. The likelihood of a wrong judgment can be expressed as a percentage, say $P = 0.05$, which means that the chance of error is 5 in 100. In other words, we can be 95 percent confident that the statement is reliable. A statistician is not content to give his best estimate of the true situation; he also arrives at a number that represents the amount of trust that can be placed in such an estimate. This number is called a confidence coefficient.

Unlike the batter's average, an approval rating for a politician in an opinion poll is no more than approximately accurate, since only a sample of the population is asked to express an opinion. There are many missing opinions. If every inhabitant of the country had been interviewed, the result might have been different, but only by a few percentage points, and the margin of error, the confidence coefficient, is usually published together with the results of the poll.

The statistics of inference allows biologists to detect the

presence of rhythms even when they are weak, or "wobbly," on the basis of sample readings of body variables which are not exhaustive, by stating the amount of trust or confidence we can have in the judgment that "this is rhythmic change and not mere random variation." Such a procedure is much less subjective than the look-see approach. A computer program matches an idealized, smooth mathematical sine curve of a specific frequency to the untidy spread of the sample readings, trying out hundreds or perhaps thousands of likely simple curves with different frequencies, until it finds the one that fits the data more snugly than any of the others. The differences between the actual data and the ideal wave are squared, and the best-fitting curve is the one for which the sum of these squares is the least. That is one way of taking care of the errors inherent in the search for rhythmicity. A fit may be called acceptable if $P = 0.05$ or less. In this way "statistically significant" patterns in time can be detected, patterns that are extremely unlikely to be the result of a fluke, a lucky throw of God's dice. An example of a time pattern appearing just by accident is the mistaken "discovery" of biorhythms. Wilhelm Fliess, a friend of Sigmund Freud's, proposed that there was a cycle of 28 days and one of 23 days which started at birth and were closely associated with sexual processes. The cycles were supposed to determine the dates of illnesses, the stages of growth and even the day of a person's death. Many attempts were made to find rhythms of those frequencies in various aspects of human life. Two of them succeeded, but that was only two successes out of hundreds of trials, and the laws of probability decree that if one hundred scientists do an experiment, about five of them might reach a wrong conclusion just by chance, so that those five results cannot be regarded as statistically significant.

Once a statistically significant fit has been obtained, the ideal curve finally chosen approximates closely not only the frequency of the rhythm in question but also its amplitude and phase. The relative prominence of rhythms of various frequencies in body temperature or cortisol or any other body function can be established. Computers are needed for this work, because even if a biologist samples temperature or cortisol every couple of waking hours, for only one month, he is faced with some 400 readings on his chart, and to fit even one ideal curve to those points by hand by the least-squares method would take him days to accomplish, because of the immense amount of arithmetic involved.

As soon as these new, statistically significant rhythms were brought to light by methods that take into account the irremovable presence of uncertainty and error in biological systems, the structure of body time began to show intriguing resemblances to that of other periodic processes in the physical world. Certainly there are important parallels to be drawn between the spectrum of frequencies in rhythmic body functions and the spectrum of musical sounds. In fact, the physics of sound can help explain the biology of time, and the biology of time offers a new perspective on the physics of sound. They both speak the same language. A tuning fork emits a simple sound that has just one frequency (an ideal terrestrial version of what Kent Cullers seeks somewhere else in our galaxy), and when the vibrations of the fork are traced out on a sheet of paper, the result is a sine wave. But sounds made by musical instruments and the human voice are much more complex than that. A female soprano voice is more complex than a tuning fork, and a loud male voice is more complex than either, while a symphony orchestra lies almost at the extreme edge of complexity of organized sounds. The waves produced by a musical instrument are made up of many component simple sine waves of different frequencies and amplitudes. A noise, say, the clanking of a radiator in a concert hall while an orchestra is playing, also consists of various sine waves, but the difference between music and noise is that in a musical sound the sine waves are in an orderly relationship with one another, whereas in noise they are combined at random. All the simple component sounds that go to make up a complex musical sound, such as a note struck on a piano, have frequencies that are exact, whole-number multiples of the lowest frequency present in the sound. The component with the lowest frequency is called the first harmonic. A second harmonic is the component whose frequency is twice that of the lowest frequency, the third harmonic is three times, and so on. The sound of the human voice may contain as many as twenty harmonics, even though only ten of them are needed for a listener to identify the sound spoken.

Certain harmonics in a sound will be stronger than others, so that the complex waveform will not be the same for a flute or a violin as for a piano, even when a note of the same pitch is played on both. When a note of the same pitch is played on different instruments, the fundamental frequency is the same in both, and so are the other frequencies in the spectrum of harmonics. But the quality of a sound depends on the relative strength, or amplitude, of the various frequencies in its spectrum, and the pattern of relative strengths will change from instrument to instrument. The flute, for instance, has its strength in the fundamental and the lower harmonics, whereas the violin is quite weak in the lower harmonics but strong in the middle and upper ones, and the oboe has stronger high harmonics than any other member of the orchestra, which accounts for its squeaky tone.

Harmonic relationships are found widely in nature. The English composer Peter Maxwell Davies, who lives on a remote, peaceful island called Hoy, in the Orkneys, off the coast of Scotland, finds that he can tell from which direction the wind is blowing in the morning by listening to its harmonic content. Ocean tides pounding a granite cliff near his house generate a deep fundamental infrasound with a frequency much too low to be heard by the human ear, but the harmonics of that frequency are present in the air, audible at least to the composer, and they are distributed differently according to the point of the compass from which the wind is coming.

Planets vibrate at one or more natural frequencies, and so, at the other end of the scale of size, do atoms. Just as the harmonics in a musical sound are exact multiples of the fundamental frequency, so also an electron "wave" in the atom of an element vibrates at certain restricted frequencies which are multiples of the lowest one. Recently, scientists have written out the spectra of frequencies of chemical compounds as sequences of notes on the musical scale in order to help blind students of chemistry to identify the substances. The notes are played as a chord, and each chord is unique to the chemical compound it represents. Ethanol sounds like "an extraterrestrial dance tune." Polystyrene "modulates from a dark, demented fugue into an icy tinkle." The physicist Victor Weisskopf once tried to play the "chord" of hydrogen on a piano, and said: "It sounds terrible, but then it's not music for our ears."

The spectrum of frequencies, with their harmonic relationships, plays an important part in the response produced in one vibrating body to the vibrations of another body. This effect is known as resonance. A familiar example of resonance is the shattering of a wine glass when a certain note is played on a violin. The glass is set into motion because it has a natural frequency of vibration, and if a sound of that same frequency occurs in its vicinity, the glass responds by oscillating. The structure of the glass is such that it absorbs the energy of the particular waves of sound sent out by the violin rather than that of any other sound waves, because the energy is transmitted at the glass's own natural frequency. A violin also resonates with itself, which is why the quality of sound in each individual instrument is so variable. The wooden body of a violin, being capable of vibration, has its own natural frequencies, which are different in different parts of the body. When a violinist plays a note that has a frequency close to one of these natural frequencies, the instrument responds loudly, but if the note does not come close, the response will be inferior. Often it happens that one of the harmonics of a note played on a violin string coincides with a natural mode of vibration in the violin body, and then the harmonic in question sounds more loudly and the quality of its tone is altered. In the future, when a violin is sold, the buyer may be given a graph showing the strength or weakness of that particular instrument's response to sounds at every frequency in the spectrum of the sounds it can make.

Responses at certain frequencies are remarkably different from responses at other frequencies.

Resonance is extremely important in wave physics. Almost all physical systems have their own natural resonance frequencies. Even an underwater air bubble, jolted by a burst of intense light, will "ring," radiating sound waves into the surrounding water. The bubble expands and contracts under the impact of the light, vibrating at its resonance frequency. A typical bubble oscillates at about 50,000 cycles a second, an ultrasound frequency which is outside the range of the human ear. Larger bubbles vibrate at lower frequencies. The tuning circuit of a radio receiver will respond only when it is exposed to incoming waves of the same frequency as the one at which it is set to oscillate normally.

Franz Halberg emphasizes that the body responds to events in a spectrum of natural frequencies that are in a harmonic relation to one another. That is part of the temporal identity of a species or an individual. It is the physiological counterpart of the violin responding selectively to sound waves of various frequencies produced by the strings, according to its own natural modes of vibration, which are not the same from one violin to another, so that violins, as well as living organisms, have an "identity." As in the case of the harmonics of a musical sound, the component frequencies of a spectrum of biological rhythms may vary in amplitude; some may be "louder" than others. Often, but not always, the circadian, or about-24-hour, component has a larger amplitude than the infradian, or longer-than-24-hour, components. A given body rhythm has a normal spectrum of frequencies, and in sickness the spectrum can be distorted, just as there is a normal and abnormal spatial structure of the cell.

It may be that resonances across the various frequencies of the spectrum of the body's rhythms play a part in the integration and coordination of the living system, that they "hold the body together." William Hrushevsky, of the University of Minnesota, says: "The time structure of the body is put together by means of an interacting web of clocks which all affect one another, the way things in the universe all affect one another. Their resonances keep them in relative check, although they are not fixed in any way. All the rhythmic systems in the body are in contact with one another across many frequency ranges. When you bend one of the systems in some way, it is pulled or pushed back into shape by its resonant relationship with the other systems. Planets are held in place by virtue of the resonance that exists between the periods of their orbits around the sun. In the body there are rhythms with a spectrum of frequencies, ultradian, circadian and infradian, and if you insult the body in some way, with a challenge, such as jet travel across time zones, these rhythms will tug and push at one another until the body is back into synchrony. After jet lag you can watch the various rhythms come back into line in their own time, some sooner, some later. Resonances across multiple frequencies hold living organisms and the universe together."

Resonance is a kind of sympathy, a tuning-in, that exists between vibrations of a certain frequency, and lack of resonance results in a sort of indifference, a tuning-out. In the first case the vibrating system is open to influences from another vibrating system, as the shattered glass is open to influences from a violin playing at the glass's own natural resonant frequency. Thus, in the spectrum of frequencies, circadian, ultradian and infradian, of a given body function, most components of the spectrum are modulated and influenced by resonant effects from other spectral components, both within the same function and from elsewhere in the body. A similar effect occurs in music. If the key on a piano an octave below middle C is pressed gently so as to open the string silently, and then middle C is struck firmly, the frequency of middle C being about 260 cycles a second, the same as the frequency of the first harmonic of the lower string, the lower string can be heard to resonate. In the same way, a circadian rhythm in, say, body temperature might be open to influences from an infradian rhythm whose frequency is a harmonic of the circadian frequency, but be unreceptive to one that is not in a harmonic relationship. This selective attunement of oscillating systems may explain why body clocks are entrained by environmental oscillators, such as the day-night cycle, whose frequencies are compatible, or resonant, with the frequencies of the clocks in the living organism.

In Franz Halberg's view, a central feature of biological time structure is the harmonic relationship that exists among the various component frequencies. A striking aspect of this relationship is that the components themselves appear to be harmonics or subharmonics, multiples or submultiples, of seven, a number that has played a disproportionately large role in human culture, myth, religion, magic and the calendar.

Rhythms of about seven days, called circaseptans, are turning up in many functions of the body. When Lawrence Scheving locked himself up in a special apartment completely cut off from external time cues, doing his own cooking and having no social contacts, in order to monitor certain operations of his body as part of a study of hypertension, he found circaseptan frequencies even though he had not especially looked for them. The results of the experiment are still being analyzed, but so far rhythms of about seven days have been found in a number of variables, including the ratio between two important neurotransmitters, norepinephrine and epinephrine, the volume of Scheving's urine, his systolic and diastolic blood pressure, and his heartbeat. The rhythms were of low amplitude, compared with the circadian rhythm. Each one free-ran at a frequency that was slightly different from the rest. But all of the rhythms showed a statistically significant fit to an ideal curve with the frequency of 7.89 days. "It is surprising to me that about seven-day rhythms were there," Dr. Scheving said. "Our knowledge of such rhythms is still at a very early stage, probably about 25 years behind our understanding of the circadian frequency."

Circaseptans appear as part of the body's defense against harmful intruders, and manifest themselves only when the intruder appears. The reason for this may be that the arrival of an interloper is a stimulus which synchronizes already existing rhythms of the body that have an intrinsic frequency of about seven days, but are too weak to be detected, resulting in a collective weekly rhythm with a larger amplitude than any of the individual rhythms. In itself, the stimulus may contain no seven-day information, any more than a single flash of light conveys 24-hour information to the circadian system, yet it can entrain that system. Again, the amount of "knowledge" of time generated internally is much greater than the information about time coming from the environment. This seemingly magical conjuring up of something unsuspected and latent is one of the curious properties of resonance. As K. C. Cole has noted, "resonance can make things seem to appear out of nowhere—like rabbits out of hats. The music broadcast by your local radio transmitter seems to spring out of thin air when you tune your receiver to a sympathetically vibrating frequency."

A circaseptan rhythm manifests itself in response to an endotoxin. Like a flash of light, an endotoxin is a single, arbitrarily timed challenge to the body. It is a poisonous substance, which is present, for example, in the walls of the typhoid fever bacillus. Lewis Thomas, a biologist and essayist who studied endotoxins for more than 35 years, speaks of them as a kind of once-only message, full of bluster and braggadocio, synchronizing various webs of the immune system: "The center of the puzzle is that endotoxin is really not much of a toxin, at least in the ordinary sense of being a direct poisoner of living cells. Instead it seems to be a signal, a piece of misleading news. When injected into the bloodstream, it conveys propaganda, announcing that typhoid bacilli in great numbers (or other related bacteria) are on the scene, and a number of defense mechanisms are automatically switched on, all at once. When the dose of endotoxin is sufficiently high, these defense mechanisms, acting in concert or in sequence, launch a stereotyped set of physiological responses, including fever, malaise, hemorrhage, collapse, shock, coma and death. It is something like an explosion in a munitions factory."

Sometimes, when circumstances change, a circaseptan doubles into a harmonic of itself, with a frequency not of seven days but of three and a half days. Normally, the spectrum of rhythms in the surface temperature of the female breast consists of frequencies of about 24 hours, about 7 days, about a month and about a year. If the span of measurements is extended even farther, a frequency of about 12 years can be detected. When cancer strikes, however, the spectrum is altered drastically. In place of the weekly and monthly frequencies, 3½-day and 42-hour components appear, as well as a circaseptan and a free-running circadian rhythm of 21 hours. In the 1950s, biologists would have simply noted that the disease had caused the circadian rhythm to become disentrained from the day-night cycle and left it at that. Today, thanks to Fourier analysis and the statistics of inference, we know that much more is going on. The weekly and monthly frequencies in the spectrum of surface temperature rhythms in the healthy breast have multiplied into harmonics of themselves, and each has a harmonic relation to the other. There has been a transposition from the monthly and weekly regions of the spectrum to the three-and-a-half, two-day and circadian regions. Dr. Halberg calls this a "spectral compromise." He says: "The system does its own reshuffling."

Mermaid's wineglass, the five-million-year-old giant alga, has an about seven-day growth response to shifts in the light-dark schedule. When the nucleus of this alga is removed, however, presto! The rhythm doubles its frequency and becomes one of about three and a half days, a circasemiseptan, which is a harmonic of the original frequency. The new rhythm results in reduced growth.

Circaseptan and circasemiseptan rhythms are not arbitrary, even though they seem to lack counterpart rhythms in the external environment. There is rhyme and reason to their existence. They are not directly adapted to the motions of the earth, moon and tides, but they are compatible with other rhythms in the body that are adapted to those cosmic motions. The seven-day cycle is a harmonic, a multiple of the month, and a division, a subharmonic, of the day, so that by frequency division or frequency multiplication circaseptans are in a definite mathematical relation to the major periodic-

A BIOLOGICAL RAINBOW 133

ities of the earthly environment. They are a sort of biological artifact, evolving internally by means of the harmonic laws. It is as if three or four components of the spectrum of the sound of a musical instrument were "given" by nature and the rest had to be provided somehow by the player of the instrument in order to make the sound quality richer, more interesting and more complex than it would have been otherwise. Probably, in the course of evolution, living organisms found that the timetabling of body processes could not be done by internalizing the limited range of rhythms of a day, a month and a year which was all the environment supplied. Extra frequencies had to be "invented," without the aid of the environment, not out of whole cloth, but by using material already in place.

Various cells, tissues and organs perform routine tasks essential to life, and these include maintenance, growth, defense and reproduction. Such tasks do not necessarily go on with the same intensity all the time, but are intermittent, continuing for a while and then lapsing. Some tasks last longer than others. Metabolism, the building up and breaking down of chemicals in cells, can be a relatively fast affair, but other activities such as growth and reproduction occur over more extended spans of time. For this reason, a particular function of the body may have a spectrum of rhythms with a dominant frequency that is very different from the dominant frequency of the spectrum of rhythms in another function, perhaps widely separated in space. Yet no matter which frequency component is the primary one in any given function, all rhythmic systems of the body probably possess an innate circaseptan frequency, so that when they cooperate to perform a specific task which is body-wide, say, an immune reaction, the reaction occurs on a weekly schedule. That schedule is a compromise between too much time and too little. A day and a night, which is the dominant frequency in the spectrum of many routine body chores, would not be long enough to complete the complicated array of chemical and other activities that compose the immune defense reaction, and a month would be too long.

"I think the big revolution in biologic time structure came rather early in evolution, when the living organism said to Mother Nature, 'Mother Nature, I would rather do it myself,' " Dr. Halberg says. "In other words, the primitive form of life was saying, 'I would rather defend myself, or at least respond to novel situations with my own schedule.' Even at the stage of the single cell, the organism evolved its innate circaseptan and circasemiseptan frequencies in response to stimuli from the environment as well as to disease. It 'chose' its own period of responses with a period of about a week and about half a week."

Clearly there are mechanisms in living species that are responsible for doubling or halving a rhythm's frequency, transposing from one part of the spectrum to another, "reshuffling" spectral components according to harmonic laws. Exactly what those mechanisms are, and where they are, nobody knows. Dr. Halberg thinks the task of identifying and locating them will require the pooling of ideas and effort by molecular biologists, physical chemists and mathematicians. When they have completed their work, we shall understand a great deal more about the temporal organization of the body. These hidden processes, in their importance and mystery, Dr. Halberg believes, represent "the frontier of biologic spacetime."

Only a few years ago the great prize, the Holy Grail of chronobiology, was to establish the existence of a single master clock. For a while, the SCN were thought to be that clock, but now that dream has faded: there may be many clocks with claims to primary status. It is significant, and highly revealing of the new attitude toward biological time, that today one of the most alluring quests is not for a simple, single control center in the brain which regulates all the rhythms in the body from the top down, but rather for the workings of a mobile system of spectral components, harmonically related, each of which contributes in a cooperative way to the symphony that is the temporal organization of living things.

SEVEN

Light and Time:
Together Again

Certain kinds of biological clocks are tuned to receive special types of messages about time, so that they can keep the body in step with cycles of change in the environment. Of all these messages, the most dependable, the most invariant, is light, so that light is the chief external means of manipulating the internal clockwork of biological time.

This amounts to an irresistible invitation to science to step in and manipulate the clockwork in its own way and for its own purposes, because artificial light of various kinds can be used as a substitute for the natural light of the sun. But that is not all. Modern science has at its beck and call substances that have an effect on the nervous system similar to the effect of light, so that a chemical synthesized in a laboratory can masquerade as light and become a message about time, once it is introduced into the body. Man can now act as the messenger and hence the manipulator.

In Einstein's theory of relativity, time, space and light

135

are linked. In fact, light is more basic than either space or time. What space is depends on how light moves in it, and time depends on how long it takes for light to travel a certain distance. Light takes precedence, because in the new picture of the universe the unshakable fact that the measured speed of light cannot exceed a constant upper limit, a fact that flew in the face of common sense, has forced science to relinquish accepted ideas about time. Light was elevated to an entirely new importance, a sovereign status it had never possessed before, and time was toppled from its throne.

In the living system, time and light are related, but in quite a different way. Light is the most important, universal agent that entrains biological clocks, resetting and adjusting, delaying or advancing them so that they run in synchrony with the 24-hour cycle of day and night. If the clocks are not modified in this way they continue to run, but at their own rate, so that biological time becomes out of step with world time, and can give no useful information about it to the body.

A light signal has a corrective effect on the tau, the intrinsic, natural period of a primary circadian clock. Most of the light striking the retina of the human eye is used for vision, but some of it is encoded as a time signal and is transmitted to the brain. The light-time message passes along the tract from the retina to the SCN clock in the hypothalamus. It is interesting that the hypothalamus, as part of its homeostatic control system for maintaining a constant body temperature, contains a receptor organ for heat, a temperature "eye," analogous to the retina of the actual eye, which is a receptor organ for light. In fact, the optical eye and the thermal "eye" for temperature are derived from the same part of the brain, the bottom of the third ventricle. During evolution, the optical eye moved outward to the surface of the body and was thus able to receive impressions directly from the world outside.

For the circadian system, light is not a source of energy in the sense of providing motive power for the clocks in the system. The clocks are self-sustaining. The energy to keep them running is supplied internally, within the body. Light is information about local time, and it modifies the circadian system in two ways. It adjusts the period of the system, making it exactly 24 hours, and it sets forward or sets back biological clocks in such a way that a periodic event in the body, say, waking up, coincides with a specific periodic event in the world, such as the dawning of a new day. The circadian system is then said to have a correct phase relation with the cycle of light and darkness.

What is remarkable about the relationship between light and biological time is that a very brief light pulse, almost nonexistent as far as ordinary human vision is concerned, can have important and long-lasting effects on the circadian system. A single flash of light, lasting only 1/2000 of a second, is enough to alter a circadian rhythm in a species of fruit fly so drastically that the rhythm is shifted eight hours out of phase with local time. Farmers who use artificial light to lengthen the "days" of chickens kept in windowless sheds in order to trick the birds' hormonal systems into thinking it is spring, causing them to lay more eggs, find they can do so with short burst of light rather than with continuous illumination, saving energy. Little external information is needed, because the circadian system already possesses its own, very rich internal structures for providing the body with information about time, and these do not need continual input from the outside world, but only periodic correction. It might be thought that the full 24-hour day-night cycle would supply an animal with a great deal more information about local time than a single, brief flash of light, but this is not the case. Exposure to the full cycle does not give the body much more information about time than the single light flash, although it helps to stabilize the phase relation between events in the body and events in the world.

This striking disparity between a trivial input in the form of a light pulse and a disproportionately large change in the circadian system, a resetting of body clocks that lasts for hours, has parallels elsewhere in the biology and psychology of the living system. For instance, a theory of modern linguistics proposes that the reason why children learn their native language so quickly and "naturally," merely by being exposed to fragmentary and incomplete information about it in the form of casual adult speech, is that they possess inborn mental structures which predispose them to acquire a correct theory of the language and to reject incorrect theories. The effects of information in the form of light on the circadian system are also great, beyond all comparison with the amount of that information, because of the rich biological time structures that preexist in the body.

What is more, light has different effects on biological clocks, depending on "what time it is" by those clocks. The circadian system of an animal or a human being has what is called a "subjective day" and a "subjective night." This internal day-night cycle is normally synchronized with the external 24-hour day-night cycle, but it need not be, if the organism is cut off from information about local time. When the circadian system is disentrained from external time cues and free-runs, subjective day and night together occupy more than 24 hours, or less, depending on the species. The body is programmed to function in a certain way at each stage of the circadian cycle. A nocturnal animal will go to sleep during its subjective night, and a day-active one will venture out during subjective day. When the circadian system is dislocated by a change of local time, as in the case of jet travel across time zones, a person may find himself wanting to sleep during the objective day, because as far as his body is concerned it is subjective night.

The Harvard chronobiologist Martin Moore-Ede points out that in certain groups of people subjective day and night may not coincide with actual day and night. These groups include not only jet travelers, shift workers and exam-haunted students who sleep on erratic schedules, but also people who live in a more or less constant environment, where the periodicity that is such a striking feature of the normal world is absent. One of the most notorious of these nonperiodic environments is the modern hospital, especially intensive care wards, where light, activity and noise are virtually continuous. The lack of cyclical time cues in a hosptial can disrupt a patient's circadian system within a few days of admission.

In normal circumstances, however, light adjusts circadian clocks by setting them forward or setting them back, thus keeping events in the temporal program of the body in phase with events in the world. A light signal may advance the program, so that, for example, sleep begins earlier than it did before the signal was received. Or it might delay the program, in which case sleep begins at a later time. All other functions of the body under the control of the circadian system are similarly pushed forward or back with respect to local time.

Light does not have the same effect on the system all through subjective day and night. The same input can have quite different results, depending on what time it is subjectively, on the stage that has been reached in the internal circadian cycle when the input occurs. Again, this anomaly is due to the fact that the body has an internal time structure that is related to external time, but only indirectly; it operates according to its own special logic. The fact that a light cue during subjective day does not have the same effect as during subjective night is an essential element of that logic. In general, a light signal sets the clock back if it occurs during the individual's subjective night, and advances the clock if it occurs during his subjective day. The switch over from a delaying effect to an advancing effect may come quite suddenly, in the middle of the subjective night. The same light signal may result in a large corrective adjustment or a small one, depending on what time it is by the circadian clock. Two main rules apply here. One rule is that external time signals elicit a weak response during the subjective day and a strong response during the subjective night. The second rule is that delays tend to occur at the end of subjective day and early in the subjective night, and these are followed by advances in the late subjective night and in the early subjective day.

There is a fairly long segment of the circadian cycle, during the subjective day, when light has little effect on the clocks. This is understandable, because daylight by itself does not give much information about local time. The clearest time cue comes during the transition from dark to daylight in the morning and from daylight to dark in the evening. If a tau is longer than 24 hours, as it is in humans, the program needs to be advanced, and the adjustment is made chiefly in the morning. If the tau is shorter than 24 hours, a delay is required, and this will be done in the evening.

Light is not the only carrier of information about external time. Some blind people have a normal, entrained relationship with the external 24-hour cycle, presumably because their circadian systems respond to other periodic events, such as meals, social routines, bedtimes and even the alarm clock in the morning. But light is the most potent and the most general time cue. It is also the least "noisy" message, since sunrise and sunset are regularly occurring events and entirely predictable, being caused by the rotation of the earth. The changing length of daylight as season follows season is equally predictable, thanks to the stability of the orbit of the earth around the sun. Light modifies the circadian system, and because the effects of light are amplified and differentiated by the internal structure of the system, it can be used as a probe to discover the details of the structure itself.

In fact, light can be a means of disentangling the component elements of the circadian system, teasing them apart as one takes a machine to pieces to see how it works. Body clocks can be desynchronized, so that they break away from one another, by subjecting them to an artificial light-dark cycle using lamps of unusually high intensity, and then gradually lengthening the period of the cycle, so that entrainment is progressively more difficult for the circadian system.

There are good reasons for believing that the human circadian system can entrain only to light-dark cycles of between 23.5 hours and 26.5 hours. That means that in a normal 24-hour day, the human clock can be reset forward by only an extra half an hour. Advancing it any more than that would exceed the 23.5-hour minimum. The clock can be set back by as much as 2.5 hours, however, without going beyond the

26.5-hour maximum. Presumably this is why people adjust to a new local time more quickly after a flight across time zones in a westward direction, where local time is shifted backward, than after an eastward flight, where every time zone moves local time forward an hour. The westward trip requires the circadian clock to be set back, while the eastward flight requires that it be set forward. British researchers have noticed that among passengers displaying symptoms of acute psychiatric disorder at London's Heathrow airport, those arriving on flights from the United States tend to be more manic and unnaturally active, while those traveling from eastern Europe tend to be more depressed.

Such limits on the range of entrainment, restricting the synchronization of internal time and external time to light-dark cycles of a certain length, are one feature of the circadian system that seems to display a clear trend in evolution. The range narrows progressively up the scale of life forms, from the simplest to the most complex species. In primitive, single-cell organisms and plants, the range is extremely wide, but it shrinks to twelve hours in insects, and is most restricted of all in vertebrates. The human circadian system is unique among mammals in the peculiar way in which it responds to the light-dark cycle. Light of normal intensity is not as effective in entraining the human body clocks as it is with other animal species, and the range of entrainment is markedly restricted.

Recently, however, it has been found that the range widens considerably for humans when light of high intensity is used. This is another lucky break for science, since the quality and strength of artificial light can be endlessly and precisely varied. Apparently, the circadian system of humans has an exceptionally high threshold in its response to light. Perhaps this is because we are a day-active species. Ground squirrels, which are creatures of the daytime, are about nine times less sensitive to light than rats and hamsters, which are nocturnal, and humans are much less sensitive than ground squirrels. The clocks of most primates are also relatively unresponsive to light, though more so than the clocks of people.

Another reason for the high human threshold to entrainment by light may be that we have been using fire as a way of circumventing the constraints of biological time for 100,000 years or more, and we need to distinguish between light as a time signal for our biological clocks, and light as an artificial adjunct to a fuller and more civilized life. It is interesting to note that Thomas Edison, who more or less banished darkness from the face of the earth by introducing electric light on a grand scale, professed to despise sleep and the obstacles it placed in the way of unfettered, round-the-clock activity, though he took discreet daytime naps.

Using very bright artificial lights, humans can now be entrained to a light-dark cycle of as long as 30 hours, a remarkable expansion of the supposedly narrow range. Also, by systematically lengthening the period of the cycle, perhaps by ten minutes every day, the limits are gradually exceeded, so that the effect of disentrainment on body rhythms can be studied in a precise, controlled fashion. Since different rhythms appear to have different ranges of entrainment, what happens is that as the period of the light-dark cycle is stretched, rhythms disentrain, but not all at once. The rhythm of deep body temperature, perhaps controlled by the proposed strong X pacemaker, usually breaks away first, whereas the sleep-wake rhythm may remain locked in to the light-dark cycle even if the period of the cycle is extended to 28 and even to 32 hours. A particular rhythm of mental performance, speed of computation, remains synchronized with the temperature rhythm in a 28-hour "day" and with the sleep-wake rhythm in a 32-hour "day." Evidently different rhythms have different ranges of entrainment.

The use of light to manipulate circadian clocks and "dissect" them, as physiologists dissect the spatial structure of the body, is a technique that is only in its early stages and may produce remarkable results in the next few years. Another promising approach is the use of chemical substances to entrain or disentrain the clocks.

Because light and chemistry are so closely associated in the living system, time and chemistry are also related. That is part of what it means to "biologize" time.

Messages about time course continually around the circadian system, keeping it synchronized both internally with itself and externally with the world outside. Both body and brain are made up of millions of cells, and this great multitude of relatively independent units must be able to communicate so that the organism functions as a harmonious whole. One medium of communication is by hormones, the chemical messengers which are carried in the bloodstream to distant parts of the body and influence the activity of cells that are specially equipped to receive them. Another medium is the nervous system, in which nerve cells, or neurons, communicate with each other by secreting specific chemicals called neurotransmitters, which cross the tiny gap between two cells and produce a change in the electrical activity of the target cell. All brain activity is mediated by these substances, which is why they are of intense interest to those who study the chemistry of the brain. Certain kinds of nerve cells convert a message carried by a neurotransmitter into a message carried by a hormone, and these are called neuroendocrine transducers. Most neuroendocrine transducers are located in the hypothalamus.

One link in this chemical-electrical information network, where light, chemistry and time all come together, is a distinct and special pathway leading from the SCN to a little gland called the pineal. Information about day and night is communicated along this route, which runs by way of the spinal cord and the sympathetic nervous system. The pineal gland is a neuroendocrine transducer which changes nervous system messages into hormonal messages by synthesizing and secreting a hormone called melatonin. The molecular structure of melatonin closely resembles that of serotonin, a neurotransmitter which also holds intense fascination for chronobiologists; the SCN regulates the rhythmic increase in a specific enzyme in the pineal that helps to convert serotonin into melatonin. Humans secrete most melatonin at night, between 11 P.M. and 7 A.M., and suppress it during the day. Exactly what melatonin does to earn its keep in the human body is not known for sure. There is some evidence that it may act like a sedative, inducing drowsiness in certain individuals when given at times when they are normally wide awake. But it is intimately and intriguingly connected with light and darkness, and therefore with time. Darkness is a signal to the pineal gland to start making melatonin, and light is a signal to stop making it, so that the amount of melatonin present is a chemical message to the body about night and day, about time. The melatonin signal sent by the pineal to other parts of the body is a wave whose frequency, phase and amplitude can all be altered by light, and this wave is a means of coordinating in time various complex body functions. In rats, melatonin can be suppressed by the dim light of the moon or a candle flame, but for humans, with their relative insensitivity to illumination, it takes light ten times as bright as an ordinary room light to have the same effect.

Luke Thorington, the creator of Vita-Lite, a fluorescent lamp that emits the entire spectrum of colors found in natural daylight, from infrared to ultraviolet, and has been used to show that bright light does suppress melatonin in humans, suggests that ordinary indoor lighting, the sort most of us work and read by, may be interpreted by the human pineal as darkness, or what Thorington calls the twilight zone. "Maybe we're all living in a melatonin haze," he says. Perhaps this could explain why some artificially lighted workplaces have a deadening or mildly soporific effect on the people who spend their working days there. The light may not be rich enough or strong enough to suppress melatonin.

The Greeks thought of the pineal gland as a sort of memory valve, which controlled the flow of memories stored in the brain and occasionally got stuck—which is why a person sometimes strikes his head with his hand when trying to recollect. In 1640, René Descartes declared that the pineal was the seat of the soul. Claude Bernard's curmudgeonly teacher, François Magendie, suggested the gland was a valve that regulated the flow of spinal fluid. As late as the mid-1960s, the pineal was called the "last great mystery in the physiology of mammalian organs." Only recently has it become a glamorous and highly respectable organ to investigate. Today, books are written about the pineal and conferences organized around it.

The pineal gland is so ancient, and so different in size and role from one species to another, that studying it, the biologist S. D. Wainwright has said, is like "seeing evolution in progress." The pineal is nearly universal in vertebrates, but its function varies. It is only about a quarter of an inch long in humans and weighs less than one hundredth of an ounce. Located near the center of the brain, it is not part of the brain itself. The pineal is directly sensitive to light, and seems to operate as a kind of third eye in some species, such as frogs, but it is not directly photosensitive in mammals. Wainwright believes that the different relationships between the pineal gland and biological rhythms in various animals alive today reflect a continuing evolution of both the gland and the relationships.

How active the pineal is in a given species depends on the gland's size. In animals that live in harsh climates, and therefore must reproduce at highly restricted times of the year, linked to the season, the pineal tends to be large and presumably plays a role in synchronizing the animal with local time, under conditions that drastically constrain the "when" of mating behavior. In arctic muskox and lemmings, the gland is fairly large, and as for elephant seal, half of all the tissue inside a baby seal's head is pineal tissue. A large pineal, therefore, means that time is important to the organism. Breeding, for example, is prohibited at a season when it would result in young being born in cold weather. The evidence suggests strongly that in some animals the pineal can "sense" what season it is from information about the length of daylight; the number of hours of daylight, varying in a regular fashion throughout the year, is the most reliable indicator of season. Animals with very small pineals usually live in the tropics or near the equator, where the day is roughly the same length all the year round. Nocturnal animals tend to have smaller pineals than day-active ones.

Melatonin, the hormone secreted by the pineal gland, may be a chemical messenger communicating information not only about time of day but also about the season of the year. In certain species of animals it synchronizes seasonal changes in body weight, coat color, and torpor, independently of the reproductive system. The full effects of melatonin are only beginning to be appreciated. It probably regulates the changing responses of the body to light over the course of 12 months as well as over a period of 24 hours. If behavior needs to be tightly constrained and synchronized with the seasons, melatonin provides very useful information. Birds migrate in spring. But what message from the outside world tells them that spring has come? The critical message is not the rising temperature, because that is a noisy signal, fluctuating and capriciously varying. Lengthening daylight is the clearest and least ambiguous information that the environment can give. Melatonin keeps the body informed of day length, which may be how the reproductive system knows what time of year it is. It is now known that animals cannot interpret day length correctly without a daily rhythm in melatonin. Somewhere in the brain there must be a summing device which converts information about hours into information about weeks and months. In the absence of a pineal, animals reproduce at unsuitable times of the year, and turn up their body thermostats at irrelevant hours of the day.

In every species studied, melatonin is secreted for the longest time in winter and for the shortest time in the summer. In humans, who are not markedly seasonal creatures, the function of melatonin is elusive since a small number of people manage to get along with little or none of it in their bloodstream. Most of us, however, show a definite daily rhythm of melatonin. It may be linked to sleep and rest, and to reproduction, but in rather ambiguous and special ways. Reproductive behavior is accompanied by alterations in the waveform of the melatonin signal; changes in its strength may help to regulate the ovarian cycle in women. Taken orally, melatonin results in a state of well-being, contentment and tranquility. Given intravenously it promotes sleep, but not in every case. It is possible that melatonin plays a role in synchronizing the various stages of sleep, including the stage of vivid dreams.

There is intriguing evidence that in a harsh climate, such as that in which animal reproduction is tightly constrained to certain times of the year, humans also conform to seasonal rhythms. Joel Ehrenkranz, a doctor at the Morristown Memorial Hospital in New Jersey, studied the church records of births in Labrador while working there as a general practitioner, delivering babies and making house calls by row boat and helicopter. Ehrenkranz transcribed the dates of more than 3,700 births between 1778 and 1940, and found a striking pattern. Yearly birthrates peaked in March and fell by 50 percent to reach a minimum in June, July and August. They rose again during September and October and then fell 30 percent in December and January. Identical trends were seen in two separate communities. The seasonal variation was four times greater than that recorded elsewhere. Even during periods of great cultural and religious disruptions in the lives of the people in these communities, the annual cycle remained stable. Ehrenkranz took blood samples from four Inuit hunters at intervals of three months to see if the amounts of melatonin changed during the year. He was especially careful not to expose the men to artificial light. The samples, when analyzed, showed that the circadian pattern of melatonin secretion did vary from season to season.

In winter, almost all the melatonin was secreted at night, but in summer roughly equal amounts of the hormone were in the blood during the day and night. It has been shown that when melatonin is present in the daytime, its usual effects on the body no longer occur. Melatonin secretion during the day appears to cancel the hormone's action altogether. Dr. Ehrenkranz found that conceptions among the Inuit peak in June and July, the time of year when melatonin was neutralized, so that births crest around March. "That shows, at the very least, that the pineal gland is an active organ in adult humans, and it may be regulated by light," he cautiously concludes. "Also, there is a correlation between the observed annual cycle of births and the observed annual cycle of pineal function. Such a correlation is provocative, though it does not prove a definite cause and effect relation between melatonin and reproduction in humans. Of course, there are other well-known links between seasonal changes in the body and the effect of light. For example, children tend to put on height in the spring and put on weight in the fall, but in the blind this periodicity breaks down and no such pattern is observed."

The role played by light in these time-determined funtions of the body is complex and of great interest. Light, especially dawn light, entrains the circadian clock, and the changing length of daylight provides information about time to the seasonal system of rhythms. Yet in the case of the pineal gland, light can also change the relationship between a rhythm and the clock that controls it. And darkness also seems to be a message about time. Darkness acts as a sort of shutter which must be closed in order for melatonin to be secreted, and the shutter is closed tighter in winter than it is in summer.

In fact, if light, time and chemistry are closely related in the circadian system, so too are darkness, time and chemistry. Recently, it has been learned that a part of the brain called the lateral geniculate nucleus, or LGN, which is connected by a pathway to the SCN, is also involved in modifying circadian time. If that nucleus is stimulated, it will reset the circadian clock in the SCN. but the changes in the clock's timing are not at all like the changes produced by pulses of light. They resemble much more effects that might be produced by pulses of dark. The coupling link from the LGN to the SCN involves a neurotransmitter which was discovered in England as recently as 1983, called neuropeptide Y, and it seems to be of great importance, since it is found all over the brain. Biologists refer to neuropeptide Y playfully as "a two-year-old neurotransmitter." If this substance is introduced directly into the SCN, it will shift the clock just as if a dark pulse had been received from the environment. It seems that there is a hitherto unsuspected "negative world" in the brain where dark pulses manipulate biological time against a background of light. Thus pulses of dark and their chemical mimics are likely to become part of science's arsenal of instruments for manipulating biological time.

Scientists who try to unravel the puzzles surrounding the temporal organization of the body are taking a new look at the circadian systems from the perspective of important and surprising new principles—namely, that light is in a sense a drug, that certain drugs behave like light, and that darkness may have its own messenger in the nervous system.

"The most powerful handle we have on biological rhythms today is that the system is controlled by light, and we can use that fact to manipulate the system and understand it better," said Thomas Wehr, a psychiatrist at the National Institutes of Health. "The way in which I would like to understand the system better is in terms of the chemical messengers and nerve cells in the brain that regulate rhythms and the responses of those rhythms to light. I would like to use pharmacological probes, photomimetic drugs that imitate the action of light."

Dr. Wehr and his colleagues have achieved some exciting results in the treatment of depression by shifting the timing of sleep in relation to the circadian cycle. Depression seems to be a time disease, and it is connected with disturbances in the temporal patterns of sleep. Depression has been described as a sort of abnormal biological rhythm in itself, spanning weeks, months or years. Light and darkness, time and chemistry, are elements of this mood disease, and manipulating them is a way of attempting a cure.

Depression has a daily and a seasonal aspect. The sufferer tends to awaken early in the morning, and his mood is worse in the morning than it is in the evening, which may mean that circadian clocks controlling the sleep-wake and mood rhythms are out of phase with the day-night cycle. Scientists have noted the autonomous nature of depression, how it resists "feedback" in the form of efforts by other people to cheer the patient up, and, as we have noted, biological rhythms do not operate on a feedback principle. Depression may also fluctuate with a yearly rhythm. It is most common in the late spring and in the autumn. Emily Dickinson cap-

tured this calendar-dependent lowering of spirits wonder-
fully in the lines:

There's a certain Slant of light
Winter Afternoons,
That oppresses, like the Heft
Of Cathedral Tunes.

The seasonal nature of some mood illnesses could be
influenced by the "darkness shutter" that regulates the secretion of melatonin. Patterns of suicides have been studied
for more than a century and they show a peak in spring, both
in the northern and in the southern hemispheres. Exposing
depressed people to bright, rich artificial light early in the
morning, extending the natural span of daylight, seems to
improve their condition. Researchers who used this procedure knew that bright light is "biologically active" in humans, suppressing melatonin, whereas ordinary room light
had no effect on depression and does not suppress melatonin.

It may be no coincidence that illnesses affecting mood
are closely linked to time of day and time of season, to light
and darkness and to sleep. All these are related to the system
centered around the hypothalamus, the SCN, the LGN and
the pineal gland. Daily and seasonal rhythms are coordinated
here. Nor may it be an accident that the medium of communication in this network of biological timekeeping is the
neurotransmitter, or rather many different types of neurotransmitter. It is well known that drugs which cause or treat
depression and mania have an effect on the manufacture and
release of neurotransmitters, which fluctuate with a circadian
rhythm.

It is a worthwhile hypothesis that in the depressed patient, who often wakes up at 4 A.M. or 5 A.M., and lies moping
and brooding, unable to get back to sleep, the circadian
rhythm of rest and activity, under the control of the so-called
weak Y pacemaker, may be abnormally advanced. The improvement in such a person's mood in the evening could be
a result of a delay in the rhythm of hormone and temperature

changes that is in the domain of the strong "X pacemaker." But why is the rest and activity rhythm advanced? One reason may be that these unlucky people are unusually sensitive to time cues, such as light. The perceived brightness of a light signal may shift the setting of the body clock. For a day-active species, such as humans, whose tau is longer than 24 hours, the effect of a strong time cue on the phase of a circadian rhythm is all the more marked.

Dr. Wehr has put the same patients on two different sleep schedules, one early, the other late. He found that when their sleep was shifted earlier, they felt better. In fact, many had remissions. When sleep was shifted later, however, they did not improve. "Now that says several things to us," Dr. Wehr explains. "First, whatever depression is, it seems that sleeping and waking play an important role in the disease. Something happpens during sleep that is responsible for depression. If you interrupt sleep you interrupt depression.

"Secondly, the fact that one sleep schedule was successful and another was not suggests that it is not sleep by itself or the amount of sleep that matters, but the timing of sleep, the phase of the circadian cycle in which sleep occurs. So it looks as if depression is sleep-dependent and circadian-phase-dependent as well. It makes you think that perhaps there is an interaction of sleep with some specific sleep-sensitive interval during the circadian cycle that plays a critical part in causing this disease."

Yet shifting the timing of sleep in relation to circadian rhythms in an effort to cure depression is a cumbersome and inconvenient procedure. People do not really like to go to sleep at odd times of the day or night. It might be better to come at the problem from the opposite direction, and instead of altering the time of sleep, shift the phase of the circadian rhythms and leave the timing of sleep unchanged. Because light is a sort of drug, and light shifts the phase of a circadian clock, scientists believe that ultimately they will be able to manipulate biological time with drugs. That means that the message from the environment about local time, usually in the form of daylight and darkness and changing day-length, can be mimicked by a pill. We may be able to swallow a time cue with a glass of water once the circadian system is more fully understood.

A pill that resets circadian clocks could be used to treat many kinds of time disorder. Shift workers whose body clocks are out of synchronization with a changing work routine could take it, entraining them at once to a new schedule. Since seasonal depression responds to light treatment, it would be amenable also to a light-imitating pill. Jet-lag would be an inconvenience of the past. A passenger flying the Atlantic from New York to London would take a pill at a precisely calculated moment during the flight. Instead of adjusting slowly, over several days, to the change in schedules, the correction would be made in the air. When this passenger of the future lands, his biology is no longer on New York time. It is on London time.

It is known already that a number of drugs alter the relationship between internal body time and external local time. Coffee is one of these agents. Martin Moore-Ede thinks "we may already be wantonly treating circadian phase disruption on a massive scale with Monday morning cups of coffee; the methylxanthines, including theophylline and caffeine, have been shown to alter the phase of circadian rhythms." The effect of coffee seems to be weak, however. It has been estimated that a passenger would have to drink fifty or more cups of coffee to reset a biological clock on a transatlantic flight.

Among the substances that are possible candidates for a jet-lag pill are cholinergic, nicotinic agonists which are similar to acetylcholine, carbachol and nicotine. Carbachol, for example, simulates the effects of light pulses and can be used to shift the timing of the "Y pacemaker" in the brain. A drug that blocks the effect of GABA, a brain messenger which inhibits the activity of nerve cells, eliminates responses to light signals in the circadian system that set the clock forward, but leaves unaffected responses that set the clock backward. This suggests that there may be two populations of nerve cells in the human nervous system, one mediating advances of the clock and the other one mediating delays. And it could all be happening in the retina of the eye.

One type of chronobiotic drug, given early in the circadian period, sets body clocks ahead, enabling airline passengers flying eastward across time zones to adjust to the new local time of their destination. These drugs include ACTH, the hormone that sends messages from the pituitary to the adrenal cortex, nomifensine, an antidepressant, theophylline, which advances or retards a biological clock depending on what time it is given, and chlordiazepoxide (Librium), which lessens the disruptive effects of phase shifts. A second type of drug works in a different way. It intensifies the modifying effect other time cues have on the period of a pacemaker clock. Some of these drugs, like deuterium, lithium, amphetamine, clorgyline and ethyl alcohol, lengthen the circadian period and make it easier for airline passengers on westward flights to reentrain with the day-night cycle. Others shorten the period and would be helpful to eastbound travelers; these include the tricyclic antidepressant drugs, melatonin and estradiol.

What is new about all this is that the connections that link clocks, that make them truly a system, is beginning to be mapped. A breakthrough in understanding the wiring of the circadian system as opposed to its various clocks has come about rather suddenly. There is intense activity. "It is going to lead to rational pharmacological interventions which were not possible before," says Dr. Wehr. "That is the exciting prospect now. The interesting thing about these substances is that they may not resemble other drugs used to treat depression, because they have come out of a completely different way of looking at the disease. A few years ago, we left the pharmacological treatment of depression and tried other techniques, such as light and the alteration of sleeping and waking schedules. Then we got interested in the chemistry of sleep and wakefulness and of light and dark.

"We hope this will lead us to a treatment of depression which is quite new, and very unlike today's approach to the use of drugs. Today the tendency is to take a drug and keep trying to obtain variations on the same drug. There is a certain amount of wheel-spinning in that approach. Sometimes you have to go away from a problem and then come at it from a completely different angle. There is another important point. The really novel aspect of these agents which mimic light is that the timing of their administration will be critical to their effect. If you give them at one time of day they will produce an effect which is exactly the opposite of the effect produced if the drug is given at another time of day."

Psychologists are talking a great deal nowadays about "metacognition," by which they mean that human beings do not just know or learn or remember in a state of innocence; they know about knowing, learn about learning and have an intuitive or perhaps even a well-informed grasp of how memory works. The mind also has a metasense of time, applying rational and objective corrections to the subjective distortions and illusions that make an hour "seem" like fifteen minutes; we shall examine this phenomenon later on. Something like this process may be going on today in our relationship with biological time. Metacircadian elements are crowding into our experience of circadian rhythms as more becomes known about them. Air travelers can adjust their behavior to reduce the severity of jet-lag; shift workers can adopt schedules that minimize disruption of their sleep-wake and temperature cycles, taking advantage of the fact that because a human tau is about 25 hours, people adapt more easily to later working hours than to earlier ones when they change shifts. Sufferers from mood illnesses can expect to have their biological clocks manipulated by light or drugs. Retiming the body's timers is a metaprocedure that will become increasingly widespread in the 1980s and beyond, and this too is a product of an advanced stage of evolution, in which conscious knowledge can be applied to the unconscious and hidden structures that subtly influence our lives.

EIGHT

Timetable
for Simple Minds

One of the chief distinctions between homeostasis and the body's network of biological clocks is that the first makes an individual more free in time than he would have been otherwise, and the second makes him less free. But that is a convenient oversimplification. There are degrees of freedom in body time structure. One type of structure may have evolved in such a way that it binds an animal tightly in the grip of time, while another type holds it in a looser embrace. It is possible to imagine biological timekeeping systems as a continuum, running between two extremes. At one extreme, every state of the body and every behavior of the organism would be rigidly scheduled at a definite time in each 24-hour cycle, or in each month or year, leaving an animal no freedom of choice as to when to perform a certain act or when to be in a certain biological state. The clocks would be the "cause" of all actions and all changes of state. At the other extreme, the individual would be at liberty to be and to act in its own time, so that behavior would be either random or the result of a deliberate decision. In the first case, the creature's existence is programmed, and in the second case it is spontaneous.

In reality, most species are a mixture of the programmed and the spontaneous, with some leaning more in the direction of temporal freedom and others more in the direction of temporal constraint. Time is important in some respects and unimportant in other respects, and it is the tension, the balance between these two factors, that helps to determine the "nature" of a species, its hidden limitations and assets, its way of life, its relationship with the environment. Even human nature, perhaps *especially* human nature, cannot be defined without considering the ways in which we are constrained and the ways in which we are free in time.

Biological timekeeping is surprisingly rigid in some respects and surprisingly plastic in others. In matters of reproduction, which has a very high priority in evolution, it can be fairly dictatorial. In many species there is a right time and a wrong time for mating and producing offspring, so that these activities are strictly clocked and synchronized with appropriate times and seasons. The clocks are the "cause" of reproductive events. It seems to be of some importance, for example, that certain mammals are born at just the right hour of the day.

Quite recently, scientists have found that an unborn mammal, still inside the mother's body, "knows" what time of day it is outside. The mother uses her own circadian clocks to synchronize those of her litter with local time. A clear, entrained daily rhythm has been found in the metabolism of glucose in animals still in the fetal stage. There is intriguing evidence suggesting that the time of day at which birth takes place is determined by a biological clock. Since in most animals it is the fetus that initiates birth, the mother must first set the clock of the unborn so that initiation will come at just the right phase of the day-night cycle. The mother, being in touch with the environment, apparently provides information about time of day, and this is the cue that entrains the fetal clock. At the critical moment, the fetus might feed back information of its own to the mother, saying, in effect, "Okay, the time of day is right and I'm ready to go. Let's do it." Then begins the cascade of events that culminates in birth at the appropriate time. It seems to be essential that such an animal emerges into the world at a time of day that is least dangerous to its survival, and the best way to communicate information about local time is while the creature is still in the fetal stage, because the relationship with the mother is more intimate then than it is after birth.

Even the lengthy process of reproducing, from conception to birth, is known now to be more strictly organized in time than was once thought. While the overall rhythms of reproduction in mammals may have periods as long as a month or a year, specific events in the reproductive process are switched on and switched off by the circadian system so that the temporal constraints for these events are quite specific as to time of day.

Restricting a creature's freedom in time need not mean that an animal is less successful than it might have been otherwise; in fact, temporal constraints may enable the animal to be highly successful in a particular, specialized way, so that it can act more effectively in its world. A built-in program, assigning biological or psychological events to definite times and ordering them into specific sequences, drastically reduces choice. It closes off many options and makes alternative forms of behavior inaccessible. But the fact that such a program was selected for during evolution makes it likely to enhance survival. The options that are closed off tend to be those that are ineffective or dangerous, and the actions that are programmed in tend to be those that are effective and safe. So the animal can engage in successful actions without having to learn from experience or experiment, without needing to try out actions that may or may not be successful, and could be harmful. The creature is "free" in the sense that correct behavior is ruled in and incorrect behavior is ruled out, rather as the Soviets argue that whereas people in the western democracies are at liberty to starve, their own citizens are provided with the freedom to work.

Time constraints can be a means of guiding animals not

bright enough to have extensive knowledge of the world to function as if they did have such knowledge. Simple minds can sometimes operate like wise minds in nature because choice has been eliminated and "wise" behavior of a particular, restricted sort has been programmed in. A complex world is drastically simplified for such beings, since they do not have to decide when to do this or that. Time slots are already provided for them by programs under the control of biological clocks. A program tells them, in effect, "act now," or "don't act now." Such a mechanism could mislead an innocent observer into thinking that the animal had made a conscious, intelligent decision.

One of the most intriguing examples of stupidity masquerading as brilliance is the honeybee, whose brain weighs only one thousandth of a gram, hardly the equipment of a mental giant. Bees can learn and remember certain things with astonishing ease and accuracy. They can distinguish a range of flower colors from yellow to green and blue and are especially adept at spotting the ultraviolet at the center of a flower, which is invisible to the human eye. They retain the information in memory. Bees can also remember the landmarks near a particular flower, so that it will be easier to find at a later time.

Honeybee learning, however, is subject to a peculiar limitation. It can be done only at certain critical periods, within quite narrow time windows, and these windows are determined innately, by means of built-in temporal programs. Bees can recognize one particular scent out of several hundred. Yet they learn that scent only during the time when they are actually standing on the flower. Since bees collect nectar from only one species of flower at a time, it is important for them to remember the color of that species, but they are programmed to learn that color only in the final two seconds before they land on the flower. Even odder is the fact that a bee, while it has both short-term and long-term memory, can learn the landmarks around a flower only in the critical, brief time period when it is flying away from the flower, and at no other time. The bee may see the landmarks, but it does not remember them unless it actually departs from the food source under its own wingpower.

Another critical period of learning for a bee is during the first flight out from the hive each morning. Bees need to remember large objects near the hive, such as a bush or the side of a house, because they are very shortsighted and are hopelessly inept at finding the small entrance to the hive without extra information in the form of bigger and more striking features of the surrounding landscape. Only during the first flight out in the morning, however, and at no other time, can a bee commit these objects to memory. If the hive is moved during the night, the bees suffer no inconvenience. They simply store the new landmarks in their little memories first thing in the morning. Total confusion results, however, if the hive is moved during the day, even if it is only rotated by 90 degrees so that the entrance is facing another way. The change throws the bees into helpless disorientation. They may spend hours hovering around the hive, flying away a short distance and then returning, as if to say, "Where's the damned entrance?" Often one bee will find the entrance by accident, land on it, begin to secrete a chemical substance that attracts the rest, and at last the muddle is sorted out.

In one sense, these time constraints are a deficiency for the bee. They determine the limits of the animal's knowledge of the world, and the limits are surprisingly strict and narrow. In another sense, however, they are an important asset. They suppress mental activity most of the time, and permit it only at the most appropriate moments. They ensure that remembering takes place only when the bee really needs to remember, so that precious space is not taken up in its lightweight brain with superfluous circuitry. The narrow time windows are sufficient under normal circumstances, but do not allow for very unusual contingencies, such as a scientist moving the hive during the day.

Professor James Gould of Princeton University has made some calculations in an attempt to discover how much space in a bee's brain would be needed for various methods of memory storage, and it is clear that without the time-determined strategy a bee has for remembering, it would run out of storage room. "It must be very expensive for a bee to memorize everything," Professor Gould said. "If you watch a bee after it has found a new source of food it studies the landmarks for many, many seconds. It needs to spend quite a bit of time really getting it down. Probably this has to do with the small size of the bee's brain.

"Restricting mental acts to specific times, and suppressing them at other times, saves space in the brain. Each time is remembered separately. A bee can learn only one thing at a particular time of day, and finds it impossible to remember two things that have the same time tag attached to them. It can learn that food is on a blue triangle with an orange scent at ten o'clock in the morning, and that food is on a green square with a peppermint scent at ten-thirty. But you cannot teach a bee that both are satisfactory food sources at the same time of day."

Time constraints such as these enable bees to conduct their lives efficiently by narrowing the focus of mental activity. Bees are specialists, tending to harvest the same kind of flower as long as it is available, and then switching to some other species at a different time of day or a different month of the year. Flowers open and close during the day at such predictable times that the eighteenth-century Swedish naturalist Carolus Linnaeus created a flower clock, consisting of different species of flower arranged in a circle, each one representing an hour of the day. When the spotted cat's ear opened, it was 6 A.M., when the passion flower opened it was noon, and the evening primrose told the hour of 6 P.M. Given this predictability, the bee's memory works like an appointments calendar. There is an entry for 11:15 A.M., one for 11:30, another for 11:45, and so forth. Bees concentrate on one species of flower at a time, treating each one as a separate challenge, which means that once familiar with a certain flower they work on it very rapidly and deftly, and as automatically as a person tying a shoelace.

Gould calls bees "carefully tuned learning machines," learning only what they are programmed for, when the program tells them to do so, and this happens under exactly defined circumstances. Other animals, too, are able to acquire important information only during brief, clearly specified time slots. That is the case with some species of bird. The white-crowned sparrow, for example, can memorize the song of its parents only between the tenth and fiftieth days of its life. At the end of that critical period, no amount of teaching will make the young bird learn the song. If the song is stored before the fiftieth day, however, it is retained safely in memory until the lengthening hours of daylight in spring release certain hormones in male birds which have the effect of starting the program of song. At first, the young birds can produce no more than fragments of the song pattern, but through practice they learn to match what they are singing with the memorized version of the parent's song stored in their brain. Once this stage is reached, nothing can stop them from singing.

Sometimes what appears to be spontaneous in an animal's behavior is really clock-determined, but with flexibility programmed in. The animal is not free not to be flexible in the timing of its actions, and this rule is strictly defined. Such is the case with many European birds which travel long distances in autumn. Migration, like reproduction, has a high priority in evolution and therefore, in many species, it tends to be confined to definite time slots. These European birds with their programmed flexibility vary their speed according to the changing nature of the terrain along the path of flight. They fly fast across the Mediterranean sea and the Sahara desert, which offer few resting places, but more slowly over Europe and central and southern Africa. These changing patterns of speed appear to be under the control of a biological clock which operates on a yearly time scale. Another bird, the chiff-chaff, which migrates in autumn to the Mediterranean and northern Africa, but later in the season than some other species, returning north much earlier in the spring, needs to be more wary of bad weather at that time of year, and cannot afford to be rigid in its departure dates. The chiff-chaff has a half-yearly clock which is a great deal less punctilious in its program than other birds, which fly south earlier in autumn and return later in spring.

An animal with limited intellectual powers, which must come to grips with a complex world almost as soon as it is born, needs some built-in guidance. It must specialize in knowing those things that are relevant to its own way of life, and get to know them quickly, without being distracted by trying to learn irrelevant information. The animal must be especially effective at processing information that is important, when it is important, rather than trying to learn and remember everything all the time. So evolution has found a way to restrict the learning and remembering of important information to certain critical times and contexts. This should not be mistaken for high intelligence. A bee acquires "knowledge" rapidly even with its dinky little brain, because its learning is extremely selective and time-dependent, confined to certain moments and certain circumstances. A bee learns as it goes along, but such learning is like putting data into a computer program.

It might be thought that human beings, who have large brains and are the leading generalists of the animal kingdom, and who, rather than having to cope with the world straightaway, spend years in the safe, protected cocoon of childhood acquiring knowledge by experience and trial and error, would be at the opposite end of the continuum that runs from being completely programmed to being completely spontaneous. If that were the case, there would be no windows of time for us, no intervals of the day when certain kinds of behavior, specific mental acts, are best performed. Such behaviors would not be clock-determined. We would be perfectly free to act at any time of our own choosing. The more we discover about biological and psychological time, the more questionable that assumption becomes.

In fact, it has been said of human beings that when they are tightly constrained by time, when time becomes important for them and a "cause" of behavior, they are existing at a rather low level of consciousness, displaying the less highly evolved side of their nature. Malcolm Muggeridge suggests that there exists an ascending scale of states of human consciousness, each one progressively freer of time. Low down on this scale of being is the will, a relatively primitive faculty, located "in the dark dungeon of the ego," a prisoner of time, with its schedules and programs and deadlines for action. Higher up the scale is the imagination, a window for the mind to gaze out of and dream of breaking free from temporal restrictions. At the highest level of all Muggeridge places mystical experience, which provides a complete escape from time. It is as if Sartre's mermaid had grown another head. C. S. Lewis talked about time as a constraint that, by restricting our mental focus, enables us to have some limited consciousness of things our minds are not designed to know directly and in their entirety. Time, Lewis said, is a sort of lens by means of which an individual can look at freedom, as if through the wrong end of a telescope, and see small and clear what would otherwise be too big for him to see at all.

Human beings are partly free in time and partly unfree. It is possible to read Shakespeare's *Hamlet* as a play about the privileges and penalties that accrue to a highly evolved species like man, who may, within limits, treat time as unimportant. Time is not a cause of behavior in *Hamlet*. A human being need not act immediately when he encounters a stimulus, as a frog strikes the instant a flylike object moves across its field of vision. Behavior is not strictly confined to narrow time slots during which a built-in program says "act now." To a being with a sense of the past and future, the present does not seem as important as it might for a species that lacks such a broad temporal perspective. The appearance, in the play's first scene, of the ghost of Hamlet's father (a famously punctual character, it should be noted) on the battlements of Elsinore, is a stimulus to action, the avenging of the father's murder by the usurper Claudius. Yet the whole play is a commentary on Hamlet's postponement of the act of revenge, while he soliloquizes, speculates, summons up theatrical representations of his dilemma in the form of play scenes, and lets his mind drift back to past events and forward to the "undiscovered country" of the afterlife, reflecting constantly on his own indecision and delay. Among many other things, *Hamlet* is a play about the glories of the human intellect, which makes man "the paragon of animals," but by granting him the luxury of ignoring time turns him inward and cripples his ability to intervene in the real world. In *Macbeth,* by contrast, the actions are a great deal more time-dependent, because the prophecy of the three witches in the first scene must be fulfilled on schedule, and Macbeth himself often seems to be a man in the grip of a program written by those sinister hags. It may be no coincidence that Macbeth is a much more primitive character than Hamlet.

Yet human beings are subject to time constraints, many of them so subtle, so far below the surface of consciousness, as to be imperceptible. When biologists or psychologists discover the existence of submerged programs of this kind, it comes as a surprise to find that our lives are restricted by these unsuspected schedules which make states of body and states of mind different at different times of the day and night, and even at different seasons of the year. Our mental powers do vary systematically during a 24-hour period, including the power to learn and remember, so that there are faint and probably very ancient analogies in the human brain of the windows of time that restrict the cognitive life of a honeybee. There are also critical periods during human childhood when knowledge, especially knowledge of language, is more easily acquired than at other periods.

"We are blind to our own blindnesses," says professor Gould. "The times when we are most susceptible to particular sorts of information may be more restricted than we like to think. Even the malleable learning we as humans pride ourselves on may have ineradicable roots in genetic programming, although it may be difficult for us to identify the program."

Such restrictions are part of the concept, once despised but now making a return to the mainstream of ideas, of "human nature." How free we are in time, and how bound, must enter into any description of human nature. Knowing what man is capable of is not enough. We should also know at what times he is most capable, and at what times he is least capable, and recognize that the temporal separation between most and least is a legacy of evolution, part of the innate structure of body and mind.

Modern philosophy is much preoccupied with the question of freedom and how to define it satisfactorily. Many thinkers have stressed the uniqueness of freedom in the human makeup. Rollo May points out that freedom is unlike every other element of man's being, because it alone does not have a definite nature. May says:

"Every other reality in human experience becomes what it is by its nature. The heart beats, the eyes see; it is their nature to do what they do. . . . What, then, is the nature of freedom? It is the essence of freedom precisely that its nature is *not* given. Its function is to change its nature, to become something different from what it is at any given moment. Freedom is the possibility of development, of enhancement of one's life; or the possibility of withdrawing, shutting oneself up, denying and stultifying one's growth. This uniqueness makes freedom different from every other reality in human experience."

The science of life time is showing that humans may be less free than we think, because "human nature" is a more complicated, and more restricted thing than we might suppose. If freedom is the power to become something different "at any given moment," then we may possess it less abundantly than we imagine. The reason is that in human biology, as well as in human psychology, one moment cannot necessarily be regarded as the same as all other moments. There is no complete escape from temporal constraints that make us different, in body and brain, at different moments. The constraints are part of the special logic of our existence and history as a species, just as those of a honeybee are part of its logic. We are temporally organized in part because the world is temporally organized, and we must adapt to the world or perish. Being organized, in space or in time, means that certain options are closed and certain other options are open.

Not everything is possible. Freedom is sacrificed so that we are more effective in our environments almost whether we want to be or not. For good reasons, evolution has seen to it that bees and humans, both in their particular ways, are caught, albeit unawares, in the net of time.

NINE

The Clocks
of Sleep and Dreams

Of all the clocked programs that suppress selected types of human cognitive activity at certain times and release them at other times, the most dramatic, as well as the most puzzling, is the cycle of sleep and wakefulness. Sleep is a condition of the body, and it is associated with rhythmic changes in thousands of biological funtions, from core temperature to hormone secretions to the operations of the bladder. But it is also the setting for mental experiences which have a peculiar character of their own. The sleeping self is psychologically as well as physiologically different from the waking self. In addition, and less obviously, we are different from one phase of sleep to another, and as in the cell cycle of division and growth, the phases are not random, but organized in time. In the past few years, scientists have come to appreciate both the subtlety and the authority of the time rules governing the external and the internal structure of sleep. Sleep is not one state, but an orderly, rhythmic succes-

sion of contrasting states, each with its own special properties and presumably with its own special meaning for the
organism.

Like the honeybee, human beings are provided by nature with time slots in the 24-hour cycle during which the
brain operates in a certain fashion and other time slots in
which it operates in a different fashion. The constraints are
not nearly as tight as in the case of bees, but they are there
for sure. In sleep, we are not free not to dream, and also not
free to choose when we dream.

Sleep is not strictly programmed. If it were, it would
submerge us beneath the tides of consciousness at a fixed
hour at night and bring us to the surface at the same time
every morning. That is a poor evolutionary strategy, because
it would make an animal completely predictable, whereas
the whole point of biological rhythms is that the animal
should take advantage of the predictable time schedules of
the environment, the ones it has internalized during evolution, rather than permit the environment, including predators and competitors, to take advantage of the predictable
behavior of the animal. We are free, within limits, to violate
the time constraints of sleep and suffer no lasting injury,
which is why sleep has been called "the gentle tyrant."

Yet the greatest mistake would be to assume that the
cycle of sleep and waking, like the cycle of sleep and dreaming, is not influenced by a biological clock. The clock is not
the cause of sleep and dreaming, but it is *a* cause. It is one of
a number of elements that together govern the temporal
structure of sleep. Biological time, like psychological time, is
a multiplicity that often functions as a unity.

Exactly why nature has equipped us with such an elaborate timetable of alternating types of mental experience during sleep is still a matter for debate. The time constraints on
the honeybee seem entirely reasonable and prudent as an
evolutionary strategy, because they organize states of the
brain so as to make the most of the brain's limited resources,
activating certain mental faculties at times of importance and
meaning and shutting them down at other times. Perhaps

there is a similar explanation for the much less drastic, much less rigid time constraints on human cognition during sleep and waking.

What is sleep for? To that question, plenty of answers have been given, but all of them are speculative. Sleep occupies a third of our lives, but its function, the reasons why we are captive to it, and even why we seem unable to spend fewer hours in its anesthetic thrall are far from clear. Thus the role of time in sleep is also something of a mystery, because the time structure of sleep is connected with the meaning of sleep, its relevance to the life of the body and the brain.

In part, sleep may be a homeostatic mechanism, restoring cells and replenishing chemicals that have been depleted during waking hours, returning them to their pristine state. If that were all there is to sleep, however, its time properties would not be as stable as they evidently are. The work of restoration would be under the control of a feedback loop of some kind. Sleep would continue until the system had been replenished, and this would take longer to complete after a day of intense, stressful effort than after a day of leisure and recreation. Yet we all know that sleep does not work in that way. We sleep for roughly the same amount of time night after night, regardless of what we have been doing during the day. One person's typical sleep length may be different from someone else's, but it does not vary greatly for each individual. Presumably it is under the influence of built-in time constraints which are not feedback devices, not easily affected by changing circumstances.

The programmatic property of sleep cannot be eradicated. Its stubborn refusal to disappear as a result of manipulation has been the cause of considerable frustration to one of the most respected of American sleep researchers, Wilse Webb, of the University of Florida. A behaviorist in his early days, Webb assumed that people sleep for as long as eight hours a night because they do not sleep efficiently enough. If we could understand the laws of sleep, its underlying mechanism, Webb reasoned, we would be able to reduce our sleeping time by two hours or so a night, which over a life-time would amount to a free gift of an extra three years of waking consciousness. Webb made the mistake of supposing that there are no built-in temporal factors influencing sleep. He spent about five years trying to teach rats to stay awake between noon and 6 P.M. which is when, left to themselves, rats spend most time asleep. The rats were uncooperative. They refused to act like good behaviorists, and showed that they did possess an intrinsic mechanism which predisposed them to sleep at their "natural" time of day. "I used every technique known to man," Webb has said. "I shocked them, I drugged them, I walked them. I did everything I could to keep them awake. But as soon as I would release them— Pow!—they would go right back to sleeping the same way they had. It finally got into my thick skull that sleep is not a learned behavior, but an innate, biological system that is set to function in a particular way."

After the debacle with the rats, Webb still had not quite learned his lesson. He decided that human beings, with their superior and more adaptable brains, could be made to increase the efficiency of their sleep and reduce its length, but these efforts, too, met with failure. Webb concluded that sleep is an inborn biological rhythm, even in humans. It is a program, but a flexible one, in that it allows for some spontaneity and opportunism. "If you are in great danger, it would be disastrous to fall asleep," Webb argued. "Woody Allen put it very well: 'The lamb and the lion shall lie down together, but the lamb will not be very sleepy.' " Yet sleep also has its tyrannical side. Webb quotes Horace: "You can try to drive nature out with a pitchfork, but she always returns."

In humans, the program of sleep and waking is determined only in part by brain clocks. Habit, routine and conscious behavior also play a part. The clocks seem to be more dictatorial when waking us up than when putting us to sleep. This means that the time constraints are very complex. In a recent study at Dr. Webb's laboratory at the University of Florida, it was found that when familiar daily routines such as watching television and talking to friends are eliminated, and people are allowed to eat and sleep whenever they feel like it, sleep breaks into temporal fragments. It no longer occupies a single, continuous block of nighttime. Sleep is still under the influence of a circadian clock, to a degree, but the influence is only partial. Episodes of sleep that started after about 7:30 P.M. were much longer than episodes that started during the daytime. On the average, nighttime sleep lasted twice as long as daytime sleep, suggesting that the proposed weak Y sleep clock stayed in synchrony with the strong X body temperature clock, although people taking part in the study were isolated from all time cues. Even when the two clocks become desynchronized, sleep length depends chiefly on when it occurs in the daily temperature cycle. People usually choose to go to bed just after the temperature cycle is at its lowest point.

In other respects, however, the temporal organization of sleep in these people who had abandoned all routines was quite anarchic. The programmatic aspect of sleep broke down and gave way to spontaneity. Instead of one long episode of sleep at night, there were many episodes, spread out across the 24-hour day, and some of them were brief. Eighty percent of the episodes lasted less than 4 hours. The average waking period was 2.7 hours, and the average sleeping period was 2.99 hours. Scott Campbell, the psychologist who devised the study, thinks the results suggest the existence of two distinct components of the human sleep system. One component is sleep duration, and it is controlled by an innate biological clock. The other is sleep placement, which is regulated mainly by behavior, familiar routines, periodic social contacts, and freely chosen activities that are alternatives to sleep. When these influences are removed, the distribution of sleep is highly labile.

The fragmentation of sleep seen in the study, Campbell proposes, represents one extreme on a "sloppiness scale" of the human sleep system. Normally, people have regular habits. They eat, read, take exercise and have a single sleep episode at night. This routine amounts to a program which reduces the lability of the sleep-wake cycle. Waking activities preempt sleep. When the routine is suspended, and naps are permitted, spontaneity enters, because there is room for spontaneity built into the system, allowed for by the nature of the sleep-wake cycle itself. At the opposite end of the sloppiness scale are those laboratory experiments in which bedtimes and times of awakening are strictly enforced and naps are forbidden. Somewhere in between is real life, poised between extreme lability and strict organization. We get up at a predictable hour of the morning on weekdays, but relax the rules on weekends.

"There is a time to sleep and a time to wake, just as there is a time to sow and a time to reap," Wilse Webb believes. In humans, these times are determined largely by behavior, by the schedules we arrange for ourselves. He goes on to say: "Each animal, each organism, has its own sleep rhythm. But this time can be overriden. Biological rhythms serve mostly as background primers rather than controls. They are not like instincts, which take over and direct behavior; they modulate it."

The spontaneous element in human sleep, the freedom to choose, within limits, when to sleep in the body's 24-hour timetable, explains the ease with which man can be a creature of the night as well as of the day. Biologically a daytime species, human beings have colonized the night. The human circadian system, with its long tau and its high threshold of sensitivity to light, is adapted to a diurnal existence rather than to a nocturnal one. But humans probably began to invade the nighttime niches after the discovery of fire. Thomas Edison, who considered that to sleep eight hours a night was a deplorable regression to the primitive state of the caveman, hoped to make sleep almost unnecessary by abolishing the night through the invention of the electric light. Like the people in Scott Campbell's experiment, however, Edison took furtive naps at odd times during the day, more evidence that while sleep duration is clock-determined, its placement in the circadian schedule is optional to a certain extent.

There is more to the timing of sleep, though, than its

external properties of total duration and placement in the 24-hour cycle. Sleep has an internal structure which is programmatic, yet forgiving of those who violate the rules. It is not a uniform state, the same from minute to minute and from hour to hour, but changes in an orderly, cyclical fashion as time goes on. This inner temporal organization of sleep is quite complex, and is not easily manipulated from outside (though some parts of it are easier to manipulate than others), just as the external property of total duration of sleep proved to be unexpectedly stiff and stubborn when Dr. Webb tried to alter it. Normal human sleep is different at different times of night. It consists of four stages, which occur in sequence. In addition to these, there is another state with unique characteristics called rapid-eye-movement (REM) sleep, in which vivid dreams occur and intense activity in the nervous system is accompanied by virtual paralysis of the muscles of the body. REM sleep was first identified in 1952, at the laboratory of Nathaniel Kleitman at the University of Chicago.

Stage one sleep is an interlude, a drifting state of passage between wakefulness and oblivion which lasts from one to seven minutes and may seem more like daydreaming than sleep in the accepted sense of the word. One of the most famous stage one dreamers in English literature is Maurice Allington, the bibulous host of a fashionable coaching inn near London in Kingsley Amis's novel *The Green Man*. Mr. Allington experienced visions in this halfway house between sleep and waking that were grotesque and puzzling, but had little power to terrify: "As against the times when an unremarkable profile suddenly turns full face and glares in lunatic rage, or becomes quite inhuman, there are the rarer times when something beautiful shows itself clearly, in a small flare of soft yellow light, before fading into nothing, into the state of a vanished fiction." It is thought that in stage one, the brain weaves fantasies, trying to organize into acceptable images the severely impoverished information that reaches it from the outside world.

Stage two is sound sleep. It occupies between 40 and 60 percent of the total sleeping time. Stage three, deep sleep, accounts for only 3 to 12 percent, while stage four, the deepest of all, makes up 15 to 35 percent. The sequence of stages is: 1-2-3-4-3-2-REM-2-3-4-3-2-REM, and so on. Normally, REM periods occur every 90 minutes or so. The first REM period is only about 5 or 10 minutes long, but the average person returns to REM sleep between four and six times a night, and for increasingly long stretches of time, up to 30 minutes or more. The sequence of stages is organized in time so as to be remarkably independent of accidental circumstances, whether internal or external. It is autonomous, with an integrity of its own, and it persists in a form not greatly altered whether a person sleeps much or little. The system clearly does not operate, as homeostasis does, on feedback principles, but depends on intrinsic timing mechanisms. All stages are rhythmic and periodic, and the stability of the rhythm is one of the most impressive features of sleep. A boy of seventeen was once kept awake continuously for 264 hours, yet when he did fall asleep at last, the sequence and timing of stages were not altered in any drastic way. The amount of REM sleep experienced during a whole night of sleep varies from one individual to another. Each person has a typical proportion of REM sleep which is reasonably constant over months and years, just as one person's "natural" duration of sleep may not be the same as another's. This is part of temporal identity.

Stages three and four of sleep are known collectively as slow-wave sleep, because that is how the electrical impulses of brain cells appear when they are amplified and charted on an electroencephalogram, or EEG. Many nerve cells fire in synchrony in these stages. Slow-wave sleep is extremely ancient, and is thought to have evolved in its full mammalian form about 180 million years ago, perhaps as much as 50 million years before the appearance of REM sleep. Slow-wave sleep still has certain prior rights when sleep has been disrupted. If a person stays up all night, slow-wave sleep increases during the next night, and only on the night after that does the person experience REM sleep in larger amounts than usual.

Clearly, the internal structure of sleep, which includes the nightly dream schedule, defies a behaviorist explanation, and must be based, at least in part, on timing mechanisms that are built into the biology of the sleeper. Allan Hobson and his colleague Robert McCarley at the Harvard Medical School have proposed that, in addition to the sleep-wake cycle, there is also a sleep-dream cycle, under the control of a timing mechanism in the brain which they describe as a sleep-dream clock. The clock consists of an oscillator in the brainstem which releases different kinds of chemicals in a rhythmic fashion. The preponderance of now one sort of chemical, now another, accounts, in this view, for the regular shifts between waking, sleeping and dreaming.

Three groups of cells in the brainstem are of critical importance here because they produce neurotransmitters, the chemical messengers that cross the gap between neurons, allowing cells to communicate with one another. Such communication is the basis of thought and emotions, so that the human brain in effect is controlled by these messenger substances, in all their spectacular variety. The three neurotransmitters involved in the sleep-dream clock are the same ones that play a role in that part of the circadian system, centered in the hypothalamus, that converts light signals into time signals and regulates the pineal gland. They are norepinephrine, serotonin, and acetylcholine.

As Hobson and McCarley see it, groups of cells on one side of the brainstem produce norepinephrine and serotonin, substances which have the technical name of biogenic amines. For that reason, this side of the clock is called aminergic, a word which means "driven by the amine force." The effect of the transmitters is to inhibit certain events from happening in the brain. On the other side of the brainstem is a special group of larger cells called giant pons cells, which produce acetylcholine, and are therefore cholinergic, driven by the choline force.

In the waking state, Hobson and McCarley believe, aminergic forces rule. Serotonin and norepinephrine are released steadily as if by the tick of a clock, and all day they inhibit the pons cells on the other side of the brainstem, which would otherwise, if allowed to be active, generate the dream state. During the nighttime phase of the circadian cycle, the production of these transmitters declines, setting free those brain circuits that were inhibited by aminergic forces during the daytime. Among the circuits released are those in the thalamus and cortex that facilitate the entry into sleep. As the aminergic force declines still further, the intrinsic excitability of many other brain cells begins to increase as the restraints fall away. This rising level of excitability, Hobson and McCarley propose, is what incubates the REM period of sleep. The brain is in a ferment of activity, even though it is asleep and cut off from contact with the outside world. The rapid eye movements, which give this stage of sleep its name, are stimulated by acetylcholine.

One of the nice paradoxes of this procedure is that among the brain systems that are released from inhibition and switched on is one whose function is to inhibit, to damp down. It is this system that immobilizes the muscles of the body by bombarding the nerve cells that control them with a chemical that blocks a response to all the excited buzzing in the brain that characterizes REM sleep. Motor command centers and pattern generators in the brain all rage away, but the commands are not acted upon. Without such massive inhibition, the sleeper would flail about in bed with his legs and arms, acting out in bodily motion the mental experiences of his dreams.

The brainstem oscillator that times the sleep-dream cycle is presumably coupled to a pacemaker clock elsewhere, perhaps the so-called weak Y clock in the hypothalamus. So there is a clock-determined alternation of states in the brain during sleep which is remarkably independent of anything going on outside the body. It is programmatic rather than being simply a feedback mechanism, and seems to be beyond the control of consciousness. The question then arises, of what use are these autonomous, clocked contrasting states? Why do different psychological domains cycle in a beautifully reliable, rhythmic succession, one after the other throughout the night?

An emerging theory of sleep holds that sleep stages do serve an ingenious and important function by servicing different parts of the body and brain at specific times during the period of sleep. According to this theory, slow-wave sleep, the first to appear in evolution, restores the body, promoting recovery from physical exhaustion. Growth hormone, which helps to rebuild tissues, is released during slow-wave sleep in a circadian rhythm that seems to be under the control of the weak Y pacemaker. During slow-wave sleep, mental activity is in abeyance.

During REM sleep, most of the brain is turned on, as circuits previously inhibited are set free from inhibition. But there are certain very important exceptions to this rule. Not all systems are active. Some are shut down selectively during periods of REM, or are damped, while the others are running at normal speed. Motor output, the movement of muscles, is blocked, as we have noted already. In addition to this, the cognitive functions of learning, attention and memory fail. It is impossible for a person to learn while asleep, and it is a remarkable property of dreams that they are experiences in which the dreamer's attention is weak and diffuse. The dreamer does not scrutinize events while they are happening, observing them closely and critically. What is more, insight is also greatly impaired. We dream night after night, but very seldom are we aware that we are dreaming and not awake. The most bizarre and improbable situations are usually accepted as if they were really occurring.

Another mental activity that undergoes a radical alteration in the dream state is memory. It is true that information is accessed from long-term memory, because it is out of such information that the dream story is synthesized. Access to remote memory is good or even enhanced in dream making. But this is a one-way traffic. The newly synthesized mental product, the dream itself, cannot be re-stored in memory; at best it is held for a few moments in short-term memory, which has a lifetime of seconds, and then is lost to consciousness. To put it crudely, the aminergic system is needed to "tell" the cells of long-term memory to make a copy of a mental experience, and in dreams the aminergic system is shut down. This might explain the very powerful amnesia that follows dreaming. At best, less than 5 percent of dreams are remembered, and most people retain less than 1 percent. Even on awakening directly from a vivid dream experience, the dreamer often feels it slipping away as it fades from fragile storage in short-term memory.

Another missing element in REM sleep is a clear concept of time. The absence of such a concept may be due to the fact that attention and the psychological experience of time are closely related. In dreams, space does not seem to have the continuity and integrity it possesses in real life, and time is often shattered into fragments. Sleep researcher Ernest Hartmann has noted that time and space are less well organized in the dream. The edges of visual space are fuzzy, and our awareness of time is diminished.

It is tempting to speculate that learning, attention and memory, which are in the domain of the aminergic system, are shunted out for servicing and checking during REM sleep, when the sleep-dream clock releases the pons cells from inhibition by reducing the output of biogenic amines. One of the big puzzles of early sleep research was the fact that brain cells are buzzing all the time during REM sleep, which made scientists wonder what on earth could be resting. The new theory suggests that one of the systems being rested is the aminergic system, keyed to memory and learning. Rest periods are determined by an internally generated timetable, and all this happens as part of the very mechanism by which sleep is made.

Specific systems of the brain can be made to operate "off-line," to use a computer analogy, at times when they are not needed, so that they can be run through and checked without interfering with the active, waking life of the organism. It is a way of testing circuits without cost. The amnesia that follows dreaming may be due to the fact that the brain is trying out short-term memory circuits, independently of permanent memory, making sure that they are working properly "with the clutch out." It is not necessary to transfer the information to long-term memory. "You don't need to remember the material," Allan Hobson points out. "All you want to do is to be sure that circuit A is working, circuit B is working, circuit C is working and so forth. The short-term memory system must be tuned and ready to act. The purpose of rehearsing information in short-term memory is to guarantee the efficacy of the system, not to retain the information for any length of time."

The various time slots that are part of the internal structure of sleep have different meanings for the organism. In fact, it is the psychological and physiological contrast between one stage of sleep and another that has been one of the most surprising discoveries in sleep research over the past thirty years. It seems to be important that each stage is kept separate in time, because each performs a special role in sleep. Deep sleep, for example, the stage four type which shows up on the EEG as high-voltage delta waves, is very different from the REM sleep stage. Delta-wave sleep seems to be a homeostatic process, replenishing the body and playing a role in metabolism. Growth hormone is secreted in the largest amount during this stage of sleep, which tends to be present in larger amounts after starvation and when total sleep time is forcibly reduced. What is more, delta sleep appears to be less clock-determined than REM sleep. It is more loosely coupled to the circadian pacemaker. How much of a whole night's sleep is delta wave depends on how long the sleeper stayed awake before going to sleep. It is not an absolute. People who normally sleep only about six hours a night have shorter periods of stages one and two and REM than those who sleep eight hours a night, but their delta periods are the same length as they are in the longer sleepers. Scientists who manipulate the inner structure of sleep find that the delta stage can be shifted to a different part of the night experimentally more easily than REMs can be shifted.

It seems clear that non-REM sleep, especially the deep, delta part of it, has a physiological function, providing a block of time in which the body's metabolic needs are attended to, while REM sleep has a psychological function, servicing the brain in some fashion not fully understood yet, perhaps by testing circuits, consolidating information, replenishing brain cell chemicals, or integrating knowledge that has been acquired during the waking day. Biologically, non-REM sleep is more fundamental than REM, but REM sleep is important for the psychological fitness of the organism, and it seems to be more clock-determined than non-REM. Some psychologists think REM dreams are a way for the brain to deal with emotionally charged experiences, a task for which there is not enough time during the waking day. People who are in the throes of a stressful divorce, for example, have long, complex dreams which tend to be more interesting than the dreams of those who lead placid lives.

The systems of the brain that are shut down for replenishment of neurotransmitters during REM sleep are presumably those that suffer the greatest wear and tear during waking consciousness. There are two or perhaps three major classes of neurotransmitters in the human nervous system. One such class, which plays a role in the sensory-motor functions of the brain, is fast-acting and assists in the rapid transfer of information that arrives from the outside world and goes out into the world. This reflex system of sensing and acting is not in continuous operation. It reacts to events as they happen, and does not plan or predict. If, by accident, you put a hand on a hot stove, you pull it back swiftly, in an unthinking, once-only movement. That is the reflex system at work. The neurotransmitters chiefly found in these fast-acting brain circuits are acetylcholine, which excites cells, and GABA, which inhibits them. These two kinds of chemicals are released in packets at the endings of nerve cells and they are rapidly broken down. A second class of neurotransmitters are involved in those functions in the brain that have to do with attention and learning. These functions are not once-only responses to external happenings. In the waking state, they keep track of recent messages in order to anticipate messages that have not arrived yet. Clearly, a system that anticipates and remembers, monitors past events and future possibilities, needs to be in a continuous state of moderate arousal, and for this a different type of nerve cell is required. Such a cell need not be large and it need not conduct messages rapidly, but it must be reliably persistent. It should be working all the time and never be switched off completely, because then it would require an enormous amount of energy to get it started again. These attention and learning circuits need to run at an idling speed below which the system should not be allowed to fall. The circuits are based on relatively slow-conducting, relatively dependable nerve cells which fire in regular pulses, like the tick-tock of a metronome, and are found all over the brain. They tend to use the biogenic amine neurotransmitters like norepinephrine and serotonin, which are not as quickly broken down after use as those in the sensory-motor systems, and in this way help the brain to maintain a constant level of activity.

Brain systems of this second type display many of the characteristics of what William James called, in his celebrated phrase, the stream of consciousness, whose chief property is to be continuously flowing. As we have seen, James emphasized that consciousness must be kept up and running nonstop in order to provide thought with its sense of continuity in time and preserve an element of sameness in psychological experience, the "law of constancy in our meanings." The stream also serves as a time context for the ideas and thoughts that float in it, a medium linking present, past and future. This continuity ensures that we usually know, if not in detail, where our thoughts have come from and where they are going.

"Every definite image in the mind is steeped and dyed in the free water that flows round it," James wrote. "With it goes the sense of its relations, near and remote, the dying echo of whence it came to us, the dawning sense of whither it is to lead." James also noted the need to restore the system periodically. He went on to say:

"We all of us have this permanent consciousness of whither our thought is going. It is a feeling like any other, a feeling of what thoughts are next to arise, before they have arisen. This field of view of consciousness varies very much in extent, depending largely on the degree of mental fresh-ness or fatigue. When very fresh, our minds carry an im-mense horizon with them. The present image shoots its perspective far before it, irradiating in advance the regions in which lie the thoughts as yet unborn." In states of extreme "brain fag," the horizon contracts almost to the span of the immediate present.

The cells of this "stream of consciousness" system in the brain would be especially prone to exhaustion, because they are running all the time during the waking day, firing contin-ually. By being rested selectively during REM sleep, these cells are able to replenish stores of neurotransmitters. Noc-turnal repairs of this kind are not so important for the fast-acting sensory-motor cells, which as reflex elements only act when they are called upon to act, and thus have plenty of time in which to rest during the waking day. Time, therefore, is central to an understanding of how REM dreams are made and what role they play in the life of an individual.

Not all of the stream of consciousness cells are shut down altogether in dreams. Some are silent, but others con-tinue to work at a low level, "just going putt-putt in a desul-tory way," in Dr. Hobson's words. "In sleep, the bottom does not fall out. A minimum level, a floor, is maintained. Activity declines, in some cases quite dramatically, but it doesn't stop completely."

The fact that certain circuits are kept in an idling state during sleep might explain why we sometimes retain partial memories of dreams, and have a fleeting insight that all is not as it should be as the dream unfolds. There is a sense that what we are experiencing cannot be real life. But such in-sights dissolve rapidly. Occasionally there is a fugitive sense of being in control of events, but that, too, lasts only for a brief time.

"When awake, you need to be oriented to the world and keep up a kind of linguistic dialogue with yourself," Hobson says. "This is very speculative, but it would seem to make sense to suppose that by guaranteeing the constant updating of memory and the persistence of attention, the aminergic system maintains both orientation and self-awareness. It is significant and clear that these functions are lost in sleep and most particularly are lost in dreaming.

"One of the most striking features of dreaming is not knowing where you are or what time it is. Orientation is violated every second by the fusion of people and places. In a sense, the stream of consciousness is taken over. It becomes a kind of maelstrom, over which one has almost no control, whereas when awake you can govern the stream of consciousness, or if you wish withdraw the governance and let it run a bit free. But in waking life the stream never runs away to the same extent as it does in dreams. It never gets completely wild and hallucinatory and out of control."

The witticism that time is a device for preventing everything from happening all at once applies with particular force to the domain of sleep and dreams. The sleep-dream clock separates in time certain states of body and states of mind which are very different from one another. It may be that REM dreams, as has been suggested, are a sort of "return to the basics," a reactivating of mental functions that are a relic of man's primitive past, with their highly emotional, vivid, sensory-motor qualities, their illogic and absence of insight. Normally, during waking life or in slow-wave sleep, these supposedly ancient mental processes exist simultaneously with the more highly evolved activities of consciousness, but are in abeyance, being expressed only in a modified and censored sort of way. In the sleep-dream cycle, they are expressed more directly, as a result of being placed in separate time compartments by the sleep-dream clock.

Yet the separation in time of these mental states may not be as sharp and distinct, as inevitable and failsafe, as was once supposed. The boundaries may be permeable. In some people, it has been found that dreamlike fantasy is not confined to REM periods, but appears also in non-REM sleep. If a person is deprived of REM sleep, some REM characteristics can "escape" into other stages of sleep or into wakefulness.

The neurologist and sleep researcher Elliot Weitzman,

shortly before he died in 1984, declared that the brain is not a single system but a complex society of systems which are both coordinated in time and separated in time by clocklike mechanisms. Such an idea has rather startling implications. It means that states which normally follow one another in a distinct, time-determined sequence may not do so if the clock functions are disrupted. Being awake and being asleep are usually thought of as being two separate and mutually exclusive states which succeed one another during the 24-hour cycle. One can sleep or be awake, but one cannot be both at once. But this seemingly axiomatic rule may not be inviolable.

When the circadian system is working as it should, and is entrained to the day-night cycle, the temporal structure of sleep shows the standard, predictable sequence of ascending and descending stages. If the circadian system is disturbed, however, the EEG chart may detect temporal anomalies. The pattern known as the spindle, which normally signifies the presence of stage two sleep, may occur simultaneously with rapid eye movements, the defining feature of REM sleep. Or alpha waves, typical of waking drowsiness, may coincide with delta waves, which are associated with the profound sleep of stage four.

"It is quite clear that the brain is not a single system," Weitzman said. "I think that chronic sleep-deprivation studies indicate that the brain may be asleep and awake at the same time. Therefore our measurement, be it EEG or behavior, may actually be five different states. Under normal and entrained conditions these are temporally structured, such that they produce the basic concept. If you disrupt that, you then begin to see fragmented components occurring simultaneously. Our measurement is really the composite of a very complicated, multidimensional, regional, physiological, temporal system."

We are compartmentalized in time, but the compartments are not always fully sealed and tightly closed. Ideally, we cycle through a sequence of contrasting states, and each state has its appropriate time, yet because of the complexity, the manysidedness, of the biological clockwork system, the program does not always separate the various states completely, so that in a sense a person can indeed be said to be asleep and awake at the same time.

TEN

The Gates
of Day and Night

If sleep is not a single, uniform state, but a rhythmic, clocked cycle of contrasting psychological states, can the same be said of wakefulness? Such an idea sounds improbable on the face of it. We do not think of ourselves as having different types of mental experiences, one after the other in sequence, according to an orderly, predetermined timetable during the day. If such sequences do exist, it would mean that there are hidden time constraints of which we are quite innocent. Alertness, competence, mental sharpness, are usually thought of as gradually declining during the day as exertion takes its toll, just as sleep used to be regarded as a smooth, continuous process in which the body was gradually restored to freshness and vigor, a process with no structure, no clear distinctions between mental states at different times of the night. Today it is recognized that there is a nighttime schedule clocked in such a way that there are cycles which repeat about once every 90 minutes, and the state of body or of brain depends on "what time it is" in a cycle, rather than showing a steady improvement as the night goes on. Dramatic shifts can be observed both between cycles and within cycles. If a similar sort of schedule were to exist during the daytime, then waking life, which appears to be a process of steady deterioration as tiredness increases, would also cycle through distinct, contrasting states according to a built-in timetable. Such a possibility was suggested by the discovery that sleep has an internal as well as an external temporal structure.

When University of Chicago physiologist Nathaniel Kleitman, together with a graduate student named Eugene Aserinsky, firmly established the existence of REM periods by attaching electrodes to the skin around the eyes of sleeping infants, it caused a sensation in the scientific world. The brief report of their observations, in the journal *Science*, in September 1953, scarcely did justice to the importance of the discovery, which was little less than momentous. Here, in the midst of presumed sleep, were interludes that resembled waking consciousness more than they resembled sleep, and they occurred in a regular, sequential fashion.

As more and more became known about the REM and other stages of sleep, Kleitman became possessed by an intriguing idea. The cycle of REM and non-REM periods spanned the night, oscillating with a frequency of about 90 minutes. Would it not make sense to suppose that this rhythm also continued throughout the entire 24 hours, so that even during wakefulness relaxed periods alternated with episodes of high alertness? Kleitman called this day and night sequence the basic rest-activity cycle or BRAC. The theory was based on the observation that infants show a regular rhythm of wanting to be fed and not wanting to be fed, and this rhythm consists of four cycles during the day and five cycles at night. Kleitman described it in a paper given at a sleep congress in London in 1960, but his report was barely noticed. Three years later, however, he published a book, and BRAC began to catch the imagination of other scientists. It was natural that Kleitman should espouse such a hypothesis. He had devoted most of his scientific career to the study of sleep, at a time when interest in the subject was sparse. Ideas about the nature of sleep abounded with superstition and guesswork. A highly professional researcher, Kleitman brought the enthusiasm of an amateur to his work. He was adventurous and eager to try new approaches. He once stayed awake continuously for 180 hours to test the effects of sleep deprivation, and his fondness for playing bridge and watching marathon bicycle races arose from long periods of enforced sleeplessness. He and a fellow researcher, Bruce Richardson, lived for over a month in Mammoth Cave, Kentucky, trying to adjust to a day longer or shorter than 24 hours; and as a result of a two-week stint aboard a submarine, the U.S.S. *Dogfish,* he recommended a new system of shifts which avoided putting men on watch when their circadian temperature cycle was at a low point.

Paradoxically, Kleitman decided to specialize in sleep because he was interested in the waking brain. He believed that sleep, far from being the opposite of wakefulness, is complementary to it, "the one related to the other as the trough of a wave is to the crest." There was only one dimension of consciousness, he thought, and it runs through sleep and wakefulness, rising and falling like mercury in a thermometer; under certain circumstances, a person could be more critically alert during REM sleep than when "awake."

As he started his career in Chicago, Kleitman reasoned that sleep would be a window onto wakefulness, a fresh vantage point from which to view it, because sleep represents a low level of activity, a sort of natural biological baseline from which other states emerge. Since a reciprocal relationship exists between waking and sleeping, knowing more about one state would surely result in knowing more about the other. In his laboratory in Chicago, Kleitman had noticed that babies display REM cycles around the clock, and he decided that adults must do so as well, although in a more subtle and covert manner, making it difficult to observe.

Today the notion of a continuous 24-hour cycle of rest and activity is regarded as an oversimplification, not to be taken too literally. Nonetheless, it may contain the germ of very important insights into the extent to which human behavior is time-determined.

There does seem to be, for example, some sort of waking version of the REM stage of sleep. One of the interesting features of the internal structure of sleep is that while a sleeper may dream during both the REM and the non-REM segments of the sleep cycle, the dreams are of a different sort. One kind of dream experience is typical of the time slots reserved for the physical restoration of the body, while another kind occurs within the time slots used for servicing the brain. Possibly, this difference, or a modified version of it, persists during the waking day.

Contrary to a belief that prevailed in the early days of dream research, we do have mental experiences during non-REM sleep. The first scientist to investigate such experiences in a thoroughgoing way was David Foulkes, who wrote a doctoral thesis on the subject in 1960. Foulkes found that in non-REM sleep there is a greater amount of conceptual thinking, less emotion, anxiety, hostility and violence, diminished visual content and less physical movement than there is in REM sleep. In the dreamlike episodes of non-REM sleep, the cast of characters is less numerous than in a REM dream. Often there is only one other person present apart from the dreamer, and fewer changes of scenery occur in the narrative. Non-REM dreams also correspond more closely to recent events and thoughts in the dreamer's life than REM dreams do. The former have an everyday quality about them, and make more sense on the surface, without the need for elaborate interpretation of hidden meanings. One man in Foulkes's study, an Internal Revenue Service agent, had a non-REM dream in which he was thinking over a point of tax law, which stated that a taxpayer had to provide over half a person's support in order to claim him as a dependent, while another man, an undergraduate majoring in English literature, saw Tolstoy's Anna Karenina sitting at a table. In REM sleep, the IRS agent dreamed that a girl rang him up in the middle of the night to tell him it was time for his "35-day

evaluation," while the undergraduate dreamed he was standing on a street corner, holding his bicycle and talking to
someone about a girl who wanted to be a striptease dancer.
The IRS agent was actually on a 90-day probationary period
in his job, but he was being paid $35 for his participation in
the dream study.

Non-REM dreams, Foulkes concluded, are more like rational thinking, while REM dreams are fanciful and unrealistic, "thinking that does not labor under nearly so many
constraints imposed by external reality or by inner standards
derived from external reality." In non-REM dreams memories are less distorted and more plausible than in REM
dreams, although there are some distortions; and thought
plays a larger role, even though the dreams themselves are
hallucinatory, dramatic episodes. Non-REM mental activity
influences REM content and is influenced by it, sometimes
taking up and developing a theme that was developed in
REM episodes.

Kleitman and others have suggested that the waking
equivalent of the REM dream is the daydream, a fantasy experience which is beginning to be studied in a serious way
by psychologists. Daniel Kripke, director of the Sleep Diag-
nostic Laboratory at the San Diego Veterans Hospital in California, and David Sonnenschein devised a study in which
people were kept in isolation for about ten hours in a room
without time cues or any diversions such as books or television. Once every five minutes, when a whistle blew, each
person wrote down what he or she had been thinking during
those five minutes. Electrodes were taped to their head so
that both eye movements and brain waves could be recorded.
Some days later, the people went back to the room and
looked through the accounts of all their thought episodes,
rating them according to how "dreamlike" they were. A significant cycle was found, 90 to 100 minutes in length, in the
intensity of these daytime reveries. One person, who was
merely restless and aware of his surroundings at 9:20 A.M.,
noted at 10:10 A.M.: "I was just hearing a Beethoven symphony in my head." At 11:00 A.M. he wrote: "Thinking about

eating dinner tonight over at my friend's house." And at 11:50 A.M.: "Imagining how it would be to be in an anechoic chamber . . . I felt like I was actually in one, screaming, but no one could hear me." Another was "wiggling my feet to stay awake" at 12:10 P.M., fantasizing about lying on a secluded, warm beach at 12:40 P.M., reliving the previous evening's experience listening to a "great pianist" at 1:10 P.M., and conjuring up thoughts of being in a Jacuzzi bath watching a meteor shower at 2:10 P.M.

At the crest of a daydreaming cycle, highly emotional and sometimes bizarre images were noted, whereas at the nadir of the cycle, real problems and plans for the future were more in evidence. Peaks of fantasy were not usually accompanied by rapid eye movements, however, which weakens the case for considering daydreams as the waking counterparts of REM. Fantasy crests were associated with alpha waves on the EEG, a sign of relaxed, waking drowsiness.

Another time, Kripke and Sonnenschein gave eight students small tape recorders and let them lead ordinary routine lives, going to classes, mixing with other students, reading and exercising as usual. A beeper on their belt sounded every ten minutes and when it did they dictated their thoughts into the tape recorders. The results showed few examples of intense daydreaming, but a rhythm with a frequency of between 90 and 100 minutes was still present.

Daydreams do resemble nighttime REM dreams in certain respects. A daydream appears unbidden. It is spontaneous and probably beyond the control of consciousness. It may even come as a surprise to the daydreamer. And while a daydream does not delve into the past life of the individual as deeply as nighttime REM dreams do, it often serves as a medium for exploring previous experiences, though in a more direct and less bizarre fashion than REM dreams. People restructure their own past half-consciously in this way. They experience fleeting, spontaneous "replays" of episodes that happened earlier in their lives, distorted pseudomemories which are often a kind of autobiography that has undergone radical editing or reinterpretation. John Caughey, an anthropologist who has collected and analyzed the day-dreams of friends, students and colleagues at the University of Maryland, finds that in these narrative treatments, loosely based on actuality, ruptured love affairs are given happy endings, old slights are repaid and old hurts healed. Friends or loved ones who have died are brought back to life, and meetings are arranged between characters who have never seen each other in real life. A woman basketball player "remembered" that her father coached her in the driveway of her house and later sat at the dinner table, lecturing her on her dietary habits, although he had died seven years before these incidents were supposed to have taken place. In the woman's memory-daydream, her father actually changed his clothes as scenes shifted. He wore a gray sweatshirt emblazoned with the word "coach" while standing in the driveway, and a blue alligator shirt while sitting at the table. The experience was less a deliberate recall and revision of the past than an un-contrived drifting into an inner world of the half-real.

The approximately 90-minute cycle of daydreaming, like the REM cycle, is an ultradian rhythm. Ultradian rhythms are those that have a period of much less than 24 hours. They range from the fraction-of-a-second period of electrical oscillations in the brain, to the one-second period of the heartbeat, the six-second breathing rhythm, and the 90-minute sleep-stage cycle. These rhythms are called "ultra" because their frequencies are high. They complete more cycles in a given time period than circadian rhythms do.

Ultradian rhythms are the neglected Cinderellas of chronobiology. Many are still a puzzle to scientists, because they are less robust, more variable and elusive than circadian rhythms. In fact, some researchers prefer to keep clear of them altogether. One man who has taken these Cinderellas to the ball, however, is Peretz Lavie, a Tel Aviv University psychologist who spent the summer of 1984 in Allan Hobson's laboratory at Harvard. Far from shunning ultradian rhythms, Lavie believes they may help rather than hinder knowledge of biological time once their mechanisms are more fully understood.

Ultradian rhythms are striking evidence of the multiplicity that underlies the partial unity of biological time, in Lavie's view. It is certainly no use hoping for an easy way out through the saving simplicity of a master clock or a single rhythm. Ultradian clocks are most definitely a democracy. There is no temporal dictator to keep them marching in step. "I see the ultradian system as a very heterogeneous one," Dr. Lavie said. "There is not one ultradian cycle, but a family of cycles or rhythms."

One clear ultradian rhythm is in the ability to fall asleep. If a person is asked to fall asleep every twenty minutes, a definite rhythmic pattern is observed in what is called "sleepability." This rhythm corresponds to the REM–non-REM cycle, though it is not exactly the same. Sleepability is a predisposition to fall asleep, and in the course of a night it varies in a predictable way as the cycle of sleep stages unfolds. If a sleeper awakens during the REM stage, he finds it more difficult to go back to sleep for the next hour than when he awakens from non-REM sleep, His sleepability is at a reduced level. What is more interesting, sleepability fluctuates in this regular, clock-determined fashion during the daytime as well as throughout the night.

"Perhaps it is too strong a statement to say we have a waking cycle during the day in the same sense that there is a sleeping cycle at night," Lavie said. "But there are certain gating mechanisms during wakefulness, intervals of time when it is easier to enter the sleep state than at other times, and these operate in a cyclic way. About once every 90 minutes, the sleepability gate is open. This certainly applies to the period from 7 A.M. to 2 P.M. or 3 P.M."

In the morning and until about three o'clock in the afternoon, the sleepability gates open about every 90 to 120 minutes. They are narrow gates. The time slots in which it is easy to fall asleep are brief. But after three o'clock, a very wide sleepability gate opens, one that changes the frequency characteristics of the rhythm. This is the siesta time, the post-lunch dip when mental efficiency slumps, not as a result of eating lunch, but because there is a clock-determined, uniquely wide sleepability gate offering an easy transition from wakefulness to sleep during the day if a person should choose to make use of it. It is a built-in, biological time slot that divides the day into two halves. The evolutionary origins of the siesta gate may lie in man's remote beginnings in a warm climate, when there was a need to have a quiet period at the hottest time of day. Primates on the African plains have two peaks of activity during the day, one in the late morning and another, more pronounced, in the late afternoon.

During nighttime sleep, the REM periods operate in a seemingly complementary fashion to the sleepability gates of the waking day. They function as exit opportunities out of sleep, gates that enable the sleeper to ascend into wakefulness. Whereas the daytime gates are times when it is easier to sleep, nighttime REM stages are "wakeability" gates, times when it is easier to wake up. This suggests that nighttime REM periods arose in evolution as a sort of sentinel device, a monitor in case of danger, and there is evidence to support such a speculation. REM sleep is a state that is vigilant, in the sense that, even though it is a stage when information already in the brain is being reprocessed in some way, and circuits are tested "off-line," new or threatening information coming from the outside world may be noticed by the sleeping brain during the REM state. On the first night spent in a sleep laboratory, sleepers often show disturbances in the cycle of stages. The first and sometimes the second REM period fail to occur. As in the case of the IRS agent who dreamed of the number 35, the amount in dollars he was being paid to undergo the sleep experiment, these novices frequently experience REM dreams that contain open or hidden references to their new surroundings and to the people they have just met.

One intriguing hint that the REM stage is an exit gate from sleep is the fact that it is much easier to awaken from REM sleep than from non-REM sleep. When people are left to themselves, 80 percent of their wakings up are from REM, not from slow-wave sleep. Asked to rouse themselves, unaided, at a predetermined time, relying on internal clocks, these people somehow manage to do so during REM sleep, nobody knows just how. That is not the case, however, when they are given an alarm clock. Then, a sleeper wakes up during REM sleep only 30 percent of the time, suggesting that the REM stage, in its role as a sentinel device, is based on internal rather than on external information about local time. There is firm evidence that if a person wakes up from REM sleep, he is better able to orient himself in time than when he awakens from non-REM sleep. The same is true for orientation in space. The right hemisphere of the brain, which is superior at dealing with space, works more efficiently on awakening from non-REM sleep than the left hemisphere, which is verbal and sequential, although there is no good reason to believe that the right side of the brain plays the leading role in dreams.

"In evolution, REM seems to have been selected as an end point of sleep," Dr. Lavie said. "It is a period of sleep when all the systems are activated very close to waking levels except for muscle tone. It is easy to terminate that state of sleep based on internal time cues, and when you awaken, you know immediately where you are, both in time and in space. This is just one function of REM sleep, of course. It is a multifunctional process. Why we dream, and whether it has any relation to the gating mechanism of REM, is another question, to which I do not know the answer."

So the brain seems to have two complementary cycles, one making easier the transition from sleeping to waking during the night and the other providing a similar series of gates for the changeover from waking to sleeping during the day. The cycle is very flexible, however. It can be modified by level of arousal, by motivation and by many other factors. This is perfectly logical, since it is only to be expected that a program which gives an easy transition from waking to sleep would be inhibited during the day. The transition takes place only when a person is free of social constraints and lacks a good reason to stay awake.

These gates offering periodic entrances and exits, which are part of the internal structure of sleep and wakefulness, play a role in the entrainment of a biological clock, presumably the weak Y pacemaker, by the external time cues of daylight and darkness. Each day this pacemaker clock, with its "natural" period of about 25 hours, must be corrected so that it is in synchrony with the day-night cycle, with its period of 24 hours. This means that the clock is put back by an hour or 90 minutes every day, the period of the ultradian sleepability-wakeability rhythm. The clock could be reset, therefore, by a process of jumping from one gate to another. REM episodes may serve as "holes" in the system that enable such resetting to take place one short, specific step at a time, by vaulting one gate, and this applies to both the sleeping and the waking parts of the cycle. If the sleeping and waking states were to consist of two continuous, homogeneous blocks of time, which is how they appeared to scientists until quite recently, before researchers began to discover hidden structure not suspected until then, such resetting one short step every day could not take place.

In this way a new and more sophisticated version of Nathaniel Kleitman's basic rest-activity cycle hypothesis is beginning to take shape. The original suggestion that there is a rhythm of actual, physical activity, a period of moving around interspersed with intervals of physical rest, is clearly not realistic and cannot be taken literally. If the idea is treated as a metaphor, however, in which "activity" is a heightened state of the nervous system, and "rest" is an increased propensity to sleep, then it fits well with observation. There is a daytime rhythm of about 90 to 100 minutes in arousal and sleepability, but such a rhythm is very loose and flexible and is easily preempted. The cortex of the brain, the center of thinking and planning, criticism, and traffic with the outside world, can override an ultradian rhythm of arousal and nonarousal in the brainstem, which is a much more primitive and ancient part of the brain. The brainstem evolved more than five hundred million years ago and resembles the brain of a reptile. In human beings the cortex is in control. Since the rhythm of daytime sleepability and wakeability is tuned to internal rather than to external time signals, information coming from the outside world can have a rudely shattering effect on a delicate ultradian cycle. That is why the "basic rest-activity cycle" is clearly observable only in certain special and rather unrealistic circumstances, for example, when a person is lying down quietly with nothing to do, cut off from the normal time cues provided by the environment. It is probably not very apparent in, say, the waking day of the average midshipman at the Annapolis Naval Academy.

Another complicating factor is that ultradian rhythms of arousal and nonarousal, as well as the REM–non-REM cycle in sleep, may be constructs, systems made up of several different oscillators that are normally coupled together but under certain conditions may be uncoupled. If the strong X and weak Y pacemaker clocks of the circadian system can separate from each other after several days of isolation from external information about time, it is very likely that the much more labile and "noisy" ultradian rhythms can also desynchronize. It may not be too fanciful to imagine that the component parts of the BRAC are assembled and reassembled every 24 hours. While the rhythm of arousal and nonarousal during the waking day seems to be continuous with the previous night's REM–non-REM cycle, it is *not* continuous with the next night's REM–non-REM cycle. The process has to be restarted. In fact, it seems likely that the first non-REM period at the beginning of a night's sleep is a time cue which "sets the clock," initiating the first REM period of the night, but only the first period, while subsequent REM episodes are under the control of an independent oscillator and drift in time, continuing into the waking day. The BRAC is interrupted by a delay, a distinctive pause, at the outset of sleep, which precedes the appearance of the first REM episode. This may represent some kind of entraining or phase resetting mechanism which synchronizes the various component oscillators that together make up the REM cycle. If the pause does not occur, its absence spells trouble for the individual. The ultradian rhythms, and the circadian sleep-wake cycle itself, are thrown into disarray and temporal chaos; the human system is no longer constrained to behave appropriately in time, and lapses into sleep when alertness is called for, as in the disease of narcolepsy.

Thus it appears that the REM stage of the basic rest-activity cycle is not one thing but is assembled out of various rhythmic elements. Among these are fast brain waves, regular changes in blood pressure and metabolism, and loss of muscle tone. Under normal circumstances, all the components are synchronized, but on occasion they can desynchronize and go their separate ways. The EEG might show the familiar REM brain waves, but the rapid eye movements will be missing, or else the muscles of the body will not be immobilized. This is what led Elliot Weitzman to say that the body can be asleep and awake at the same time, and only when the constituent biological rhythms come together as they should are the two states fully separated in time so as to form a clearly defined sequence of different states.

Possibly daydreaming, the waking equivalent of REM sleep, is also assembled by synchronizing various components, but the parts are put together in a special, different way. In sleep, vivid dreams typical of REM periods occur when the brain is making theta waves, which emanate from the hippocampus and are also present in the waking brain at times of anxiety, when a person is unable to take in a great deal of information from the outside world. Most important, during REM sleep the nervous system is in a state of high arousal. But at peaks of daytime fantasy, and when theta waves emanate from the brain, the nervous system is in a state of low arousal. It is as if the clock of arousal had been reset, so as to synchronize with different components of the cycle.

Some ultradian rhythms behave as if they ought to be linked to form a system, because they show the same period of about 90 minutes. In fact, however, each rhythm may be independent. During sleep as well as in wakefulness, there are episodic hunger contractions which have the same period as a REM episode, but do not appear to be related to REM sleep. The luteinizing hormone rhythm, which persists during the day, is weakly coupled to the REM cycle, but it seems to be under the control of an autonomous oscillator. The kidney is regulated by both circadian and ultradian oscillators, but the ultradian rhythm, which governs the kidney's fluid balance, is independent of the circadian rhythm. More than half a century ago, it was found that there are 90- to 100-minute cycles in contractions of the stomach, but this rhythm is not closely coupled to that of REM sleep. The property that all these various rhythms share in common is that of having a period of about an hour and a half. Lavie calls this "the 90-minute enigma," and speculates that it may have something to do with the rate of protein synthesis in cells, or the time taken to transport a biochemical substance from cell to cell or from a cell to a target organ somewhere else in the body. If that is true, the ultradian elements of biological time could simply be a matter of biochemical distance. Cats, with their smaller brains and bodies have a REM cycle of about 20 minutes, and that is also the approximate frequency of certain of their ultradian rhythms.

How many different oscillators control this democratic, heterogeneous population of rhythms, weak and strong, dependent and independent, is a question to which no final answer can be given yet. Research into ultradian rhythms is at least 30 years behind the fast-moving and self-confident circadian rhythm industry. There is no clear evidence that rhythms in the ultradian system have the same period and little that would suggest they are all run by the same master clock. Immense complexity reigns.

"Possibly the newest aspect of our knowledge of ultradian rhythms is that, as far as we can tell, every single system in the body seems to have its own rhythm and is running in an ultradian fashion or a circadian fashion or both, but not necessarily in conjunction with anything else," said John Gertz, a graduate student in neurophysiology at the University of San Francisco. "And you can disentangle the rhythms, tease them apart very easily by experimental techniques. We have been able to show that, on the average, there is definitely an ultradian rhythm in every parameter we have measured in the EEG of humans. But in virtually no single individual were the various parameters running in harmony with one another as they would be if they were under the control of the same pacemaker. They are out of phase. They are not coherent, in a statistical sense.

"Almost everything we can measure in the body and

brain is rhythmic. It is very rare to find something that is not. Yet as we measure more and more, hoping it will all become more parsimonious and simple, the universe of biological rhythms gets more and more complex. You want to throw up your hands like Einstein did and say, is God really playing at dice? Nothing seems to be in synchrony with anything else. Sometimes it appears as if there are 100,000 different clocks ticking away inside the body, and that strikes a scientist as absurd. It just doesn't intuitively appeal to anybody."

PART THREE

TIME AND INFORMATION

ELEVEN

Time and
the Biological Brain

Since measurable rhythms are everywhere in the living system, from the ebbing and flowing tides of fluids and chemicals in the body to the waves of electrical activity in the brain, it is not easy to draw a line and say, here, on this side, is time biologized, and there, on that side is time psychologized. The two overlap and influence one another, so that, for example, a disorder of mood and mental state, such as depression, and the impairment of a physiological process, such as sleep, may have a common basis in the disruption of circadian timing mechanisms and may be treated with a drug that sends the nervous system a message about time. REM dreams may be psychological in their effects, but the dream state appears to be programmed into the structure of sleep by a clock that is as "biological" as the heart or the liver. In principle, the clock can be seen and touched. One of the famous blind alleys of dream research is the mistaken assumption that dreams are exclusively psychological, that they occur because, unconsciously, we want them to occur, and when we cease to need them, they will go away. This was the trap into which the psychologist Alfred Adler fell. He underestimated the biological element of dreams, proposing that dreams are a form of self-deception, appearing extensively in the sleep of people who are too immature to know themselves as they really are. Such people, Adler thought, unconsciously create for themselves masks of false appearances, and their dreams reflect these masks. When self-deception is outgrown in adult life, dreams are also outgrown. They become occasional experiences or disappear from sleep entirely. Today we know that such an idea is quite erroneous. The occurrence of dreams is biologically determined and is as inevitable as the tick of a clock. Nature has put them on our nightly schedule and we have no choice but to experience them. Yet dreams can also have important psychological meaning, which if interpreted correctly can, paradoxically, make us understand our waking life more fully than before, and thus be more free in our choices because we are less "blind to our blindnesses."

A complex and subtle relationship exists between rhythms of the body and rhythms in the efficiency of the brain as it performs its normal tasks. It was once believed, for example, that the brain is better or worse at manipulating information, crunching numbers or juggling words, depending on which stage has been reached in the 24-hour body temperature cycle. Common sense suggested that as temperature went up, mental performance improved, and as the body cooled down, efficiency deteriorated. As it happens, however, common sense, once again, has let us down. There is a relationship between biological rhythms and mental performance, but it is neither simple nor obvious. There are many different kinds of rhythm in the silent orchestra, and more than one way in which the brain approaches even the most routine of its daily chores. Brain styles and rhythms interact in various combinations.

Since the nineteenth century, psychologists have wanted to know why it is that people use their brain more successfully at certain times of the day than at other times. The most boring explanation was that efficiency declines as tiredness increases. Hermann Ebbinghaus, the pioneer of memory research, found that lists of nonsense syllables could be learned quite rapidly in the morning, but more slowly later in the day, which seemed to fit the tiredness hypothesis. Nathaniel Kleitman, whose importance as a scientist should not be obscured by the fact that he was sometimes wrong, proposed in 1963, without being adamant about it, that mental efficiency and body temperature rise and fall in synchrony with one another.

It is clear today that while biological clocks do have an influence on mental performance, the influence depends on which clock we are talking about and on the particular strategies the brain is using to perform its various tasks. The brain is not simple or single, even when it is engaged in an activity to which we give a single name, such as "remembering." There is more than one type of memory, and the types are related to one another in a sort of family, just as there is a family of biological rhythms in the ultradian system.

Different kinds of mental operations have their own schedules, peaking at different times of day, and their own special correspondences with biological clocks. Efficiency at simple tasks may indeed be linked to the circadian cycle of body temperature, rising and falling in step with it, as long as no heavy demands are made on memory and only slight burdens imposed on the ability to reason. If someone is asked to search through a list of letters of the alphabet, printed at random, and cross out all instances of a particular letter, say T, a cycle of performance is usually noticed which matches the cycle of temperature. But when heavier demands are made, and the tests involve memory and verbal reasoning, efficiency seems to be under the control of a different clock, one with a period of about 21 hours.

Routine, repetitive manual exercises, such as fitting pegs into holes using first the right hand and then the left, are usually more successfully performed in the late afternoon or evening, when body temperature is high. Activities that perplex or challenge the brain tend to be better done during the late morning or at midday. Remembering and reasoning, applying rules and formulas, manipulating syntax and coordinating ideas, are mental acts that improve or worsen across the day in a rhythm that is probably linked to the sleep-wake cycle. The peak of efficiency in intellectual operations such as these is usually reached during the first half of the day, and they seem to be more strongly influenced by what happens during sleep than tasks that require deftness with the hands. This suggests that REM episodes have a tonic effect on reasoning and on certain types of memory. When people are waked up from REM sleep they are usually more agile mentally and more talkative than when they are roused from slow-wave sleep. Hearing and seeing, on the other hand, as well as hand and eye coordination, tend to reach a high point of efficiency later in the day.

Jet travelers, after flying across several times zones, tend to be clumsy when doing things with their hands, and are also duller than normal mentally. They tend to recover their wits sooner than they get over their clumsiness, however. It appears that manual dexterity depends on the strong X pacemaker, which is "stiffer" and more resistant to change than the weaker and more plastic Y clock, which seems to play a part in rhythms of verbal reasoning. For that reason, the X clock takes longer to adjust to the new day and night cycle than the Y clock does. The jet traveler's capacity to steer his way through ideas and conversations is more quickly restored than his ability to steer a car through traffic in the city of his destination. The same is true of workers on rotating shifts in a factory. Piloting a supersonic flight simulator is a psychomotor skill that needs up to 10 days to adjust to the effects of jet-lag. Different kinds of tasks appear to be under the control of different kinds of biological clocks.

Similarly, there is no point in asking whether memory is better or worse at particular phases of the circadian cycle without being sure which memory it is we are talking about. As it happens, "memory" is not equally good throughout the waking day. But short-term memory and long-term memory differ in important respects, and the differences are the result of definite strategies of adaptation during evolution, related to the survival of species. The human nervous system has a rhythm of arousal which crests at about 8 P.M. and long-term memory improves as arousal is heightened, peaking late in the day. Other rules apply, however, to short-term memory, which is better in the morning, reaching its zenith at about 10 A.M. or 11 A.M. Short-term memory is about 15 percent more efficient in the morning and about 15 percent less efficient in the evening, almost a mirror image of the behavior of long-term memory. The reason for this may be that when the nervous system is in a high state of arousal, as it is in the evening, more attention is paid to important and interesting information. Long-term memory usually works better when it is dealing with information that has special relevance and meaning for the rememberer, whereas short-term memory is a more automatic, meaning-free mechanism. At nine o'clock in the morning, importance has an almost negligible effect on short-term memory, but at three in the afternoon, it plays a more potent role. Scientists at NASA's Biomedical Research Division, whose job it is to schedule the work of American astronauts, say that a football coach instructing his team in a tricky new series of plays should do so at 3 P.M. At 9 A.M., the difference in the efficiency with which the team remembers the plays is as great as if they had slept for only three hours the night before.

The influence of time of day on long-term memory also depends on whether information is being stored or retrieved. Schoolchildren who listened to a story being read at 9 A.M. were better at immediate recall of its contents than children who heard the same story at 3 P.M. That was only to be expected. It turned out, however, that when they were asked to remember the story seven days later, the children who had heard it read in the afternoon recalled more details than those who had heard it in the morning. Evidently the storage of the information in long-term memory is more efficient in the second half of the day than it is in the first half, but time of day does not matter much in the case of retrieval, since it made no difference to the results whether the memory test, as opposed to the story itself, was given in the morning or in the afternoon.

How well and quickly human beings process information depends on many factors, some internal, some external, including the influence of circadian and other clocks. Another factor is personality, but personality cannot be considered by itself, apart from temporal identity, which has a specific, biological basis in the body's clocks. One of the few personality traits researchers agree is easily identified and consistent from one study to another is the extent to which someone is introverted or extroverted. And an especially reliable component of this complex trait is how impulsive one is. An impulsive person tends to act on the spur of the moment, not stopping to think, has a relatively fast tempo and is more inclined to take risks. He is more extroverted than people who are not impulsive, and he is also more likely to be an owl, at his best late in the day. People who are not impulsive tend to be larks, morning people.

Stress induced by caffeine is a hindrance to the mental efficiency of nonimpulsive people in the morning, because they are already in an aroused state, and a help to them in the evening, when they are less aroused. For people who are impulsive, however, the reverse applies. Caffeine is a help in the morning and a hindrance in the evening. Yet the temporal contrast between these two types is not inflexible and tightly programmed. As larks and owls practice certain mental tasks day after day and get better at them, the morning and evening difference is not so clearly marked. Also, when larks and owls are in close visual contact, the larks lose their "natural" time-of-day advantage. In general, our performance rhythms are less consistent, day in and day out, than our physiological rhythms. They are more easily affected by factors outside the circadian system proper.

Human biological rhythms are not like instincts. They do not exercise supreme authority, taking over behavior and directing it in specific ways, as birds instinctively build nests and ants organize themselves into colonies. Nor are they on the same level as civilized consciousness, which has more and more separated itself from the basic instincts during evolution. Biological rhythms modify behavior. In Wilse Webb's words they serve mostly as "background primers" rather than controls. And there is a two-way traffic: behavior can modify the rhythms, suppress them, test their limits, defy their authority, override their schedules, set their clocks. A mental decision can regulate a biological mechanism.

Some rhythms are more obedient to the influences of the conscious mind than others. Ultradian episodes of sleepability in the daytime, for example, can be overridden by the cortex. A large cortex is the hallmark of highly evolved creatures, and it is the cortex that preempts cycles of arousal in the more primitive brainstem. Evolution does not eliminate existing mechanisms when it modernizes a species. It adds new structures onto old ones, as if an engineer were to "evolve" a new automobile by building the engine of a Lincoln Continental on top of the chassis of a Model T Ford. Because nature operates in this fashion, ancient brain systems have to coexist with the more up-to-date ones, sometimes competing in a sort of rivalry. François Jacob has compared evolution's patchwork methods with those of a tinkerer, who builds a gadget or utensil out of odds and ends, using what is close to hand—pieces of cardboard, lengths of string, fragments of wood or metal. A tinkerer works on what is already there, either transforming an article so as to give it a new function or combining various parts into a new and more complex whole. Nowhere is this more true than in the case of the brain, where a comparatively recent structure, the neocortex, center of intellectual and cognitive activities, has been added to the older "visceral" brain, which deals with emotions and drives. "This evolutionary procedure—the formation of a dominant neocortex coupled with the persistence of a nervous and hormonal system partially, but not totally, under the rule of the neocortex—strongly recalls the way the tinkerer works," Jacob says. "It is somewhat like adding a jet engine to an old horse cart. No wonder accidents occur."

Under such an arrangement, outmoded time schedules

of body and behavior, developed long ago in evolution, still remain and are under the partial rule of the higher and more modern brain centers. The "cause" of the wide sleepability gate after lunch, for example, is a biological rhythm, which schedules the propensity for a nap at that time of day. The cortex, however, does not know that the rhythm exists (unless it subscribes to *Psychology Today*) and assumes that the cause of the feeling of drowsiness is an external one, namely lunch. In many cases, it is able to override the rhythm, by a command to go on working at full intensity and ignore the sleepability gate, although some people find this easier to do than others. As wartime prime minister of Britain, Sir Winston Churchill always went to bed for at least an hour as early as possible in the afternoon, postponing cabinet meetings so that he could enter the ancient sleepability gate as soon as it opened. It is possible that in more highly evolved beings of the future, whose cortex might be even larger than that of present-day humans, the afternoon nap would be more easily suppressed.

The relationship between biological rhythms and the work of the brain has another surprising feature. Mental operations, decisions of the conscious brain, can act as a *zeitgeber*. They may entrain and shift the phase of the rhythms. There is, for example, an ultradian rhythm of efficiency in the sort of task that requires good mental focus. If someone begins this kind of task, there will be an improvement in his work, followed by a deterioration, leading to a second improvement and then another deterioration. The period from peak to peak of this cycle is always about 90 minutes, the biological time unit of the REM cycle in sleep. The rhythm is quite predictable and programmed. Yet there is no obvious environmental cycle which could act as a time cue for such a rhythm. How, then, is the clock that controls the rhythm reset, so that, no matter at what time of day the task is performed, the peaks of efficiency occur at those particular phases of the cycle? Apparently the mere act of starting the task acts as a time cue, resetting the phase, rather as a flash of light can entrain a circadian rhythm—say, the rhythm of core temperature, resetting the phase so that the peak occurs at the same time every day. Once a person begins a demanding piece of work, a cyclic process is set in motion, with a biological time unit of about 90 minutes, and it rises and falls in a predictable fashion, which is just the way an entrained biological rhythm is supposed to behave.

In fact, it is possible to synchronize a whole group of people so that their peaks of improvement and troughs of deterioration occur at the same time. The people entrain each other. "Factory workers were tested in Cardiff, Wales, and their performance was found to be synchronized in about 90-minute cycles," Peretz Lavie said. "The period and the phase of the rhythms matched one another. You can phase-shift people along the 90-minute cycle, entrain them to this biological time unit. That is my belief."

This tension between the higher, thinking centers of the brain and its lower levels has a parallel in cybernetics. There is a principle of systems theory known as system dominance, which states that the larger and more complex a system becomes, the more constancy is required of those basic functions that are essential to the survival of the whole. A tug exists between forces working for predictability and programmed change and forces working for spontaneity and deviation from a fixed schedule. In the human system, the conscious brain often breaks out of the temporal confines created by its more archaic regions.

Conscious behavior can act as a correcting influence on circadian rhythms that have been disrupted by air travel across time zones. One day, scientists will develop a jet-lag pill to shift the phase of biological clocks, acting on the nervous system much as light acts to synchronize them with the local day-night cycle. But even without drugs, various dodges are being devised to reduce the effects of jet-lag. Allan Hobson recommends that a passenger planning a transatlantic trip from Boston to London should try to reset his circadian clocks to London time three days before departure by getting up progressively earlier each morning: at 6 A.M. on the first day, 5 A.M. on the second, and at 3 A.M. on the day of the flight. Once on the plane the passenger should resist the temptation to drink cocktails, eat dinner and see a movie, which is what the airlines, in their irrational way, try to make him do, ruining any chance of a good night's sleep. A meal can act as a time cue, entraining biological clocks, and booze suppresses REM episodes, leading to early awakening. Ideally, airlines should serve a meal at the terminal before takeoff, show no movie, and keep the cabin dark and quiet throughout the entire journey. Flights should leave in the late afternoon, when it is easier to go to sleep. Whether they will ever do so remains to be seen.

Charles Ehret, a chronobiologist at the Argonne Laboratory in Illinois, prescribes a diet program with its own rhythm of feast and fast to advance the circadian system for eastbound flights or delay it for westbound travel. Fasting lowers the level of glycogen in the liver and so reduces the body's supply of energy, preparing the circadian clock for resetting, while a high-protein meal signals the system that a new time frame has begun. On the day of departure, two or three cups of black coffee or strong tea should be drunk between 7 P.M. and 11 P.M., because these help to shift the phase of the sleep-wake rhythm. There is a whole laundry list of items that act as *zeitgebers* for human biological clocks: light, high carbohydrate or high protein meals, coffee, tea, cocoa, chocolate, exercise such as aerobic dancing, social contacts and intellectual activity. Dr. Ehret recommends that more than one of these clock-resetting agents should be brought into play simultaneously, so as to have a strong synchronizing effect on the pacemakers of the circadian system. Ideally, a passenger flying from New York to London should eat a high-protein breakfast at 2 A.M. and carry on a witty conversation with her neighbor at the same time. His program has been followed by the U.S. Army's 82nd Airborne Division to reduce the disruption caused by long-distance flights to the Middle East for exercises, by executives of the Control Data Corporation, and by some members of the Minnesota Vikings football team when they played against the St. Louis Cardinals in London. On that occasion the Vikings won by 28 to 10.

Dr. Hobson points out that the universal ritual of the morning coffee break may be more than just a mindless indulgence of a pampered age. Like the jet-lag pill, the effect of coffee seems to depend not just on how much is drunk but also on *when* it is taken. The stimulating effect of coffee may act on the trough of an ultradian cycle of alertness. It may serve as a time cue to a biological clock. By drinking coffee at periodic intervals, Hobson wonders, "could we be paying homage to an archaic force, a kind of physiological appendix no longer necessary to ensure our survival? Does our morning consist of two ultradian cycles linked together by a coffee break?"

Conscious control of internal time schedules is not always effective, however, and even if it succeeds for a while, biology is apt to impose a penalty in the end. Harvard physiologist Martin Moore-Ede and his colleagues describe an incident at the Los Angeles International Airport one night when air-traffic controllers were shocked to see a Boeing 707, which had filed a flight plan to land, overshoot the airport at 32,000 feet and head west over the Pacific ocean. Every member of the Boeing's crew had fallen asleep. The aircraft, on automatic pilot, was 100 miles out over the sea before the Los Angeles controllers managed to rouse one of the crew by triggering a series of alarms in the cockpit. Part of the reason for incidents of this kind, the Harvard group believes, is that pilots do not adjust completely to their shift-work schedule. Also, local time keeps changing for a pilot as he flies across time zones. Senior pilots often choose the most disruptive schedules, because these are the ones that carry the most generous vacation allowances. A pilot's rest and duty time should be computed not only on the basis of the number of hours he has flown, but also on the phase of his circadian system; and flights that are scheduled during the crew's subjective night should be avoided. The accident at the Three Mile Island nuclear power plant in 1979 happened at 4 A.M., in the middle of the night shift, with a crew that had been working at night for only a few days and had been on weekly rotating shifts around the clock for six weeks before the accident.

It is clear that rhythms of knowing and rhythms of body function are not separate and distinct. They coexist and exert influences on one another to an extent we are aware of only vaguely if at all. Thus even our cognition, our mental performance, is constrained in complex and perhaps quite unexpected ways by inner timetables that often, but not in every case, correspond to timetables in the external physical world. By making time important for the body, these built-in structures, a product of evolution, make time important for certain types of mental activity as well. It makes a difference when we think as well as what we think and how we think, which is of course a limit on our mental freedom. Temporal identity is psychological as well as biological. All this is entirely compatible with the view of some cognitive psychologists today that the mind is restricted as to what it can know by specific mental structures which are innate, and therefore must be regarded as part of human biology as well as human psychology.

One school of modern linguistics holds that the human brain is not a universal instrument, open to all possible kinds of knowledge, but a biological organ, not too different from other, more mundane organs such as the lungs or the heart, which evolved to perform specific, limited functions, namely breathing and pumping blood. They are no good for hearing or seeing. Similarly, the brain's peculiar structure, acquired during evolution, enables it to pick up certain kinds of knowledge easily and quickly and other kinds only with difficulty or not at all. Among the world's invariant structures that the brain has internalized, according to this theory, are the universal, unchanging principles and rules of language, "universal grammar," as it is called. Universal grammar is not a grammar but a theory of grammars, one that makes it possible for young children to acquire rules of their native language quickly and surely, without needing to be exceptionally intelligent. It is almost as natural for children to master their native language as it is for them to dream at night. Noam Chomsky, the author of the theory, talks about the existence of a "language organ" which is restricted as to the kinds of linguistic knowledge it can acquire, because it has internalized only certain kinds of structures and not others. The organ is receptive to one particular theory of language by virtue of the special way in which it is built, rather as pacemaker clocks of the body are sensitive only to certain types of external time cues. A child, therefore, possessing this organ as a birthright, masters the rules of its language for reasons of biology, of human nature. The abstract principles which govern the structure of human languages, Chomsky believes, "are universal by biological necessity and not mere historical accident."

The concepts of scientists like Chomsky and Roger Shepard open the way for a theory of biological rhythms that integrates brain and body, because the brain is seen as a biological organ which has internalized specific, invariant features of the world, including its time patterns. This provides the brain with information about such time patterns, often at an unconscious level, so that it knows them by biological necessity, not by luck or special cleverness. A being from another planet, who had internalized different patterns, including a different period of rotation of his planet on its axis, would not entrain well to our 24-hour cycle of day and night, nor might he even perceive our world in the same way that we do, because his brain would be a different biological organ from ours, still restricted, but restricted in different ways.

Our extraterrestrial visitor might be unable to master other time patterns in the external world that are of great importance for human beings, perhaps more important than the rhythms of day and night—namely, those that underlie the forms of speech. These patterns are rapid, complex, rhythmic sequences in which temporal order is the key to meaning, and communication depends on our ability to perceive that order, not by deducing it in a deliberate act of thought, but by biological necessity. The difficulties facing the traveler from outer space might include not only a failure to entrain his circadian system to our day and night, but also an inability to tune in to the time structure of our speech.

TWELVE

Chemical Clocks and a Biological Miracle

Man-made clocks in the modern world are often designed for purposes other than simply telling the time of day. They are essential to the technology of communications. Clocks are indispensable components of machines that encode, transmit, receive and process information. An electronic computer has a master clock built into the control section of its central processing unit, and it has more authority over the machine than any one of the biological pacemaker clocks has over the body. The computer's master clock is not for keeping local time. Whether or not a computer is synchronized with the "real time" of the world outside, the machine must conform to time constraints which are intrinsic to its own operations. The clock provides the fundamental beat which dictates the internal time behavior of the computer itself.

This master clock is an organizing device which restricts the computer's freedom in time. It is part of the temporal anatomy of the machine, ensuring that certain events occur at appropriate times, just as circadian pacemakers are part of the temporal anatomy of the body. Information, as it is processed by the logic and arithmetic circuits of the computer, is represented by a code consisting of only two symbols, a 0 and a 1. If a switch in the computer is in an "off" position, it signifies a 0, and if the switch is in an "on" position, a 1. Each change of state from 0 to 1 or from 1 to 0 has a definite and unique meaning. So far, so simple. But a message in a computer circuit is carried by an electromagnetic wave, which is much like a wave of light, and though it travels very quickly, the wave takes a certain amount of time to move through space. At the fastest, it covers about a foot in a billionth of a second, so that there is a minuscule delay between the time it enters a complex logic circuit and the time it appears at the output of that circuit, ready for the next step in the procedure. During that brief interval, the information that comes out of the circuit is untrustworthy. The circuit is making an adjustment between its present input and its previous input. Voltages are apt to fluctuate before reaching their final values: a high, stable value for 1 and a low, stable value for 0. Instability may result in a chance value, high, low or in-between, that is random and therefore probably wrong.

Always, there is a certain randomness in the speed at which a particular switch responds. Such tiny discrepancies create very brief, transient repsonses to signals, and these responses are unreliable. They may be nonsense rather than intelligible messages. For example, a string of 1s and 0s that is supposed to encode the letter H might be changed by the random fluctuations in the circuit so that, for an imperceptible fraction of a second, it encodes the letter L. Some way must be found to introduce constraints, so that the computer ignores messages during the intervals when they are unstable and pays attention to them only when the circuit is in a trustworthy state. This is accomplished by means of the master clock, which generates regular and extremely rapid pulses. A computer clock, often a crystal oscillator, ticks four

million times a second or faster, and it is only on a tick that
anything meaningful happens in the computer. In between
ticks, the computer is told, in effect, not to take the states of
its own circuits seriously. The clock protects the flow of information from being corrupted by random changes of state
because it separates sense from nonsense in time. There are
intervals of time when messages are reliable and intervals
when they are unreliable, and the different kinds of intervals
are clearly distinguished from each other. Contradictory information states are separated in time in the computer, rather
as incompatible biochemical states of the body are separated
in time by biological clocks. In this way, the master clock
tightly limits the freedom of the machine so that it can be
effective, like the conductor of an orchestra imposing his
own tempo on the players. Scientists talk about the characteristic "rhythm" of a computer's operations. In all digital
computers there is at least one clock, and clocks are involved
in such essential processes as sequencing, calculation and
the storage and retrieval of information. Clocks are also used
in television sets and aboard communications satellites used
by many subscribers, where a change of state occurs as fast
as 100 million times a second, and thousands of different
channels are interweaved, separated from one another by
pulses in time.

The close relationship between time and communication, and therefore between time and meaning, is embodied
in the physical device of an oscillator. Oscillators of one type
or another are found throughout nature, in biological systems
and even in lifeless chemical processes, as well as in the
machines of the modern electronic age. One of the strangest
of all natural oscillators is the chemical clock, which seemed
so odd many scientists were incredulous when they first
heard about it. The discovery of the chemical clock was made
in 1921, when a chemist at Berkeley named William Bray
spent some time studying a reaction involving iodine and
hydrogen peroxide. To his surprise, Bray found that the reaction of these chemicals was not uniform, nor did it fluctuate
randomly over time. Instead, it became a color clock. The

reaction oscillated between blue and nonblue, in a periodic, clocklike fashion. Bray published his results, but they met with general disbelief from other scientists, who assumed that he had made a mistake. The principle of entropy, which decrees that systems left to themselves tend to become more random as time goes on, seemed to forbid chemical behavior in which a reaction reversed itself repeatedly to produce a steady oscillation.

Years after Bray's experiment, in 1958, a Russian chemist, B. P. Belousov, noticed the curious activity of a solution of citric and sulfuric acids, potassium bromate and a cerium salt. The mixture changed from colorless to pale yellow and back to colorless again twice a minute. Few researchers paid any attention to this curious observation, because it was published in a highly specialized set of abstracts on radiation medicine. Some time later, however, another Soviet scientist working near Moscow, named A. M. Zhabotinksy, began to make a thorough study of the reaction, which he modified in certain ways. Zhabotinksy produced a spectacular variation of it in which chemical waves, vivid blue in color, moved rhythmically across a dish of liquid, forming concentric rings.

For a system to oscillate, there must be two separate ways of storing the energy needed to sustain periodic motion. A pendulum, for example, stores the energy of gravity as it swings upward. Going downward, it stores energy in its own momentum. In the case of the Belousov-Zhabotinksy reaction, the metal cerium, an ingredient used in the experiment, can exist in two different states, a reduced state and an oxidized state. In one reaction, reduced cerium is transformed into the oxidized form, departing from chemical equilibrium, the state in which no further net chemical change can take place in the system, and thus storing energy. This is similar to the pendulum swinging upward, departing from mechanical equilibrium. In another reaction, oxidized cerium yields reduced cerium, which is similar to the pendulum swinging down.

Chemical clocks may be used as a medium of communication. In fact, it is possible, in principle, to build a chemical radio that would broadcast messages. The physical chemist Joseph Higgins has said that "virtually all the dynamical characteristics associated with modern electronics can be generated through chemical reactions." By selecting the appropriate reactions, which act as oscillators, like those in the dishes of Belousov and Zhabotinksy, signals could be sent to distant places, much as they are by an ordinary radio, although the frequencies involved would not be too low to make it a practical device for listening to broadcast speech. Higgins assumed, however, that in living organisms, chemical oscillators must play a role in biological functions in which information is transmitted and control exercised.

A spectacular, even melodramatic example of the power of chemical clocks to communicate and control is found in the strange and marvelous congregating activity of slime molds, the Acrasiales fungi. Under ordinary circumstances, these primitive entities exist as a disorganized mass of single amoeboid cells which feed on decaying plant stuff in swampy places. During rainless months, when the swamp dries out, the molds behave in a fashion that one biologist, not normally given to romantic exaggeration, describes as "truly miraculous." What happens is that hundreds of thousands, even millions of the molds suddenly start to swim in a consistent manner and gather together in a lump. The lump develops a stalk which in turn produces a collection of spores on top, and this remains in place until the lump dries out, at which stage it breaks off and disperses into an airborne cloud of spores. The spores reproduce themselves when they encounter food in a new swamp, and form a fresh colony.

The transformation of a randomly moving mass of single amoeboid cells into a whole body with a definite form, which changes its shape in such a way as to provide the means for travel to another and better environment, thus ensuring the survival of the species, seems more like science fiction than everyday biology. What makes it possible is a signaling process which is under the control of the chemical clocks in individual cells. When food is available, the clocks are not in operation. After several hours without food, however, a few cells start secreting a chemical substance in pulses, and the pulses are periodic, even though they are not tightly regulated. The substance, one that is intimately involved with the human body's circadian timing system, is called cyclic AMP.

In the metamorphosis of slime mold, periodic waves of cyclic AMP are a medium of communication. A cell acts as a "center of attraction" by putting out pulses of the chemical, and other cells start to converge on the center in a rhythmic motion. Each cell responds to the signals sent out by the attractor by speeding up the rate at which cyclic AMP is secreted. All are producing spurts of the substance at the same moment, and by the time they have come together to form a single mass, they are in rough synchrony. The chemical clocks are ticking in unison. This cell mass is called a slug, and before it settles down to produce spores, it moves along the ground by means of rhythmic contraction. The synchronized waves of cyclic AMP, acting as a kind of pacemaker, control the contractions. In the final stage the slug ceases to move about and the uniform mass is differentiated into a base, a stalk and a round head of spores. This process is guided by the cyclic AMP which the cells are still secreting.

One of cyclic AMP's many roles in the human body is to act as a chemical messenger. Order is maintained within the living cell, with its complex temporal and spatial structure, partly by manipulating the time scales of events there. Inside the cell, many different kinds of molecules are swarming about, colliding with one another, dividing or exchanging atoms with other molecules. It might seem that in this great confusion, chemical reactions must be taking place at random and that chaos reigns. In fact, an exquisite order prevails. The reason such order exists is rather unexpected, and has to do with time. Most of the reactions of interest in a cell proceed very slowly, taking hours or even days to complete. Specific enzymes ensure that the wrong reactions do not happen and the right ones do happen by catalyzing certain slow reactions, making them go faster. An enzyme selects one slow reaction out of many, and compresses it in time. The reaction is made to proceed so quickly that there is no time for any other, slower reaction to interfere. In this way chaos is averted and only what is supposed to happen in the cell does happen. An enzyme is specially made so that it speeds up just one particular chemical reaction and no other, and it can do so only if that reaction is one that would have proceeded in any case, without the action of the enzyme, though at a much slower rate.

Cyclic AMP, in its role as chemical messenger, controls the speed at which enzymes function, as well as regulating the expression of genes. The extraordinary behavior of slime mold, which organizes itself in space and time, at the appropriate time, is due to the presence of oscillating cyclic AMP outside the cell. In humans and other animals, cyclic AMP, which won its discoverer, Earl Sutherland, a Nobel prize in 1971, is known as the "second messenger," because it stays inside the cell, where it performs its task of organization, acting on the orders of a "first messenger," a neurotransmitter, which sends a signal to the cell membrane. For example, in response to fear or other kinds of stress, the body provides a sudden rush of energy, enabling an animal to turn and flee or stay and fight. This process is actually very complex. It begins with the hormone adrenaline, which travels to the cells of the liver and binds to specific receptor sites on the outside of the cell wall. But adrenaline is only the "first messenger." By binding to the cell wall, it triggers the production of cyclic AMP, the second messenger, which diffuses throughout the cell, relaying the signal from the first messenger and instructing the biochemical machinery of the cell to transform glycogen, or animal starch, held in reserve in the liver for just such an occasion, into glucose, which exits from the liver into the bloodstream and supplies the energy needed to cope with the crisis happening in the animal's life.

Cyclic AMP amplifies, many thousands of times, the weak signal sent by the first messenger impinging on the outside of the cell. It also plays a critical role in amplifying events in time, extending them so that instead of lasting for only a few seconds they last for minutes or for hours. The second messenger transforms a momentary happening into a longlasting one. Cyclic AMP is at work in nerve cells, so that it is involved in the communication networks of the brain, opening up exciting prospects for a new understanding of brain function. Because the second messenger system works rather slowly by comparison with the speed of much nerve cell activity, it seems to specialize in mental states that are stretched out over time. The time scales of events in the nervous system range from the few milliseconds it takes for a muscle to be flexed, to minutes for short-term memory and years for long-term memory.

As well as playing a role in the immediate flight-or-fight reaction, cyclic AMP appears to mediate much longer-term events, the more continuous "stream of consciousness" systems of learning and memory that appear to be shunted out for repairs during the REM stages of sleep. Such an amplifying in time of short-term happenings into longer-lasting biochemical changes by means of this versatile second messenger may be the biochemical mechanism by which information is stored in the brain, the basis of long-term memory. That is only a hypothesis in the present state of knowledge of the brain, but it is an interesting one just the same.

Possibly, cyclic AMP is the mechanism that makes biological clocks tick, that actually generates the rhythms that organize the body in time. Cyclic AMP is part of a feedback process, and a feedback circuit is normally homeostatic. It regulates a system by bringing it back to a particular state whenever it deviates from that state. In many respects, the body is like a factory in which substances are being built up and broken down continually by means of intricate sequences of chemical reactions. Apart from the selection of correct reactions and the exclusion of incorrect ones, some form of control is needed to ensure that the substances are made in the appropriate amounts, neither too much nor too little. Often such control is in the form of a feedback mechanism. The end product of a sequence of chemical reactions may exercise an inhibiting influence on the reactions themselves, so that the process creates its own destroyer. As the process slows down, less of the product is made, with the result that the inhibition is relaxed, and the relaxation speeds up the process once again. Such a cycle of speeding up and slowing down creates an overall constancy, a steady state, because the output of the process feeds back into the process itself, maintaining it at the desired level.

This is rather like the population cycle of rabbits and foxes in a wood. The ratio of rabbits to foxes fluctuates, but has an overall constancy, because when the rabbits multiply, the foxes that feed on them also flourish and increase in number. But then the more numerous foxes eat too many rabbits, which grow scarce. Eventually, the decline in the population of rabbits creates a shortage of food. That results in hard times for the foxes, thinning out their ranks. Fewer foxes means that the pressure is taken off the rabbits, which begin to multiply, and the cycle begins all over again so a self-regulating steady state results. The effect of a decline in the rabbit population feeds back into the system, correcting the oversupply of foxes, and the effect of a scarcity of foxes feeds back, enabling the rabbits to increase.

But while feedback may be a means of control, it is also a creator of instability. The system can become an oscillator, because feedback is not a single, once and for all event, but a process that swings back and forth around an ideal set point. The population of rabbits does not always, or even often, equal the population of foxes exactly, but alternates from a high level when foxes are few to a low level when foxes are plentiful. Error and fluctuation are the essence of a feedback system, because error is the message that starts the process of correction.

Feedback is a source of oscillation, and oscillations are involved both in the measurement of time and in the transmission of information. The link between oscillations and messages was never brought home with more painful force than in the career of a famous pioneer and genius of modern communication, who made the disastrous mistake of neglecting to take the connection seriously enough. The term feedback first came into use in the early days of radio. In 1912, the American inventor Edwin Howard Armstrong, then an undergraduate of twenty-two at Columbia University, took the momentous step of feeding back the output of a vacuum tube into the grid circuit of a radio receiver, thereby reinforcing the strength of the signals to which the set was tuned by as much as a thousand times. In doing so, Armstrong discovered a new principle. Armstrong's regenerative circuit, as it was called, using feedback, made it possible to bring in broadcasts from faraway transmitters, revolutionizing radio's future as a medium of communication. The first, dramatic demonstration of a long-distance radio link took place in 1914, when Armstrong arranged for the German embassy to receive messages from Nauen, Germany, at its only wireless station in the United States, at Sayville on Long Island. Overnight, the new type of circuit made newspaper headlines and its inventor became famous as "Feedback Armstrong."

One curious effect was noticed in the regenerative circuit. Too much feedback led to hissing and howling noises in the headphones, because the tube started to oscillate. As an oscillator, the tube became more than just a receiver of signals. It was capable of transmitting signals. As had been realized much earlier, oscillations of this kind radiate electromagnetic waves into space, and the waves can carry messages. In the early days of the Second World War, many British merchant ships carried receivers with regenerative circuits, but these were quickly abandoned when it was realized that oscillations in them could send a signal, betraying the position of a ship to German submarines.

The literally fatal mistake Armstrong made was to file two separate patents, the first for the receiver functions of his feedback circuit, and the second for the oscillating or transmitter functions, a few weeks later. A patent was issued on the first application in October 1914, but the second became enmeshed in elaborate legal proceedings. The flamboyant inventor Lee de Forest, who had broadcast a performance by the tenor Enrico Caruso live from the Metropolitan Opera in

New York as early as 1910, claimed priority. De Forest filed a patent for a circuit that he maintained had both regeneration and oscillation. Armstrong referred seven times in his first patent application to the fact that when the feedback reaches a certain level, the circuit became a generator of high-frequency oscillations. Later he scrapped his second patent application, arguing that the transmitter functions were intrinsic to the circuit. Yet because he had not specifically claimed the transmitter uses in his first application, a legal imbroglio of epic proportions ensued, which ended with Armstrong losing on a technicality when the case went to the Supreme Court. In 1954, nearly destitute, Feedback Armstrong took his own life.

Because of its power to control, feedback in living organisms was widely selected for in evolution, since control increases fitness. Feedback systems have certain unusual properties which make them special and well worth studying. In a feedback circuit it is often possible to use cheap, imperfect components and still obtain overall results that are of superior quality. Feedback is a way of overcoming defects and creating a reliable whole out of unreliable parts. It enabled electrical engineers to link hundreds of amplifiers of varying quality in a transcontinental telephone line and still ensure that messages were heard clearly at the other end. Feedback is also an indispensable feature of hi-fi record players. In the human body, among its many other functions, feedback enables a heart that is deteriorating with age to beat at a rate that is within the limits needed for active life. Biological feedback is one of the most important innovations ever to emerge in evolution; it is absolutely essential for the continuation of life. One result is that oscillations are likely to arise in living organisms, and oscillations are the basis for time measurement, communication and perhaps much else besides. Engineers say one particular difficulty in designing a chemical factory where chemical reactions are taking place and feedback is present between one part of the factory and another is to prevent oscillations. The processes actually seem to "want" to oscillate, and often such behavior takes unexpected forms.

The first biochemical oscillation was detected in 1955, by A. T. Wilson and M. Calvin, in the dark reactions of the cycle of photosynthesis. In 1964, nearly continuous oscillations were seen in baker's yeast, in the process known as glycolysis, a nearly universal means by which cells break down glucose. Certain enzyme reactions in the cells of animals turned out to be oscillating in a robust fashion, producing wave patterns with the mathematical properties of a sine curve.

Cyclic AMP, the second messenger, which is opening up new horizons of understanding in the human nervous system, is associated with feedback, with oscillations, and with time. Since this chemical substance is such a powerful regulator of cell functions, the cell needs to control the amount of cyclic AMP within its borders, and in most cases control is exercised by means of one or more enzymes, called phosphodiestrases, which degrade cyclic AMP into an inert form. If the amount of second messenger becomes too great for more than a few minutes, more of the degrading enzymes are manufactured, and the cyclic AMP level declines. The decline of cyclic AMP then reduces the amount of degrading enzymes. Such a fluctuating foxes-and-rabbits situation, while it makes for overall stability, may lead to oscillations, and cyclic AMP might in that case become a sort of clock. Its level has been seen to oscillate in the pineal glands of chicks studied in the laboratory.

Another reason to suspect cyclic AMP as the mechanism of biological clocks is that it plays a role in the resetting of a clock in a marine snail called *Aplysia*, the sea hare, an unprepossessing creature much studied because it possesses the largest nerve cells in the animal kingdom. Cyclic AMP also mediates the output of a biological clock. Since the same substance is involved in the input to the clock in one system and the output from it in another, some biologists speculate that cyclic AMP may also be in the clock itself, the oscillator that makes it tick.

These ideas pervade our existence, both biological and psychological, because messages are the stuff of life. Feedback, oscillations, clocks and communication are linked concepts, in the body as well as in the information machines of the modern age. They elucidate the close connection between time and meaning. And they are also involved in movement, both physical and psychological. Slime mold is not only an internally communicating organism, but also one that moves in a periodic, rhythmical fashion thanks to the pacemaker effect of waves of cyclic AMP. As we shall see, clocklike mechanisms govern the body's movements through space and the psychological movement through time of temporally organized activities like speaking and conversing. The body and brain, like the external environment of human communication in all its many forms, from sentences to symphonies, are rhythmic. Oscillations make patterns in time, and such patterns can be the basis for sending and receiving messages. We are able to decode the meaningful sounds of speech and music in the world around us only because the rhythmic brain can tune in to the rhythmic environment, following its intricate temporal structure; and since we are masters of that structure by a gift of nature, we have access to the rich knowledge it conveys.

THIRTEEN

The Conversational Waltz

In an imaginative new approach to the puzzle of human timekeeping, Daniel Stern, a psychiatrist at Cornell University Medical Center, screened a film in his laboratory of the first round of the 1966 world heavyweight boxing title fight between Muhammad Ali, then known as Cassius Clay, and Karl Mildenberger, held in Frankfurt, Germany. Dr. Stern's chief interest at the time was in how an infant, so soon after birth, is able to respond with such a precise sense of timing to the movements and sounds made by its mother.

Ali, then aged twenty-four and unbeaten, was reputed to have one of the fastest left jabs in the history of heavyweight boxing. Before the match, a Frankfurt cab driver inquired of one of the fight's promoters: "Do you think our Karl has a chance?" To which the promoter replied, with a sigh: "To live." Dr. Stern carried out a split-second timing study of the movie. He counted the number of film frames it took for Ali to throw a left jab. He found that just over half of all such

229

punches were faster than 9/50 of a second, which is the most rapid visual reaction time observed in human beings. What is more, Mildenberger, a southpaw, threw left jabs faster than the critical speed of 9/50 of a second more than a third of the time he was in the ring.

This result raised some intriguing questions. If a punch is regarded as a "stimulus," in the jargon of psychology, and a boxer's defensive blocking of that punch as a "response" to the stimulus, then most of Ali's jabs and many of Mildenberger's were too rapid for anyone to block or dodge, because of the psychological time constant of 9/50 of a second, which places a limit on how fast a person can react to something he sees coming his way. Good boxers do not signal their punches before they throw them, and yet nearly all of Ali's blows were blocked or avoided, failing to find their target. Ali won by a technical knockout in the twelfth round.

Analysis of the film, Dr. Stern found, shows how quickly the brain can understand and predict the deliberate movements of other people. Mildenberger, in avoiding Ali's lightning-swift jab, was not reacting to an event that had already happened, but was decoding Ali's programmed sequences of behavior, anticipating them in time and space, even though they were intended to be difficult to decode. In order to land a punch it is necessary to estimate exactly where the other boxer will be thousandths of a second in the future, and blocking a punch requires a similar microaccuracy of time judgment.

The Ali-Mildenberger bout is a fair analogy to what goes on when people come together in an everyday social encounter. A person does not simply react to what another person does or says after the fact, as a tennis player reacts to the stroke made by his opponent at the other end of the court. In many cases, the actions of people overlap in time, and often the overlap is so great that an action cannot be described as a response at all, because both action and "reaction" happen simultaneously. This is not like a tennis match, where a cause leads to an effect after a lapse of time, but resembles more a waltz, where both partners know the steps and the music by heart and move precisely together, because they are generating their own rhythms, each in synchrony with the other's, like two coupled clocks. "We are forced to think," Dr. Stern concluded, "that they are following a shared program."

As we have noted already, an essential property of circadian rhythms is that they are relatively autonomous, even though they are normally entrained by an external time cycle. Circadian rhythms, and biological rhythms generally, do not depend on continual feedback from the environment. The program is set to run and it runs. The clock is a "cause" of metabolic and other changes in the body. Much the same could be said of psychological rhythms, which govern the timing of human speech and behavior when people come together to talk. One party to a conversation is not necessarily "activated" by the other, or "responding" to the other, but rather both together generate autonomous sequences of joint action.

The enigma of what has been called "the serial order of behavior" has mystified researchers for many years. The term was made popular by Karl Lashley, one of the colorful figures of American psychology, who devoted a great deal of his time to the study of how human beings can perform, at high speeds, actions that are precisely organized in time.

Lashley spent his early career studying motor skills, although his own motor skills were notoriously deficient; a colleague once said that being a passenger in Lashley's car, usually a flashy convertible, was a truly hair-raising experience, because Lashley "operated an automobile with supreme disdain, both for the traffic regulations and for the mechanical limitations of the vehicle." Lashley's tastes were catholic and his mind was receptive to ideas from many different sources. Under the strong intellectual influence of his mother, he learned to read at the age of four, and once said he was raised on "Nestlé's Food and Nietzsche." He played the cello with the Jacksonville, Florida, Philharmonic Orchestra, and was fond of Virgil, Balzac, Bertrand Russell, burlesque shows, daiquiris, pink ladies, and his dog, Till

Eulenspiegel. Lashley was strikingly thin and pale as paper. He wore a pince-nez on a black string. He did not have a very high opinion of human nature, but tried not to hurt the feelings of individuals. He was shy and bold at the same time. A close friend said that meeting Lashley was like approaching "a thing of wildness and timidity."

How did humans, Lashley wondered, perform such marvels of synchrony as hitting a ball with a polo mallet while sitting astride a cantering horse, or playing the piano, or arranging a complex idea into a properly constructed sentence? Existing psychological theories assumed that these exquisite timing skills were produced by chains of reflexes, in which one small segment of the complete action triggered another small segment, and so on until the end of the sequence was reached. This was essentially a feedback explanation, which assumed that when a movement started, it continued until stopped by the end result. A pianist, for example, was supposed to play sequences of notes one at a time, each finger movement being triggered by sensory impulses sent to the brain as a result of the preceding finger movement.

Lashley doubted this explanation. He studied pianists' performances and compared the speed of finger movements needed to play a rapid cadenza with the speed of transmission of signals between nerve cells in the brain. He found that the time interval between one depression of a key and the next was too brief to support the feedback theory. A sensory message would take too long to travel from the fingers up to the brain and back down again to the fingers. He suspected that instead of chains of reflexes, a whole series of events was under the control of a motor program, and feedback set off an entire series, not just a single action. Exactly what drove the fingers at lightning speed across the keyboard Lashley could not say for sure, but he speculated that there must be some sort of circuit or circuits in the central nervous system transmitting "waves of excitation" which predetermine the timing and the time order of speech and other skilled motor sequences, at an unconscious level. Such circuits are clocklike mechanisms which are more reliable, more independent, and more central to the organism than the sensory feedback reflexes proposed by the older school of psychology. They organize movement in time and in space. Lashley thought these reverberating circuits were always fired up and active, even in sleep.

Temporally programmed actions, Lashley said, exist even in insects, but they do not reach any degree of complexity until the appearance in evolution of the cerebral cortex. They are "especially characteristic of human behavior and contribute as much as does any single factor to the superiority of man's intelligence." Yet they also have archaic origins. A pianist superimposes two rhythms when he plays, and a fish superimposes two rhythms when moving its dorsal fin, and both rhythms are centrally maintained. The performance of the pianist is a vastly sophisticated and more highly evolved version of the performance of the fish.

In the nervous system, there are elaborate networks of interrelated nerve cells that can transmit wave patterns in time to other parts of the nervous system. Always active, they form the basis of behavior and play a role in all perception and every integrated movement, from walking to speaking. "The rhythms tend to spread to almost every other concurrent activity," Lashley noted. "One falls in step with a band, tends to breathe and even speak in time with the rhythm. The all pervasiveness of the rhythmic discharge is shown by the great difficulty of learning to maintain two rhythms at once."

In the 1980s, clear evidence has been found for the existence of oscillators that govern the rhythmic action of movement, from the trotting of a horse to the marching of a regiment of soldiers. In a landmark article in *Science* in October 1980, Fred Delcomyn, professor of entomology at the University of Illinois, declared that the long debate over whether the central nervous system does or does not require feedback from the sense organs in order to generate properly sequenced, rhythmic behaviors, such as walking, is dead and done with. The answer, beyond a doubt, is that feedback is not needed. Evidence built up over the past twenty years

234 TIME AND INFORMATION

amply supports such a conclusion. Neurobiology, possessing few universal principles that apply not just to single cells but to whole systems of neurons, has acquired a new one, the principle of the internal, autonomous and precise generation of movements which are patterned in time. It applies to all nervous systems. "Recognition of this principle will mean the resolution of a controversy nearly three-quarters of a century old," Dr. Delcomyn says. The range of species and the range of activities to which it applies are impressively broad and include creeping in the sea hare, flying in the locust and dragonfly, swimming in the eel, hopping in the rabbit, scratching in the cat, croaking in the frog, eclosion in the moth, and shell opening in the mussel. Oscillators, behavioral clocks, produce these rhythms of activity. "It seems possible that oscillators may have evolved many times in the animal kingdom," Delcomyn notes. "This suggests a significant evolutionary pressure."

The questions we should now be trying to answer concern the nature of the behavioral clocks, how many there are, and how they interact with one another and with input from the senses. Sensory feedback may help to stabilize a rhythmic movement while it is going on, and a rhythm imposed from outside the organism may reset the inner behavioral clock, entraining it and shifting its timing, just as light pulses can reset a circadian clock; but like circadian clocks, these are self-sustaining and do not depend on an outside source of energy or a continuous source of information about time. External factors—the roughness or smoothness of the terrain, the small adjustments needed, say, for each soldier in a ceremonial parade to keep his place in the line of march —modify the actual movements of the body, but the rhythm itself is generated from within, whether it drives the fingers of a concert pianist or the fins of a fish. It is a clocklike system regulating an action in which time is of paramount importance.

The actual mechanisms responsible for such timed behavior are not known. It is fairly clear, however, that we cannot speak of a single mechanism, a master clock which drives all rhythmic responses. That is because temporal organization varies so much, depending on what we are doing, how we do it, and even what sense organ we happen to be using. Specific parts of the body and even specific muscles within a given part are not under the same control. Two such simple and seemingly closely linked actions as extending a finger and flexing it to tap are separately timed. What is more, a sound lasting for a certain time seems to go on for longer than a light switched on for the same length of time, and "perceiving" an interval, just saying how long we think it lasted, is a different mental operation from producing an interval, trying to imitate a clock by striking a pencil on the table to mark the start and finish of the interval.

Much of social life, and of human communication in the broadest sense, consists of rapid sequences of actions which are timed to the split second, and depend for their success, and even their meaning, on rhythmic properties. Rhythm enables the partners in a conversation to predict important next moments in the stream of speech, based on what is happening at the present moment, the "now." It is both a clock and a guide to meaning, showing, once again, how close is the link between time and information. When people meet and talk, they often behave more like musicians who improvise together with moment-to-moment precision. Such an encounter is both spontaneous and predictable at the same time. It needs to be a bit predictable in order to be a bit spontaneous. That is what makes conversation very different from the dialogue in a play, where actors speak lines that have been written for them in advance, and which they have rehearsed together many times.

Almost any kind of human encounter tends to have a regular time pattern. Young children are more rhythmic in their actions when they play with other children than when they are playing by themselves. At a family meal, children are apt to bang their spoons on the table in time to the rhythm of the adult dialogue going on around them. When talking to each other, people often cough, sneeze, laugh and blow their noses in synchrony with the prevailing tempo of the conversation. Teachers sometimes match their own rhythms with those of the class, and sometimes require the class to match theirs.

"Ensemble," says the linguist Ron Scollon, "is what is real about real time." By that he seems to mean that in many kinds of social activity, the participants come together like players in a string quartet, who must lock onto the prevailing tempo of the music if they are to know at all times what the other players are doing and what they are likely to do next. Talk is social ensemble, and unless the members of the ensemble are linked by this "bond of immediate temporal predictability," as Scollon calls it, the encounter may go awry, just as in a chamber music concert the way in which the players are entrained to tempo can make or break a performance.

There is a stream of talk, and it functions as the stream of consciousness of William James functions, to provide a time context for the thoughts and messages that flow in it. Time cues must be continually given and received, so that people conversing know whereabouts and "whenabouts" they are in the stream. That is what makes conversation such an original, lively and creative activity, because its time structure favors decisions that are made on the spur of the moment and adjusted continuously in the light of rapidly changing circumstances. It is difficult for anyone taking part to maintain fixed attitudes and predetermined roles from beginning to end of the encounter, and this is what gives conversation a useful psychological role, apart from being an example of human communication pure and simple. But spontaneity does not happen unless a person knows, from moment to moment, "what time it is" in the conversation.

The regular pulse that tends to be set up in any encounter where two or more people are talking does much more than merely stabilize and smooth out irregularities in the stream of speech. It may determine whether the encounter makes sense, whether words mean what they are supposed to mean. The role of pulse in conversation is in some respects similar to, though in other respects very different from, the role of the clock in a computer, which separates sense from nonsense in time.

For many years, it has been known that time patterning is an essential aspect of ordinary human talk, but only recently has the rhythmic structure of conversations been closely investigated, so that it is now a topic of intense interest in linguistics and psychology. This new work is throwing up some surprises, showing us the inner, unconscious psychological mechanisms by means of which people adapt their own, personal time to the peculiar demands of the ensemble, which is the "real time" of the conversation.

Frederick Erickson, of Michigan State University, has shown that, in a conversation, listening as well as speaking involves a moment-by-moment process of sending out signals, which need not be verbal, about the state of play of the conversation now and the likely state of play in the immediate future. By means of signals of this sort, people adjust their behavior continually. In Erickson's view, listening in a conversation plays a role almost as active as that of speaking. By means of nods of the head, gestures of the body, smiles or frowns, the listener keeps the speaker informed on how his message is being received, whether it is time he stopped talking, or whether he should continue. At any stage of the encounter, speaker and listener are obliged to draw inferences about the meanings being exchanged. They must signal both their own meanings and the particular manner in which they are interpreting the meanings of the other person. They must also indicate in some way the significance of the present moment and the moment to come in the sequence of talk. All this is done spontaneously, step by step as the conversation proceeds. Two people having a conversation guide each other along like a couple of ballroom dancers who regulate the movements of their partner as they glide across the floor in a waltz. Rhythm is one of the main organizing principles of conversation, as it is of ballroom dancing. It is a social steering mechanism by means of which people are able to communicate information about this moment now, the next moment and the moment past.

The cues that parties in a conversation send each other

are timed so that they often occur with clocklike regularity. In English, which is a heavily stressed language, certain syllables are made more obtrusive than others by having a higher or lower pitch, greater volume, a change of tempo, or a purer vowel quantity. Sometimes a syllable will be stressed by means of more than one of these devices. A stressed syllable tends to be timed so that it synchronizes with the underlying beat of the conversation; so too are the body movements and gestures of a speaker and a listener, and the exchange of roles that occurs when one person finishes speaking and surrenders the floor to another. Such rules of timing are not always observed, but they are observed more often than not, and sometimes the regularity of the pulse is almost as strict as that of a metronome. Erickson surmises that the regular beat of conversation enables partners to "entrain" each other so that each can anticipate next moments. This is analogous to the entrainment of a circadian rhythm by a day-night cycle, making it possible for the organism to anticipate strategic "next events" in the environment.

Whether or not words have the effect they are intended to have may depend on whether they fall on the beat, whether they occur in an appropriate time or an inappropriate slot in the conversation. Sometimes the result of a missed beat in a human encounter can be almost as disruptive to the success of that encounter as the result of seeing a relevant landmark at an irrelevant time is disruptive to the normal activity of a honeybee. An utterance loses its meaning if it falls into the wrong time slot, in the sense that though it may be heard by the ear, it is ignored by the conscious brain.

Frederick Erickson found a striking example of lost meaning due to wrong timing when he videotaped and analyzed a test given by a special education teacher to a five-year-old child named Angie. The intention of the test was to find out whether the child had any handicaps for which extra coaching might be needed. Some of the results are remarkable for the light they shed on the critical importance of periodicity and meter in even the simplest kind of face-to-face contact between people. During this interview, Angie and the teacher sit together at a table with Rita, a child who has already taken the test.

The teacher gives Angie part of a sentence and asks her to complete it. "Listen carefully now," she says. "In the daytime it is light. What is it at nighttime?" Rita at once speaks up with the right answer, "Dark," in a loud voice, just before Angie also utters the same word, but in a softer voice. The teacher does not register Angie's answer. She reprimands Rita for speaking when it is not her turn, and then proceeds with the interview:

TEACHER: What is nighttime?
ANGIE: Dark. [*The teacher does not register Angie's answer.*]
TEACHER: [*Begins to speak in a singsong before Angie has finished the word.*] If daytime is light, nighttime is . . . ?
ANGIE: Light.

Here is a perfect example of the way in which a failure of timing can result in the breakdown of social meaning. The child being interviewed gave the correct answers, but at an incorrect time. The teacher had established a rhythm by means of stressed syllables which are placed about one second apart. These stresses form a regular metrical pattern in time, and they signal the appearance of new information. The new information required in this case is the answer to the question, "What is nighttime?" An answer is expected to fall into the right time slot. When it misses the slot, the teacher ignores the answer. Twice Angie gave the correct reply, "Dark," but it was out of synchrony with the one-a-second pulse of the interview. Apparently thinking she had made a mistake, the child changed her answer to "Light," and because this answer did fall on the beat, the teacher noted it, and marked it down as wrong. Luckily, the standard required of children taking the test was so modest that Angie passed.

In another revealing study, Erickson filmed a dinnertable conversation at the home of a family of Italian-Ameri-

cans in a suburb of Boston. The transcript of the evening's
table talk showed how rhythmic ensemble work can preserve
sense and structure even in the face of what might appear to
be sheer chaos, with more than one person holding the floor
at the same time. An underlying rhythm generated by the
group as a whole has an organizing, orchestrating effect, so
that a speaker is able to talk even while another conversation,
unrelated in content, is in progress, without seeming to interrupt or destabilize that second conversation. The rhythm
serves as a kind of master clock for the ensemble.

The company at this dinner included seven-year-old
Bobby, his mother, an older sister of Bobby's and a woman
guest. The talk was relaxed and bantering. Asked if the salad
was homegrown, the mother replied proudly: "Everything
came out of the garden." Then the conversation broke up
into two streams. Bobby began to tease, asking if literally
everything on the table came from the garden. He ran
through a list of items, one after the other, with a relentless
emphasis, like a jackhammer: "'n the *dishes*, 'n the *forks*, 'n
the *napkins*, 'n the *glasses*, 'n the *pepper*, 'n the *drinks* . . ."
At the same time, the guest inquires jokingly of the mother if
the pasta was grown on a lasagna bush in the garden. The
following exchange then took place, to the insistent accompaniment of Bobby's rhythmic monologue:

MOTHER: If we actually had a lasagna bush in the
backyard we'd have lasagna every other *night*.
GUEST: Yeah, that's one of my favorite things to *eat*.
MOTHER: An' they should have a *steak* tree.

The stressed syllables in Bobby's accompanying monologue contain a new item of information on every beat. His
rhythm is a simple anapest, ta-ta *tum*, ta-ta *tum*. Each successive noun—dishes, forks, napkins, glasses, pepper, drinks—
falls on the accent. Meanwhile, in the more complex adult
dialogue, the stress falls precisely on a boundary, marking
both a moment of new content and the end of a syntactic
clause. What is most surprising is that the stressed syllables

in each of the two separate conversations, those of the boy and those of the mother and guest, occur either when there is a moment of silence in the other conversation or in synchrony with the pulse of that conversation, so that neither adult nor child destroys the rhythmic integrity of the ensemble. Nor do they interfere with critical time slots in the other stream of talk, where new information tends to occur. The two streams overlap but do not interfere with one another, almost like a musical composition for string trio.

Erickson has written out part of this conversational ensemble in musical notation, and it shows that each speaker avoids disrupting the ensemble even when talking simultaneously, by the exact matching of stressed syllables and by the mother doubling her speed of delivery just before the next rhythmic emphasis:

We'd have la - sa - gna ev-'ry oth-er *night*.

'n the *dish-es* 'n the *forks* 'n the *nap-kins*

"Conversational partners appear to be rhythmically field-sensitive," Erickson observes. "They appear to be taking action on account of the rhythmic action of others, not only within pairs, but across pairs as well. More generally, rhythm seems to be the fundamental glue by which cohesive discourse is maintained."

Ron Scollon, a linguist whose past career as a player in a chamber orchestra leads him to take a special interest in the question of tempo in face-to-face encounters, has recorded and analyzed a wide range of talk, including several university lectures, a radio symphony broadcast announcer, Groucho Marx on the radio show *You Bet Your Life*, a baseball commentary, some third-grade reading lessons, three family members chatting while cooking breakfast, and narratives of Athabaskan tradition bearers in both Athabaskan and English. The tapes provided a wealth of new insights into the role of timing in human communication. Scollon found that a steady, fairly slow measure in 2/4, or duple time, the fundamental measure of music, underlies many kinds of ordinary talk, although many beats are silent.

One important discovery is that tempos vary not only from person to person but even in a single individual. In this particular case, tempo is defined as the speed of the succession of downbeats and upbeats in a measure of 2/4 time. It establishes a basic, regular pulse which persists like the tick of a clock, although the rate of the clock may speed up or slow down. Tempo is measured in beats per minute. It varies considerably according to context in ordinary talk, but in narrative it is more uniform. Tempo is only one aspect of rhythm, and rhythm is more complex than tempo. Fast and slow speakers may observe the same tempo, but the fast speaker compresses many syllables into one measure of 2/4 time, while the slow speaker's measure contains fewer syllables. The quick-witted Groucho Marx spoke at a very slow tempo of only 75.9 beats a minute, but squeezed an average of 4.62 words into each measure. It is as if a composer used 32nd and even 64th notes in a bar rather than quarter or eighth notes.

What surprised Scollon most about tempo was not just that it underlies talk in nearly all contexts and circumstances, but that tempo also seems to be negotiable as part of the give and take that goes on continually when people get together and converse. A speaker will time his entrance into a conversation according to the particular tempo already set by others. Often, after entering, he will speed up or slow down the beat in order to establish a new tempo. It is very unusual for anyone to begin speaking at an arbitrary rate without first matching the prevailing tempo, so that the whole ensemble moves smoothly and together to a new underlying beat. It is as if a new conductor were to come out and lead the same orchestra. Radio announcers often start speaking in synchrony with a beat that is the same as that of their introductory theme songs. In some of Scollon's recordings, after a longer silence than usual, someone will clear his throat in a gesture, as Scollon puts it, "which predicts the tempo that is to follow as accurately as a conductor's silent 'one, two,' does before the entrance of the orchestra."

The density of words in a measure and the number of silences observed tend to be fixed features of a person's conversational style, but tempo is far from fixed. It is a variable which is used to negotiate interactions between speakers. On Scollon's tapes, the mean tempo of talk was strikingly variable, ranging from 115.4 beats a minute in a radio spot announcement to 70.2 beats in another radio program.

Certain sounds that are meaningless in themselves, such as "um" and "ah," which seem to do no more than fill up hesitations, empty moments in the stream of talk, may be a way of integrating the ensemble of a conversation, since they almost always fall on beats of the measure and seem to serve a timing function. Repeated words, superfluous if we consider only the content of what is said, may play a similar role.

Several colleagues listened to Scollon's tape recordings independently, to ensure that the 2/4 measure Scollon heard was not a figment of his imagination. They found that in some parts of the samples it was easy to hear the beat, and in other parts more difficult. The tempo seemed to wax and wane. It was at the boundaries of speech events that tempo tended to weaken in its regularity. When people taking part in a conversation were negotiating adjustments of tempo, the tempo itself was hard to determine. Moments of ambiguity, however, when talk seemed to be now in 2/4 time, now in 3/4 time, were subsequently resolved when the speaker settled down to his own particular beat.

Tempo, as Scollon sees it, is the bond that enables us to move together in real time. It provides us with an account of the immediate past and a basis for anticipating the immediate future. In a musical performance, the note being played "now" gives an individual performer information about the loudness and tuning of what he himself is playing at that moment, but it is the tempo that tells all the players when to play their next notes. Similarly, in a conversation, it is tempo that keeps the talkers and listeners in touch with each other. Negotiating tempo is a way of coordinating the whole ensemble.

In human encounters, tempo is highly nonuniform, from occasion to occasion and from person to person. The same person's tempo may change considerably, depending on what he is thinking, feeling and doing. The heart rate, for example, is much steadier and more regular when the brain is engaged in a demanding and difficult mental task than during periods of relaxation and play. Such lability of tempo seems to be necessary, almost a prerequisite, for participation in human communication which takes place in real time.

The existence of an underlying beat, or measure, agreed upon collectively, automatically and without thinking about it, reduces uncertainty about the "when" of events in the stream of talk. A beat supplies an element of sameness and of constancy in conversation, a form of communication that is full of uncertainty and change by its very nature. Being spontaneous in a human encounter is impossible without a regular time base. In a play, the dialogue is determined in advance by the script, and the actors know exactly whereabouts and "whenabouts" they are in the script at every moment. It is the script that organizes the players in time. The essence of a conversation, however, is that it is not predetermined. A conversation has no script. What saves it from being disorganized in time is the existence of a common pulse, a kind of clock, whose ticks mark temporal reference points in the action. A play sounds spontaneous only because the timing is rehearsed with such skill and care that it mimics spontaneity. Just as jazz musicians are unable to improvise together without a clearly defined beat, so talkers cannot manage their spur-of-the-moment shifts of topic, their split-second entrances and exits, without the metronome of negotiated tempo.

Hesitation is uncomfortable in a conversation not so much because talk stops and starts, but because it does so in a way that makes the timing of the talk unpredictable; other partners in the conversation find it difficult to match their actions with that of the speaker. If one person departs by as little as 1/6 of a second from the established conversational pulse when starting to speak after another person has finished speaking, stability may break down for a short period and there is a competitive "tugging" at the rhythm rather than mutual coordination.

"The ways in which each person talks and listens from moment to moment become an environment for the others," Erickson says. "That is why, although we are now able to program computers to talk, we are unable to build them so as to act as engaging conversationalists. People can do that, we argue, because they are able to make sense in the immediate circumstances of the local scene from moment to moment in real time."

The pulse supplies sameness, predictability, even though the conversation itself may be full of unforeseeable changes of content, bursts of humor, which depend for their effect on surprise, and new signals of intent from the other speakers. In fact, studies of conversations have shown that the novel, unexpected parts of a speaker's message tend to fall on the beat. Stress is used to mark the moment in time when new information is given, a subject changed, a fresh point of view or a contrast introduced. The result is that sameness in time may denote difference in content or in context. Predictable time slots are reserved for unpredictable events, perhaps as a way of ensuring that the listener pays attention at a specific moment, and this is all done below the level of conscious awareness. Stress is also used to signal a fresh subject or topic, a point of view, or definite as opposed to indefinite information. This means that listeners are able to anticipate rather than simply react to conversational events. Tempo enables people to adapt to the environment of communication much as the circadian timing system enables all species of animal to adapt to the physical environment, by ensuring that the body "knows when it is" in real time, so that it is able to prepare for what the environment will do next.

Real-time activities, those in which the brain must synchronize with the time of the group and not drift into some private time scale of its own, often mystify by the speed at which they take place. Such rapidity, essential to all communication, is possible because the brain always knows where it is and when it is, and therefore predicts ahead, into the future, rather than treating each moment as a complete surprise.

FOURTEEN

Evolution and the Retreat into Time

Time is important in the lives of human beings for reasons that do not apply to the rest of the animal kingdom. All living creatures need to perceive and be aware of their environment, and different species do so in different ways. Since knowledge of the external world is always selective, through sense organs, and is never complete and exhaustive, a particular species may emphasize smell over sight, for instance, so that those aspects of the environment that can be smelled are more important for the species than aspects that are seen. The internal representation of the world created by this process is necessarily a simplified and specialized version of the real thing, a device enabling an animal to make some sort of sense out of an immense amount of incoming data. The kind of sense that is made of the data differs from one species to another. "We ourselves," said the biologist Francois Jacob, "are so deeply entrapped in the representation of the world made possible by our own sense organs and brain, in other words by our genes, that we can barely conceive the possibility of viewing this same world in a different way. We can hardly imagine the world of a fly, an earthworm or a gull."

Information from the environment bombards the living organism, but much of it is filtered out as being irrelevant, and what part does gain admission is encoded, stored, manipulated and transformed in such a way that the nervous system and brain create an analog of reality, one that is suited to the particular needs of the animal, by means of internal representations of the external world. Such a model made by the brain "is, in a way, a possible world," Jacob notes, "a model allowing the organism to handle the bulk of incoming information and make it useful for its everyday life. One is thus led to define some kind of 'biological reality' as the particular representation of the external world that the brain of a given species is able to build. The quality of such biological reality evolves with the nervous system in general and the brain in particular."

If a species of animal perceives the environment chiefly through the sense of sight, space will predominate in its internal model of the world. If another species makes greater use of hearing than of vision, however, time will be the important dimension. "Biological reality," therefore, while it is a space-time analog of what actually exists, may emphasize space or emphasize time, depending on the way of life of the animal and its evolutionary history. The biological reality of human beings incorporates time in a unique fashion, enabling them to master the extremely complex temporal structure of the special kinds of sound sequences used in spoken language.

From eye as well as ear living creatures find their way around the world by using information of both kinds, and often one supplements the other. A frog, for example, recognizes food and enemies by vision, but mating partners by hearing; the frog ascertains the species, sex and age of a likely mate by the amount of low- and high-frequency sounds contained in a croak. The rules that apply to the environment of sight are not the same as the rules that apply to the environment of sound. In the visual mode, space is more important than time, while in the aural mode, time is more important than space.

Vision does operate in the time domain, of course. We do not take in a painting all at once, in a single instant, but make discoveries about it little by little. Yet a painting stays in one place while we look at it, and all its parts are simultaneously present. Certain schools of twentieth-century painting, such as cubism, represent an object as if seen from all sides simultaneously, thereby collapsing the time it would take in real life to experience the object in its entirety, by walking around it, into an instantaneous experience in art, spatializing time in a new way. Quite the opposite is true of a piece of music, which is made up of sequences of notes that follow one another in time. The music has a precise beginning and a precise end. It does not last indefinitely. A lapse of time is necessary for the music to exist at all; it is nonsense to speak of music in terms of the sounds it is making at a given instant. Without a lapse of time, music cannot exist. Music's effect on the listener lies in its temporal structure, the way it makes patterns in time, and it also makes considerable demands on memory. In order to appreciate music, a listener must train his memory or risk missing much of what it has to offer. The crucial distinction between the visual arts as being primarily spatial and music as being essentially temporal means that an aesthetic theory that holds good for one may not apply to the other. It is one thing to purge painting and sculpture of all literary and sentimental content and make it wholly abstract, but quite another to insist on a similar cleansing for music.

The English composer and conductor Constant Lambert was caustic in his censure of critics who condemn music and painting indiscriminately for "telling a story" or representing episodes from real life and romantic fiction. That fallacy arises, Lambert said, because critics have failed to notice that a picture with a narrative element is bad, not because it has committed the crime of being literary, but because it is trying to represent time by cutting a section through it in space. It shows an instant of a span of time, and the viewer's interest is led away from the scene as it occurs in space, because he is trying to reconstruct the events that have led up to that particular moment in time. The picture is "in the nature of an uncompleted sentence." The same scene could be depicted in a novel and be perfectly acceptable, because then it would be one of a series of events in time. The "impure" painting, one that tells a story and appeals to the emotions, is closer to music than a purely abstract painting which has no literary or sentimental content, because of music's emotional appeal and its time element. Music, far from being an abstract art, is as naturally emotional as painting is naturally representational, Lambert declared, adding:

"The repetitions of a certain underlying curve in an abstract or representational picture have no dramatic content because they occur in the same moment of time—one's eye can choose which it looks at first, or take in the various statements of the same form simultaneously. But the return of the first subject after the development in a symphonic movement has an inevitable touch of the dramatic, merely through the passage of time that has elapsed since its first statement. Time, in fact, is rather vulgarly dramatic; it is the sentimentalist of the dimensions."

Hearing is a hundred times more agile than vision at detecting the fine details of temporal structure. If a light is made to flash on and off in rapid sequence, the human visual system loses the ability to perceive the light as being intermittent if the intervals between flashes are shorter than a fiftieth of a second. Then the light appears to be continuous. For sequences of sounds, on the other hand, the threshold of fusion is much higher; the intervals between sounds need to be as brief as about one five-hundredth of a second if we are to have the illusion of hearing a continuous sound. When Morse code operators are made to work with light signals instead of sound signals, they are about five times slower in perceiving the messages.

So mercurial is the human sense of hearing that even the

fastest passages in music played by a symphony orchestra do not approach the limits of the ear's power to discriminate time structure. Only modern electronic synthesizers are able to cross that threshold, using computers which generate rhythms too rapid for the lips and fingers of the most skilled human players, though listeners can appreciate them.

The ear is complex, and it craves interesting sounds, those with rich and elaborate waveforms. The instruments of the modern orchestra are the result of a sort of Darwinian process of evolution, in which the unfit were eliminated by natural selection. The fit instruments, the ones that excite and stimulate the ear's agility in coding time patterns, survived, whereas those that failed to do so became extinct, for the very good reason that composers ceased to write music for them.

Sound information can be converted into sight information, and vice versa, but curious anomalies arise as a result. Composers have sometimes written passages into their scores which the eye can detect easily, but which the ear is apt to miss when the space code of the printed score is transformed into the time code of the actual performance. The Dutch composers of polyphonic music in the seventeenth century would occasionally write two sequences of notes, one of which resembled the other played backward. On paper, a reader of the score can see the reversed sequence immediately. When musicians play the work, however, the mirror-image effect usually goes unnoticed. There is a character in Thomas Mann's novel *Doctor Faustus*, an eccentric lecturer on music who used to regale his sparse audiences with examples of visual curiosities that composers smuggled into their scores, and "wager that very few people would have detected the trick by ear, for it was intended rather for the eye of the guild." In a more rigorous fashion, Bela Julesz, of the Bell Telephone Laboratories, has shown that the ear cannot perceive temporal symmetry, a sequence of sounds played forward and then backward, if by "perceive" we mean the effortless, spontaneous recognition of pattern. Julesz suggests that this ability failed to appear in evolution because in real life time is not reversible. In the physical world, things tend to become less orderly as time goes on. Parts of a system that are separated are inclined to become mixed together, and do not unmix without intervention from outside. Time is thus a one-way street from order to disorder. Only on a movie film can the events of a day be shown unfolding from evening to morning instead of from morning to evening; on film a mixture can unmix itself spontaneously, but not in actual experience. Even though a reversed film might make visual sense of a sort, the sound track will be unintelligible. The human brain internalizes a representation of time as invariantly one-way, with an arrow that points from past to future and not from future to past. A melody sounds very different when played from finish to start instead of from start to finish. By contrast, repetition is as easily perceived by the ear as by the eye.

The psychological connection between hearing and time is so close that people who become deaf as adults, after a normal childhood and youth, often notice that life seems to lose its quality of forward movement. It grows static. The sense of the passing of time, so strong in most of us, despite the insistence of today's scientists that temporal passage has no objective reality in the physical world, diminishes when the sense of hearing fails. The psychologist Ira Hirsh believes that the auditory system is a timekeeper, perhaps complementary to the role of vision, helping us to deal with time as vision helps us to deal with space. In fact, our "sense" of time may depend in part on active stimulation of the circuits of the brain that control hearing. In an ingenious experiment, Hirsh showed that the ability to judge the duration of an event is affected by the presence or absence of a background noise, but not by a change from darkness to light. People think the event lasts longer when the sound is introduced.

In hearing, space and time come together in a curious manner. One way of understanding this convergence is to consider the fact that today's electronic engineers have not been able to design stereo record players of such fidelity that they can recreate exactly the experience of sitting in a concert hall listening to an orchestra playing. Nobody hearing a record of a concert could ever be fooled into thinking a live performance was in progress. The reason engineers are unable to accomplish such a feat of deception is that science does not fully understand yet how the human auditory system is able to construct an acoustic space, of such a kind that we can sit in the concert hall with our eyes closed and have a representation in our head of the space in which the music is being played and the location of the sounds. Essential pieces of the puzzle are still missing. If an engineer could invent and market at an affordable price a stereo player that reproduces faithfully the acoustic space of an actual performance at home, he would make a fortune.

The problem is that hearing is not just physiological but psychological as well. To ask, how does the ear hear? is to ask another, more difficult question, namely, how does the brain deal with time? Hi-fi designers do not simply study acoustics, but are interested also in the wider discipline of psychoacoustics, which requires a general understanding of the nervous system. The brain creates an acoustic space by converting the time code of sound into information about space. This means that the direction from which a sound comes has an important effect on the way in which that sound is perceived. The American composer Charles Ives, happening to walk one day around an orchestra that was playing in the square of a New England village, was astonished to find that changes of position made a remarkable difference to the sort of sounds he was hearing. John Cage also remembers experiencing "a sudden joy," at the age of seventeen, when he stood at the intersection of two streets in Seville and realized that he could hear two different sounds at the same time.

Information about space is encoded in the auditory system of the brain in part by exquisitely accurate comparison of the difference in the time of arrival of the same sound at each ear. If a sound is coming from the extreme left or extreme right of the listener, it will arrive at one ear earlier than at the other, but only 675 millionths of a second earlier.

Yet so subtle is the brain's power to analyze time that a discrepancy of just 10 millionths of a second can be detected. The main function of the ear itself is to convert sound waves in the air into electrical pulses, which are then passed on to the brain. In a coiled chamber called the cochlea, filled with a clear fluid, thousands of tiny hair cells, connected to the brain by means of the auditory nerve, are tuned to sound waves of a particular frequency, like the strings of a piano. A hair cell responds to movement of the cochlear fluid at or near its natural frequency by producing electrochemical pulses which travel along the auditory nerve to the brain. Exactly what goes on in the brain when the information from the cochlea is being processed is not known for sure, but it is thought that several different regions of the brain go to work on the signals independently but in synchrony to create the internal experience of an external space, the "acoustic space" in which the sounds are occurring. Comparison of the different times of arrival of the same sound at each ear is made in the superior olivary complex, which is connected to the hair cells in the inner ear by nerve pathways only one neuron long.

Such fine time discrimination enables the auditory system to be selective; it can suppress echoes, which otherwise would distort and confuse the perception of sound signals. In a room, a speaker's voice sends sound waves bouncing off the walls and ceiling, and these waves arrive at a listener's ears slightly later than those that travel by a direct route. The listener is unaware of these secondary signals, however, because the brain holds sounds in a temporary buffer for about a twentieth of a second, adding information on where each sound originated. By means of this brief delaying device, sounds can be compared, and only those sounds coming directly from the speaker are judged to be acceptable. Echoes, if they arrive within a twentieth of a second of the primary sounds, and from the wrong direction in space, are ignored. The ear detects them, but the brain ensures that consciousness will fail to notice them. In a cavernous hall or in a mountain gorge, a listener is aware of echoes because they travel greater distances and therefore arrive too late for the brain to compare them with the primary sound.

Good as they are, modern stereo players do not make it possible for the listener to recreate the original acoustic space of the performance because they fail to provide the right time cues. This means the space is seriously distorted. Robert Carver, of the Carver Corporation, believes that a two-channel player system can convey all the information needed for the sensation of hearing a live performance, as soon as scientists learn how to manage the time cues. Existing systems, Carver declares, would not deceive a four-year-old child. "Some very bright people are going to have to thoroughly investigate and understand exactly how it is we hear things, and then that's going to have to be translated into hardware," he thinks. "If the necessary psychoacoustic research happens, music systems 25 years from now can be breathtaking in their realism." At Johns Hopkins Peabody Institute, psychologists are recording the brain waves of people while they listen to music, hoping to unlock the secrets of how signals at the ear are converted into models of space in the head.

Using the environment of sound to obtain information about the environment of space by means of time cues is a strategy that appeared widely in evolution. A barn owl can fall upon and kill a mouse rustling in leaves in total darkness by turning its face in the direction of the sound. The difference in time of arrival of the sound signals at each of the barn owl's ears is only 15 millionths of a second, but the bird can fix the position of its prey both in the left and right direction and on the up and down axis in space. The owl is able to detect noises too faint for humans to hear. It can also store very complex sounds in memory, and these can be used later as a standard of comparison for new sounds that the bird encounters.

A different method of locating objects is used by bats, which emit high-pitched shrieks, generated in rhythmic pulses, and these sound pulses bounce off surfaces almost as light does, enabling the animal to distinguish not only the shape of things and where they are located but also their texture. A bat can tell the difference between the softness of foliage and the hard, smooth surface of a window, just by the quality of the reflected sound.

Hearing, vision and smell are distance senses, unlike touch and taste, which deal with the close at hand, the here and now, and they are distance senses with respect to time as well as to space. A sessile organism tethered to a rock on the bottom of a riverbed responds only to immediate events, those that impinge physically on the surface of its body, and only at the moment when contact is made. As soon as organisms began to develop senses that could detect things far off in space, time entered as an important aspect of existence. If the creature could be aware of a distant object, it gained a certain amount of time in which to start a pursuit or take evasive action, to plan deliberately instead of reacting impulsively to the here and now. Of course, nature endowed certain species with fixed patterns of responses, so that a specific movement in space would elicit an automatic reflex action, as when a frog strikes at a dark spot traveling across its line of vision, but more highly developed creatures had a wider choice of actions at their disposal.

Distance receptors were extremely important in the evolution of intelligence, because the brain was built on them and grew out of them. The cerebral cortex, seat of the higher mental functions of learning and thought, evolved out of a distance receptor, the sense of smell. Sir Charles Sherrington, one of the founders of neurophysiology, called the cerebral cortex the "ganglion" of the distance receptors. A ganglion is a congregation of nerve cells having a common function. In advanced vertebrates, especially in man, the mass of brain tissue that grew out of the eye, ear and nose, became larger and more complex than the receptors themselves and came to dominate the entire nervous system.

The need to analyze time as a means of creating a mental map of space may have been a critical factor in the emergence of this "ganglion," which eventually made humans the brainiest species on the planet. Human beings belong to the class of mammals. Early mammals evolved from primitive reptiles, and one of the mammal's distinguishing features as a new type of creature was the homeostatic control of body temperature. Reptiles, as one of four classes of vertebrates that inhabited the dry land after the amphibians had crawled out of the sea, are cold-blooded. Today, the reptile class includes such animals as snakes, lizards, alligators and turtles. About 225 million years ago, the reptiles introduced a momentous innovation by producing eggs that could be hatched on land.

Early reptiles relied on sight as their main distance sense. Their visual system became remarkably complex, even though their responses to what they saw were rather simple and stereotyped. The control systems for reptile vision were not high up in the brain proper, but out on the edges of the nervous system, in the retina of the eye. Here the external world was represented by a kind of map, with the retina as a grid on which the position of an excited nerve cell symbolized the location of an object or event in the space of the environment. The retina of the reptile eye was so rich in neural circuitry it has been called a sort of peripheral brain in its own right, able to analyze visual information. Both amphibians and reptiles performed most of this analysis in the "eye brain." It was the task of the brain itself to coordinate visual messages with appropriate reflex responses. The reptilian distance sense of hearing was much less elaborate than that of vision. Some lizards alive today have less than 100 hair cells in the ear, suggesting that they see more than hear their way around the world. Probably in early reptiles, ears were useful mainly for detecting vibrations in the ground or sounds of low frequency in the air.

The earliest mammals were reptilian in many ways, but one important difference was that they adopted a nocturnal way of life. They needed the protection of darkness in order to escape the attentions of the ruling reptiles, including the monstrous dinosaurs, which dominated the landscape by day. Some of the giant reptiles fed on plants, but many were eaters of meat. Dinosaurs are commonly thought of as lumbering, stupid beasts, but in fact many of them were agile and quick at snatching and consuming various unfortunate fellow creatures which attempted to flee from them. The name dinosaur means "terrible lizard," and these violent predators certainly must have inspired terror in their prey. One, *Ornitholestes*, was built for speed, and was expert at seizing a fugitive small animal off the ground with its three long fingers. *Struthiomimus* robbed nests of eggs, which it crushed in its toothless beak. The dangerous killer *Deinonychus* had a clawed toe which could be lifted off the ground and rotated through 180 degrees, enabling the animal to disembowel its prey with a sharp backward kick of the legs. All through the Mesozoic era, which began about 200 million years ago and is known as the Age of Reptiles, mammals flourished, but discreetly, under cover of darkness. They lived in trees, in holes and under logs and stones, lying low during the day and exploring for food at night. Mammals coexisted with the giant reptiles, but separated from them in time for their own safety, retreating to another part of the 24-hour cycle.

Animals inhabit ecological niches, and since the definition of a niche includes not just its physical properties but also the behavior of the species that occupies it, including the rhythms of activity under the control of biological clocks, a description of a niche purely in terms of space is incomplete; it must be described in terms of time as well. There are daytime niches and nighttime niches. A nocturnal animal is adapted to a nighttime niche by evolutionary strategies, just as a diurnal species is adapted to the temporal niche of daylight, and a crepuscular animal to the twilight hours. The circadian system can be modified in certain respects to suit the particular niche to which a given species is adapted. Nocturnal creatures usually have a shorter tau than day-active ones, their pacemaker clocks tend to be tightly coupled, and they are often more sensitive to light signals. The circadian system is organized in such a way that if a species switched from a daytime to a nighttime existence, as the early reptilian mammals did, the requisite biological changes would not be too difficult to accomplish. The level of melatonin secreted by the pineal gland is a physiological index of daylight and darkness in both nocturnal and day-active animals, so that all that was needed was for that information to be interpreted in a different way by the body. The circadian system of modern humans is adapted to a daytime niche, but the discovery of fire, and much later the invention of the electric light, extended their niche artificially into other parts of the 24-hour cycle.

For the newly nocturnal mammals, sleeping in the light and roaming about their world in the dark, the distance sense of seeing, in which space was the paramount dimension, ceased to be of much use. One way out of this dilemma might have been to develop night vision, but nature does not usually resort to such a strategy. Being able to see well in the dark is quite rare in the animal kingdom. More often, other senses are sharpened. Owls and hawks hunt the same sort of prey in the same territory, but owls hunt at night, mainly by ear, and hawks prey in the daytime using extremely keen eyesight. There were two candidates to replace the distance sense of vision in mammals. One was smell. The other was hearing. Mammals specialized in both kinds as they evolved. Sound, however, has a distinct advantage over smell as a source of information about the world. It travels in straight lines, whereas a scent meanders and curls, at the whim of the breeze. Sound, therefore, is a more reliable guide to direction than smell. But using sound cues to create a mental model of the environment means that the nervous system must convert the time code of sound into internal models of space, and this, as hi-fi engineers have discovered, is a mystifyingly complex process in which the brain is intimately involved.

Harry Jerison, a psychologist at the University of California, has presented a wealth of evidence to suggest that mammals evolved a sophisticated hearing apparatus as a major distance sense, much more elaborate than that of the reptiles, adapted to their domain of darkness, and in the process became the first creatures to construct an internal model of the external world which incorporated fully the time dimension. That was something their reptilian ancestors had not needed to do. Time was introduced as an essential element in the life of these animals and was crucial to their survival.

At first, the new auditory system of mammals may have been similar to the map arrangement of reptile vision. A sequence of sounds arrived at the ear, and the sounds would be stored and labeled, so that they could be compared with one another, giving information about the distance and direction of the source of the sounds. But as we have seen in connection with music, coding time in terms of space, or space in terms of time, is less easy, less "natural," than coding space as space or time as time. Switching from one code to the other places heavy demands on the capacity of the nervous system and requires special circuitry. In the case of reptile vision, where information about space is registered spatially, on the retina, the code is part of the structure of the eye, a relatively simple and economical arrangement. A new complexity is introduced when sounds must be processed so as to tell the animal as much about space as sight would, giving information as to the whereabouts of an object in that space, whether it is large or small, near or distant, still or moving, and if moving how fast and in which direction. It is important that the auditory system interpret sounds that occur separately, one after the other in time, as a connected, integrated sequence, creating the perception of a single complex event, perhaps a dinosaur crackling twigs as it stalks hungrily across the land. This linking process is known as "time-binding."

The extra circuitry needed for transforming time into space could not be packaged in the ear itself, as reptile vision was packaged largely in the eye. There was not enough room. The new equipment had to be installed higher up in the nervous system, and it required the evolution of larger brains. In mammals alive today, only a small fraction of the nerve cells that process auditory information are at the middle and inner ear; by far the greatest number are at various levels of the brain. "The first expansion of the vertebrate brain may have been primarily a solution of a packing problem, and may only incidentally have resulted in the evolution of intelligence," Dr. Jerison believes. Likewise, the sense of smell, which was also important for the nocturnal mammals, was a factor in the expansion of the forebrain, because it depended on the storage and integration of smells, so that a certain scent sniffed today could be compared with an identical, remembered scent which had been located and pinpointed yesterday or days earlier.

Integrating successive sounds means taking events that occur over a span of time and representing them as a single, complex event which can be held in the brain as a whole and perceived as a whole, all at once, as if it were a spatial pattern, like the printed score of a musical work as opposed to the music itself. The brain needs to encode when an event happened as well as where it happened, in comparison with other events. That ability, Jerison thinks, can add up to a time sense, and mammals acquired it. Mammals can handle time codes with great facility, whereas relatively few modern reptiles are set up to do so. Reptiles can be alert to unusual sounds, but they cannot fix the location of an object in space acoustically. Natural selection did not make the drastic modifications to the nervous system needed for space-time manipulations.

For more than 100 million years the mammals lived an inconspicuous nocturnal existence, mousy in appearance and obscure in their way of life, keeping out of the way of the frightful reigning reptiles. Few mammals were larger than a modern rat. During this dim period of retreat, however, the mammals were developing distance senses that would lead to the complete transformation of the brain. Alfred Romer, one of the twentieth century's great authorities on the evolution of vertebrates, calls this nocturnal interlude in the history of man's ancestors a time of trial and tribulation, but also a preparation for a glorious future. It was, Romer said, "a period of training during which mammalian characters were being perfected, wits sharpened. As a result, when, at the close of the Cretaceous, the great reptiles finally died out and the world was left bare for newer types of life, higher mammals, prepared to take the leading place in the evolutionary drama, had already evolved."

About 70 million years ago, the mass extinction of the giant reptiles took place, for reasons that remain unclear. The daylight niches occupied by the dinosaurs and their like fell vacant. The dinosaurs, the flying reptiles and all marine reptiles except turtles were wiped out, but land plants were spared, and so were mammals, which were now free to abandon their protected nocturnal existence and emerge as creatures of the light. At first mammals responded to the disappearance of the huge predators by developing larger bodies, since they did not need to skulk in small crannies any longer. Later, in addition to the sharper wits, certain mammalian species acquired sharper eyesight, but unlike reptile vision, where information processing was done at the margins of the nervous system, this late model seeing sense imitated the mammalian hearing sense and became organized at the level of the forebrain, the thalamus and the cortex.

Headquartered in the brain, the day-active mammalian visual apparatus probably used a time code to represent information about space. Assuming that nature is conservative, and builds new cognitive systems along the lines of existing ones, Jerison proposes that as these brainy animals evolved, sight took on "auditory" properties. In other words it was time-binding, and it integrated events taking place at successive moments in time so that they could be perceived as a unified, connected whole. At the retina, the code would still be spatial, as it had been in reptiles, but at the higher levels of the nervous system objects or other animals in space would be seen to have not only location in space but also duration in time. A significant new temporal element was added to visual experience. The lower vertebrates, Jerison thinks, construct only very simple internal models of the world, if at all. They are programmed to respond in fixed ways to specific patterns in the environment. Their "mental" world is a world of the immediate present, in which events are momentary and disconnected. Intelligent species experience their surroundings very differently, being able to recognize changing patterns of stimulation as objects that persist and maintain their identity over time even if they change their appearance. Things have a history; what is happening now is integrated with what happened before and with what might happen afterward, an essential prerequisite for planning actions rather than simply reacting to what the world does. William James regarded the ability to perceive sameness over time as the basis for the formation of mental concepts and abstract categories. A jellyfish would be a conceptual thinker if a feeling of "Holo! Thingumabob again!" ever flitted through its mind, James said.

In the mammals which were active during daylight hours, therefore, information about sound and sight was coordinated by means of space codes and time codes which represented the world as stable, and the coding mechanisms were packaged chiefly in the brain. It is reasonable to suppose that a brain which constructed space-time models of the world, perceiving it as consistent, enduring, yet always changing, also produced behavior that was more flexible and less governed by fixed responses to specific stimuli. If a frog saw the world in terms of objects, each with its own identity and history, it would not strike at a dark spot moved across its field of vision by a scientist in a laboratory; the frog strikes at the spot because its nervous system is wired to do so whenever something the size and color of a fly comes into view.

Probably in early man mental images were integrated over longer spans of time, for seconds, minutes, hours, or longer. His perceptual world was extended in space and in time, containing objects and experiences that are familiar because they are constant over time. Temporal integration, the ability to perceive events spread over time, not as separate and unrelated but as holding together in a unity, may have arisen first in nocturnal mammals. To hear a sequence of rustling noises in dry leaves as a connected pattern of movements in space is a very primitive version of the ability to hear, say, Mozart's *Jupiter Symphony* as a piece of music, entire, rather than as momentary sounds which come and then are no more, like the fly-spot at which a frog strikes, and to experience a spoken sentence as a whole whose end modifies its beginning and whose meaning depends on the way in which the parts are organized in time. Perhaps, too, the capacity to perceive a world of enduring objects and to connect past with future led to a sense of the self as a lasting entity, a characteristically human trait. In this special kind of "distance sense," time is a consideration of the most profound importance. The reality of the self as something that has constancy over time, Jerison points out, is one of the most compelling of human intuitions. And a brain that can construct a space-time model of the real world, using the senses of sight, hearing and smell, can imagine a possible world that could exist, even if it does not. This is the mermaid's human head of Sartre's time metaphor. Such a capacity for mental imagery, for modeling possibilities as well as actualities, was present in the hominids, Jerison concludes, and must have led them at an early stage to appreciate the idea that the world did not begin at the moment of one's own birth, but had a past and will have a future after one is dead. "These have become singularly human hallmarks," Jerison says, "understandable as cognitive structures with temporal as well as spatial extent. When it became selectively advantageous to construct a reality beyond any immediate sense modality, the path to hominization could be followed."

Do we have the dinosaurs to thank for the evolution of our own perceptual worlds which stabilize experience in space and time and make possible such bedrock essentials of civilization and culture as rich extended memories, imagery, a vision of the future and a sense of the past? Were these ugly brutes the start of the chain of events that culminated in the existentialist philosophers? It seems likely that the retreat of the mammals into the safety of darkness brought time to the fore as the principal dimension along which to explore space at a distance. Space became temporalized and time became spatialized, and in the process the brain grew in size. Biological time structures became a way of adapting to the environment very early in the history of life on earth. Psychological time codes, used for the same purpose, are relative newcomers; their evolution is at least as curious as that of biological clocks and is perhaps more critical to an understanding of that most advanced of all mammals, man, and the evolution of intelligence itself.

FIFTEEN

The Cough
That Floated

Time is an important aspect of "biological reality" for the human species, and part of the very nature of such a reality is that its scope is limited in some way. We have some blind spots, even if we do not realize they exist. Certain aspects of the world's reality are excluded, made inaccessible to consciousness, and for good evolutionary reasons, because the organism must not be overloaded with too much useless awareness. So there are constraints on the brain's power to detect and discriminate the temporal structure of the environment, and these constraints, as might be expected, reflect strategies of adaptation that are innate, built in. They are there whether we want them or not. François Jacob has stressed that a given species of animal perceives those features of the "real" world that are directly related to its own behavior. For humans, one of the most important of all behaviors is communication, especially in the forms of speech and music.

Speech and music are profoundly temporal, and bearing in mind the strategic retreat into the nighttime hours by early mammals, it may be no coincidence that they are also aural and can be experienced in the dark. It is the time order of speech sounds that determines meaning in language. Since one sound follows another very swiftly, in small fractions of a second, we would expect to find that human beings are specially equipped to perceive these micropatterns in time. In fact, the brain perceives the small-scale structure of time in speech and music automatically and without effort. The opposite is true of the human perception of space; we are able to detect the large-scale structures of space more easily than the small-scale structures. An unexpected discovery is that human beings seem to be able to detect microtime patterns *only* in the case of language and music, and nowhere else. The time order of other rapid sequences of sounds appears to be outside the limits of our unique biological reality.

If three different meaningless sounds—say, a hiss, a buzz and tone—each lasting for a fifth of a second, are played in sequence over and over again on a loop of tape, a human listener is unable to say in which order the sounds occur in time. Yet our brains can detect the time order of speech and music sounds easily and without strain, even though they follow one another more rapidly than the hiss, buzz and tone. When we make music, we can structure time with remarkable accuracy. It is thought that Galileo, when he was investigating the behavior of a body in free fall, used his own singing voice as a clock with which to measure the increase of speed of a ball as it rolled down an incline. Man-made clocks of the period were not precise enough to time very brief intervals, but the psychological clock which governs human music making did have the necessary accuracy. Stillman Drake, a leading authority on Galileo, has replicated one of the great scientist's free-fall experiments, believed to have been carried out in 1604, timing the speed of a rolling ball by singing the hymn *Onward Christian Soldiers* at a tempo of about two notes a second.

In the 1960s, Richard Warren and his colleagues at the

University of Wisconsin found that the time order of sounds in rapid sequences can be perceived if, and only if, the sequence has meaning, and makes sense to the listener. When meaning is lost, temporal order is lost, Warren's work suggests. Biological reality for a given species appears to include those aspects of the world that have special relevance for that species, as we saw in the case of the honeybee, which is unable to learn landmarks if the landmarks lack importance for its own existence.

The difference between perceiving sound sequences that are relevant and perceiving those that are irrelevant, when time is an important factor, is quite remarkable. In a spoken sentence, the elementary unit of speech called a phoneme, which is the shortest speech sound that can be recognized on its own, lasts, on the average, for seven or eight hundredths of a second. In music, notes lasting only five hundredths of a second can be perceived in their correct order. If the crash of cymbals in a performance of a symphony comes a fiftieth of a second late or early, the effect is noticeable at once, the audience cringes, and the cymbalist should have the grace to blush.

That is not true in the case of sound sequences that are unrelated to speech and music, however. The length of such sounds needs to be increased considerably before the human brain can identify their time order. Richard Warren found that only half the people he tested using hisses and buzzes could name the sequence correctly even when the sounds were lengthened to between three tenths and seven tenths of a second. In many cases, listeners could not tell at first even how many different sounds the sequence contained. What a difference he found when four spoken digits were played, each digit lasting two tenths of a second. Their temporal order could be detected at once.

Even more surprising was the effect of tapes played by Warren and Charles Obusek using spoken sentences in which certain phonemes had been erased and replaced by the sound of a cough, as if an after-dinner speaker were trying to make himself heard above the din of a noisy banqueting room. Listeners to this doctored tape experienced a curious illusion. They "heard" the missing speech sound, as clearly as if it were physically present, even though the sound itself had been obliterated from the tape entirely and replaced by the cough. Even when the undoctored tape was played back to them afterward, they still could not distinguish the sound they had imagined from the sound they actually heard.

As for the cough, it could not be located at its correct place in the sentence. It had no clear temporal order in relation to the speech sounds, because it lacked meaning; it was noise, not speech. The cough seemed to float free in time, coexisting with the speech sounds, but not interfering with them. A buzz or a tone, provided it was as loud as the other sounds in the sentence, had the same effect as the cough. When a brief interval of silence replaced the missing phoneme, however, listeners were able to fix its position exactly.

This constraint on the ability to recognize the time order of sounds may seem like a deficiency, a "blindness," to use Professor James Gould's term, but it is really an asset for human beings. Being able to extract meaning from fleeting sequences of sounds, which are always subject to random disturbances, is as critical to the proper functioning of *Homo sapiens* as being able to recognize and learn the color and smell of a relevant flower is to the routine activities of a honeybee. If an extraneous, meaningless noise were allowed to interfere, occupying a definite, detectable location in the sequence, it would disrupt the temporal order of the speech sounds and therefore impair meaning. By means of the biological constraint that makes it impossible for the brain to place the noise in its correct temporal order, causing it to float free, indeterminately, the speech code is protected. In a restaurant, for example, one can eavesdrop on a tantalizing conversation going on at the next table and extract meaning from it, in spite of coughs, clatter and raucous laughter going on all around. One reason for this is that the brain does not seem to perceive sequence step by step. It recognizes the overall pattern of a spoken sentence before it identifies individual phonemes. In fact, such prior recognition of pattern makes it possible to synthesize, unconsciously, a missing phoneme, which explains why listeners do not notice that the phoneme is absent.

Another example of unconscious limits on the way we can perceive the time structures of sounds is that of temporal fission, sometimes known as streaming. Temporal fission is a fancy name for something that has been around for centuries, but only in the past few years has it been given serious study by psychologists. No special equipment is needed to produce temporal fission. Anyone can experience it by listening to J. S. Bach's C Major Solo Violin Sonata on a record player. In the finale of this piece, the melody seems to break up into three separate streams, and then into two streams, coexisting in time. Yet the melody itself is written as if it were one continuous sequence of notes, and it is played by a single instrument, since this is a solo work. The same effect can be heard in another of Bach's compositions for one instrument, his Suite No. 3 for Cello Solo. Many composers of the baroque era made use of this device, and it can be heard too in the modern jazz music of John Coltrane.

When music streams, listeners cannot perceive the sequence of notes in the "correct" temporal order. Instead, they hear two or three separate sequences simultaneously in time. Because of the nature of the brain's aural system, temporal fission occurs. It is an internal, psychological phenomenon and not an external, physical one. The effect evidently reflects some innate constraint on the human ability to perceive time structure. Under special circumstances, a melody, which is an orderly series of sounds, one after another in time, seems to break apart into more than one series, coexisting in time, like the cough that seems to float free of the temporal order of the after-dinner speaker's speech.

Often we are not aware of temporal fission. That is partly because the breakdown of time order need not lead to complete chaos, which would be noticed at once. The sequences may split apart in a lawful, orderly way. Rules of time ordering still apply, but they are different rules. If sequences of speech sounds are speeded up artificially, syllables and even connected vowels may separate into two streams. For example, if the vowel sequence

uu (as in who'd), *ae* (as in had), *ee* (as in heed) and *aw* (as in hawed)

is played on a computer synthesizer at a speed slightly faster than that of normal speech, listeners often hear the word *radio* The link between *uu* and *ae* is broken, and the first vowel coexists with the second instead of preceding it. This is heard as the consonant *r*. Streaming occurs, not so much because of the faster rate, but because the transition from one vowel to another is abrupt, and the vowels are artificially self-contained. The effect is like that of two musical notes, far apart in pitch, which are too close together in time. As a result, a vowel seems to pop out of its actual time slot in the sequence. When human beings speak, vowels are not self-contained. The transition from one to another is smooth and continuous, and this smoothness is the glue that prevents sounds from flying apart in time.

Why does streaming happen? It is not produced simply by speeding up the sequence of sounds, although that is part of the reason. Temporal fission tends to occur when there is a wide distance between the pitch of successive sounds, and too brief an interval of time between the sounds themselves. Louder sounds within a sequence must also be separated by a sufficiently long time interval if streaming is to be avoided.

Pitch, loudness and time may be regarded as three distinct aspects of musical structure, as frequency, amplitude and phase are aspects of a biological rhythm. The serialist composers treated them as three dimensions of music. A limitation of the brain, part of human biological reality, is that it does not perceive one of these dimensions by itself, in isolation. If a dimension changes, the change affects the way we experience the other two dimensions. Increasing the distance between the pitch of successive notes so that a low note is followed by a high note, or vice versa, means that the time interval between the notes cannot be too brief, otherwise the sequence will stream. The same is true if the dimension of loudness is altered. This amounts to a sort of relativity theory of hearing, analogous to Einstein's overthrow of absolute time in physics. The phenomenon of temporal fission suggests that as far as the human listener is concerned, the dimension of time cannot be regarded as an absolute, divorced from other dimensions of the structure of speech and music.

Mari Riess Jones, a psychologist at Ohio State University, has made an extensive study of streaming. She believes that streaming reflects a special feature of the human brain, enabling it to perceive with ease the various dimensions of musical sounds in combination, but making it difficult to perceive one dimension on its own. If the music is composed in such a way that the time dimension is treated as an absolute, the result may be a failure on the part of the brain to detect the correct temporal order of sounds in the sequence. The rules of perception are such that changes in pitch and loudness must be proportional to changes in the time interval between notes, and if these rules are broken, the listener's normal powers of discrimination are likely to be disrupted.

This law of psychological relativity applies widely. It can explain why people sometimes "hear" subtle changes in sounds, changes that are not really there. If a speaker says "ABCABCABC" over and over again without any pauses, the sounds of all letters may seem to be exactly the same. One letter cannot be distinguished from another on the basis of acoustic features alone. The situation changes radically, however, if a pause is placed after every third letter, so that the time interval between each "C" and each "A" is longer than those between the two other letters:

ABC pause, ABC pause, ABC pause, etc.

In that case, the first sound in each group of three will be heard as more distinct than the other two, even though it is not. Although there is no objective difference between the loudness and pitch between each letter in a group, we experience them subjectively as being different, so that each "A" sound seems higher and louder than each "B" or "C."

In the brain, Dr. Jones proposes, there is a psychological mechanism that is sensitive to relationships among the dimensions of a speech or musical sequence. The brain uses the relationships to make sense of its world. A listener tunes in to meaningful temporal patterns in the environment by means of an array of perceptual rhythms in the nervous system, similar to circadian rhythms but with much shorter periods. Perceptual rhythms are among the brain's internalized representations of reality. Their time scales are a hierarchy, shorter ones nested inside longer ones, as in music the rapid oscillations of pitch are nested inside the slower periodicities of tone sequences, and both are on a shorter time scale than an entire movement of a symphony or concerto.

A perceptual rhythm in the nervous system may be entrained by a world time pattern, as long as its frequency is a rough match to the frequency of that pattern. Fast perceptual rhythms with short time periods would lock onto world patterns with a small time scale. Within limits, each perceptual rhythm is adjustable to a real world time pattern that falls within these limits, just as a circadian rhythm with a free-running period of 25 hours can be entrained by a light-dark cycle with a period of 24 hours.

Temporal fission would occur when a perceptual rhythm fails to lock onto the pattern completely. Suppose the pattern is the finale of Bach's C Major Violin Sonata. A perceptual rhythm might lock onto the sequence of notes with certain prior expectations that the rules governing the relationships among pitch, loudness and time will be observed. In the passages where streaming occurs, the listener can no longer adjust his expectancies, because the music violates the rules in too extreme a fashion. The result is that one perceptual rhythm by itself cannot synchronize with the pattern. The differences in pitch from note to note are too great in proportion to the time intervals between the notes. So the brain brings another perceptual rhythm, with a different frequency into play. The use of the two rhythms together creates the illusion of two separate melodies being played at the same time.

In the case of the ABC, ABC sequence, where a longer time interval between groups of three letters gives the impression that each "A" is more distinct than other letters in the group, the illusion arises from the fact that the perceptual rhythms that lock onto relatively long periods in a temporal pattern are tuned to expect larger differences in pitch and loudness than may actually be present, and the brain "hears" the differences even though they are not really there.

Certain kinds of modern music make their effect by stretching or breaking many organizing principles which constrained the music of the past or elevating secondary elements to the status of organizing principles. Since, as we have seen, the ability of the brain to deal with world time patterns depends to a great extent on the rules that govern those patterns, music that takes liberties with the rules tends to test the limits of the human time sense. A strong, steady pulse locks the listener into the movement of the music, providing him with predictable time cues which make him feel at home in its temporal landscape. Music that dispenses with a regular beat, or distorts it in various ways, deliberately deprives the listener of such cues, almost as Michel Siffre, the time hermit of Midnight Cave, was cut off from the normal time cycles of the environment. In fact, some musical compositions are written with that very purpose in mind. Such pieces are a kind of forced desynchronization of internal time structure from external time structure.

One example of this is Karlheinz Stockhausen's *Zeitmasze*, for five woodwind instruments. The title means "time measurement" but the work is actually anti-time, and aims at the complete abandonment of pulse, although it is occasionally present just to remind us what we are missing when it vanishes again. The time of each part speeds up or slows down independently of the others. Without pulse, the sense of the one-wayness of time, its irreversible forward direction, emphasized so much in physics and in psychology as well, is absent.

The beat, the basic unit of duration, which divides time into equal segments, is the master clock of a musical composition. It is the primary, most regular and dependable keeper of time, a pacemaker to which other temporal structures in a work are coupled. The beat is a more fundamental unit of musical time than meter, which is a grouping of beats, say in twos or threes, to form a measure. Like the doctrine of the master clock of the universe, the concept of the musical beat has had a checkered history. Until about the year 1200, European art music was a purely vocal affair, in which the beat was close to the meter of prose, specifically Latin prose, with its pattern of long and short syllables. This was a peculiarly intellectual measure, suited for the relatively loose, labile rhythms of the expressive solo voice. Only in dance music was a strictly regular beat observed, appropriate to its function as an accompaniment to physical movement. Sacred music was written for the mind and the spirit, not for the feet, though some dancing was allowed in churches. So great was the popularity of the dance, especially in the court circles of Europe of the twelfth century and afterward, that it was an important spur to the development and elaboration of secular music.

As musical forms grew more elaborate, they had to be more strictly organized in time, to make them acceptable to listeners as well as to performers. In polyphonic works, music did not rely on a single melody continuing throughout the piece to give it coherence. Polyphony is music in which two or more separate melodies are going on at the same time. Polyphony began by being sacred and strictly for the voice, but as the motet evolved in the latter half of the thirteenth century, it spread to secular, instrumental music. Melodies with contrasting rhythms ran simultaneously and were ingeniously intertwined. More was happening in polyphonic music than in earlier works scored for one voice only, so that the master clock of the beat was needed to coordinate and synchronize the various parts, in somewhat the same way that a common tempo provides reference points in a conversation. There had to be limits on the temporal freedom of the parts, providing predetermined time slots, easy to anticipate, which all must observe. Since the music contained such a large amount of change, it needed an extra ration of sameness to guide performers and listeners through the maze. The music was fitted into a predictable framework, and that framework was the stricter periodic beat. The bar line, a vertical line drawn across the staff at intervals of time, was used at first at irregular intervals, as a merely technical device to help the performers find their places in the score. It was a constraint which began by being fairly free, but became stricter and more regular as the music became more complex. Bar lines were introduced at the start of the seventeenth century and eventually enforced a rhythmical unity, making the timing of ensemble playing more accurate. By providing a new kind of temporal constraint, another organizing principle, the bar line made increasingly elaborate rhythms and combinations of rhythms acceptable to the listening audience.

When polyphonic music ceased to be essentially vocal, and the orchestra expanded its role from being a mere accompaniment to the dance to performing as an instrument in its own right, the more physical, simpler rhythms of the dance asserted themselves alongside the looser, expressive rhythms of vocal art music. The two were in competition, and remained so.

The clocklike beat of the dance was paramount at first, as instrumental music established itself in the seventeenth century. It was prescribed by conventions and sharply curbed the freedom of players and composers. Some music of this period was based on stylized dance forms which nobody danced any more, such as the allemande, the courante and the saraband (which was once banned because of its licentiousness, but became a popular court dance in a modified, more decorous form). The suite was influenced by the dance as well as by the opera, and was made up of short sections in the manner of a minuet or a gavotte, which were strongly rhythmic. The dance suite emerged in part out of ensemble playing by bands consisting of as many as two dozen stringed instruments at the court of Louis XIII, which introduced popular dancè music to royal circles, and the craze spread to the courts of England and the rest of Europe. A species of serious dance music began to evolve, not intended for people to actually dance to; and it was performed for an audience. It was the forerunner of the symphony. Between serious and popular music a rift began to grow, but the popular dance had given to serious music a basic framework of regularity, which included the regularity of a strong, clear clocklike beat. Sounds became synchronized vertically, from top to bottom, as well as being horizontally ordered in time.

All through the seventeenth century and for most of the eighteenth, the master clock of the underlying beat gave musical form a high degree of predictability. Rhythm, which is superimposed on the beat but is more free to vary in relation to it, was relatively simple in most music of this age. As the eighteenth century came to a close, however, rhythm became more complex, more like the expressive rhythms of the old days when the voice had been the chief instrument. Meter became less regular and dancelike. Previously, accents had been placed on the first beat of the measure, but now composers experimented, sometimes shifting the strong accent to a part of the measure that normally was unaccented. In Beethoven's music this tendency became more marked. The rise of the orchestral conductor as a figure of glamour and power was due in part to the need to have an external timekeeper who would impose a more personal, more flexible kind of temporal order in the absence of a strict beat internal to the work itself. The conductor became the chief pacemaker of the performance. Until the seventeenth century, timekeeping was usually the responsibility of one of the singers, who would sometimes tap on the music book with one hand or with a roll of paper called a sol-fa. Then, instruments began to acquire more independence; a violinist or harpsichord player, surrounded by his fellow musicians, took the role of conductor, perhaps sharing it with the composer. The emergence of the pure conductor who was neither composer nor instrumentalist was a product of the nineteenth century, when, under the impact of composers like Beethoven, orchestras grew in size and scores became increasingly complex.

One important result of the introduction of temporal regularity in music was that composers began to write longer works. The strain on the listener of coping with the unexpected was reduced; a work could be extended in time and still be comprehended as a whole, because aspects of it were familiar. Meter gives a listener a sense of stability in time, as the dance beat from which the musical beat was derived gives the dancer a secure sense of how and when to move his body in space. It enables the listener to form expectations, to anticipate coming events. Strict timekeeping, which guided an audience through the maze of polyphony, increased predictability. The attention span of the listener was actually stretched out. "This development is significant in that it provided the composer with a greater opportunity to create a more total experience," Wolfgang Stefani points out in his study of the psychology of musical form. "A progression of feelings could now be expressed and explored, and thus musical communication could be made more intense." The "expressive" element which Constant Lambert found in time, the effect time has of heightening the drama and emotion of a work of art, increases as the length of the work increases.

The transformation of the listener from a sprinter into a long-distance runner, by giving him predictable elements of nonchange with which to follow the time course of a composition, opened up the horizons of music. Between about 1600 and 1750, orchestral music was not expected to go on for a very long time. It ran for a few minutes and then stopped. Even operas were constructed out of short, self-contained sections, though polyphonic masses as early as 1500 went on for as long as 45 minutes. In the baroque era, however, large forms of music were established, such as the sonata da camera, later called a suite, which was made up of a series of dances in the same key; the sonata da chiesa, consisting of four movements, also in the same key; the overture, in which the key was often changed; and the instrumental concerto.

The observance of clocklike meter and clear, consistent rhythms, tied to the underlying beat, contributed to the lengthening of the time scale of such works. Perhaps it is significant that the reverse of this process can be seen in certain kinds of modern music which have abandoned strict meter and other formal elements of nonchange. The time scale shrinks and shortens, as if the attention span is collapsing. The contraction of length is quite drastic, for example, in the case of Anton Webern's work, notably in his *Six Bagatelles* for string quartet, written in 1913. Some of Webern's pieces take less than a minute to perform. Another source of regularity, now much diminished, which helped to expand the length of musical compositions in the eighteenth century, was in the art of modulation. This made it possible to extend a theme over time by repeating it, but not repeating it exactly, transposing the theme into a different key. Sameness was supplied by the underlying theme, which retained its identity under various disguises and transformations, and change was provided by the disguises. The work would start in the home key and then increase the drama by moving away from that key. A listener can predict that the music will return home eventually, but the suspense lies in not knowing just when it will return and after what adventures along the way. J. S. Bach, in particular, was a master of the art of modulation.

There is more to meter, however, than an audible beat or the movements of a conductor's baton. It is well known that when people are asked to count evenly spaced events, whether they are flashes of light or the tapping of a pencil on a table, they invariably group the events into sets of two or four each. The brain experiences the repetitions, not as a linear sequence, like a series of numbers in mathematics, but as a cycle, or wave motion, oscillating from one to two and back to one again, or from one to four and back to one again.

The mere passage of time from one repetition to the next is experienced as motion. It is not a sound wave but a "time wave."

Emil Zuckerkandl examined in detail the proposition that when we listen to music, we do not simply hear it, we oscillate to its time wave. There is an internal, psychological pulse generated in the nervous system, just as circadian and other biological rhythms are internally generated. A resonance effect occurs. As each bar of the music is played, the listener's experience is of a wavelike motion, a falling from the crest of the wave, a fresh upward motion as the trough is passed, then a rising followed by another fall. In pictorial form, the process looks something like this:

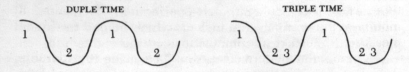

There is no need for strong and weak accents to establish the metrical pattern. It arises naturally, out of the unconscious activity of the mind. In fact, there may be no sound at all on the beat, yet we know the beat is there. An oscillation in the nervous system of the listener keeps time with the music. It is entrained by the beat, as circadian oscillations of the body are entrained by the cycle of daylight and dark.

By itself, meter is as regular as a clock, evenly spaced, uniform and predictable. Linked to the clock of meter, but observing much freer rules, is rhythm, with its irregular patterns that vary unpredictably: Donald Tovey said rhythms "have as much right to change their meaning while we listen to them as the cats of Wonderland have to grin." Rhythm is to meter what the inventive movements of a dancer are to the strict beat of a conductor's baton. In music, rhythm is constructed out of groups of long and short tones which fill out exactly the time of each bar, or measure.

The varying patterns of rhythm are superimposed on the wave of meter. Long and short tones are distributed along it, crowded together or spread far apart, while the contour of the wave remains constant. What is surprising about this new interpretation is that the wave is not spatial or physical, but psychological, and represents only the lapse of time felt as a wave. A tone is rather like a cork on an ocean wave. The ocean wave imparts to the cork a certain quality of directed motion, which alters depending on where the cork is on the wave: close to the crest, in the trough, or somewhere in between. So it is with the tones of music. Each has a certain quality of motion, not in space but in time, imparted by the wave of meter, and this quality is different on different points or phases of the wave. That is why when we listen to music we do not need to count "one" to know we are at the beginning of a bar. Because of the special way we perceive musical time, as an aspect of our biological reality, the quality of motion of the tone by itself tells exactly where we are in the measure. No further information is needed.

A number of modern composers have given considerable thought to the puzzle of the human sense of time and how music can manipulate and play on that aspect of our psychological makeup. In some cases, this has resulted in a radical restructuring of these very elements of music that help the listener to find his way around in an art form that is temporal to its foundations and does not allow an audience to stop the action, pause to look around or go back to a previous moment, as the spatial arts do. Music "takes time seriously," so that the listener must take time seriously also, and he needs assistance in that task from the composer in order to bring his own inner resources of time mastery to bear on the music.

The American composer Elliott Carter in particular is fascinated by the difference between clock time and the mind's experience of the passage of time, the second so much more complex than the first. In music there can be an interplay between the two. Thus, in his Second Quartet, a violin keeps a clocklike pulse going at the metronome rate of MM 70, which is close to the rate of the human heartbeat, or MM 140, which denotes high blood pressure, while other instruments play to different and varying meters, creating the illusion of the experience of subjective time, which stretches and shrinks according to mood, thought and context. In the opening of Carter's Cello Sonata, written in 1948, the piano percussively marks a regular tick-tock while the cello observes a time that reflects "the many different 'psychological' times—expectation, anxiety, sorrow, suffering, fear, contemplation, pleasure," irregularly speeding up and slowing down, never synchronized with the beat of the piano. Carter can give an effect of compressing time, as in his Second Quartet, or make time expansive, as in his First Quartet. He emphasizes the importance of motion in music, but in his compositions the music can seem to move backward as well as forward. The accelerating and decelerating pulses of some of his pieces have an effect that has been compared with the experience of time in dreams, "whirling out of control or running down." Carter aims to express in his music the nonsimple nature of psychological time, combining slow material with fast. Some instruments are completely out of synchronization with others. Meter is suppressed by the use of pulse rates that do not coincide. Needless to say, all this places severe demands on the audience. Carter himself agrees that the lack of recapitulation, the absence of fixed patterns and repetitions on the large scale "perhaps makes it very hard for the listener."

Charles Ives, who exercised a strong influence on Carter, also manipulated the listener's time sense, but did so in a different way. Often Ives deliberately set out to deemphasize the temporal in his music and create an impression of space. He did not take time as seriously as other composers. Ives was intrigued by the spatial aspects of music, how the physical placing of instruments in an orchestra affected an audience's experience of the sound. He specified particular seating arrangements in some of his scores. In *The Unanswered Question,* the strings are instructed to play offstage, and Ives described the passage of his *Concord Sonata* as sounding like a hymn "heard over a distant hill just after a heavy storm."

The effect of an Ives composition on the listener is often

to weaken his awareness of time, by undermining the logic of temporal sequence, of one event following another. He uses a variety of devices to suspend the sense of forward motion, cutting short a progression, leaving ideas incomplete, disappointing the listener's expectations at every turn. In some cases Ives omits bar lines altogether or places them irregularly, as they were placed in the early stage of their history (in madrigals, for example), so that they constrained more loosely than in their later, strictly regular form. His rhythms are notoriously complex and difficult. The score of his Fourth Symphony is so intricate that the conductor Eugene Goosens had to learn it wearing a towel wrapped around his head and with copious amounts of coffee at his disposal. When conducting *Washington's Birthday*, Nicolas Slonimsky gave seven beats with his right hand, holding the baton, three with his left hand, and led two beats by nodding his head. In some of Ives's works there may be twenty rhythms going on simultaneously. Rhythm, in much modern music, has lost its former function as a thread of sameness to guide the listener through the complexities of harmony and counterpoint. Instead, it has become a new source of unpredictability, an extra element of strangeness and surprise. Stravinsky, in particular, used rhythm for its own sake. Henry Pleasants argues that the attempt to make rhythm interesting again by depriving it of any recognizable regularity has been a catastrophe. "By denying the listener a tonal frame of reference, the effort to free music of tonality left the listener incapable of harmonic participation," Pleasants declares. "Similarly, the effort to free rhythm from the constraints of easily recognized patterns deprived the listener of a rhythmic frame of reference and rendered his rhythmic participation impossible."

Perhaps music must contain a certain minimum amount of temporal sameness if it is to draw the listener in and give him a sense of participating, of being at home in the timescape of the work. The circadian system depends for its successful operation on the existence of dependable periodicities in the environment which are within the range of entrainment of its own biological clocks. Subjected to man-made, artificial light-dark cycles which are outside this natural range, the circadian rhythms uncouple from the rhythms of the environment and desynchronize from one another. In a similar way, music that violates temporal regularities in its search for novelty and surprise may fail to engage the attention of those not willing to make the intellectual effort to come to terms with it; if there are cognitive rhythms in the human nervous system with natural, intrinsic periods, as some psychologists believe, music that breaks certain temporal rules, like the light-dark cycles with abnormal frequencies used by scientists to force the desynchronization of the circadian system, may be outside the range of entrainment of its audiences. That is a result of the inherent limitations of human biological reality.

SIXTEEN

Temporal Intelligence

The ability to track the internal time structure of a conversation, a symphony, a social encounter that either is spontaneous or is nothing, seems to be innate, at least in part. It belongs to that part of biological reality which is ruled by biological necessity, and it may have emerged in evolution because the early mammals became creatures of the night, listeners rather than watchers, and therefore used a time code to perceive events in space. Inborn, clocklike mechanisms enable a person to be securely located in the time stream of communication, from moment to moment, as a man walking down a street knows where he is in space. We are guided through the immense complexity of ensemble work. conversational and musical, by the constraint which ensures that certain events fall into a predictable time slot and other events do not.

Such ease, such "naturalness" in mastering temporal patterns that have special meaning for human beings is present at birth or soon afterward. The child is able to establish a "close and wise" relationship with its environment, the environment of adult speech and gesture. The world of mother and child has been called a split-second world, in which timing is all-important for communication between the two, long before the infant is capable of understanding a word the mother says. By the age of three or four months, a child can predict the timing of its mother's behavior when the two are locked together in a wordless dialogue of glances, smiles and movements of the head. One of these gestures, in itself a sort of message, is followed in quick succession by others, and the infant is already equipped to track the temporal structure of these rapid "messages," anticipating what comes next instead of reacting to it. The psychologist Beatrice Beebe suggests that the mother acts as an external pacemaker, entraining high-frequency perceptual rhythms in the child's nervous system with the rhythms of her play. One of the most important tasks for an infant is to mesh its own built-in timing systems with those of the adults around it. By the age of four months, the child responds to the mother with definite rhythms of its own, rhythms that are different from those used when communicating with the father and markedly different from patterns of responses shown to a stranger. In such infant behaviors as sleeping, making faces, making sounds, gazing, and early speech, there is a close temporal correspondence between child and mother. In early childhood, the brain seems to focus on time scales that are very brief. A baby only four weeks old can make a categorical distinction between the micropatterns of the speech sounds *bah* and *pah*. Sensitivity to fine temporal structure, very important if the infant is to master spoken language and anticipate subtle alterations of pattern in speech and harmonics in music, may be characteristic of early childhood. In later life, Mari Riess Jones thinks, attention tends to shift to longer rhythmic periods, more complex, more idiosyncratic and removed from physical dimensions. Clearly, evolution has established certain priorities in the development of a "time sense" in the maturing human individual.

Like the ability to acquire a language, the human sense of timing, in its specific function of fitting every person at a very early age for the real and important skill of ensemble work with other people, appears to be wired into the brain by instructions encoded in the genes. That kind of time cognition is special to humans, inborn and uniform across the species, being more or less the same in all individuals, regardless of how clever or stupid they may be in other ways, how well or badly taught. A defect in an infant's sensitivity to the temporal micropatterns of communication is usually a sign of pathology.

In many other respects, however, the human time sense is not special at all and is far from uniform. It leaves much to accident, intelligence and circumstances, and can be improved by education. It can be called innate only in the most limited and misleading meaning of the word. Deliberate thought, effort, worldly experience, teaching and maturity are needed to acquire it. What is more, whereas inborn discrimination of the fine time structure of communication tilts a child toward reality, so that the child is able to master just those sequences of sounds that have meaning in its life, rather than others that have no meaning, time cognition outside the micropatterns of ensemble work is often "unreal," in the sense that it is unreliable. Psychological time need not march in step with objective time, as measured by man-made clocks. It may vary in its apparent length from occasion to occasion and from mood to mood, now stretching out, now shrinking. That is what Virginia Woolf was thinking of when she remarked on the peculiar variability of time, "when it lodges in the queer element of the human spirit." The brain's sense of longer passages of time, as opposed to the split-second timing used in communication, is not a dependable guide to the world. Even within the confines of the moment-by-moment time of communication, different rules apply to very brief and not so very brief intervals. Time segments of about half a second or less are judged very accurately by means of "absolute timing." In absolute timing, a person may underestimate or overestimate an interval of this length by a few fractions of a second, but the size of the error will be roughly the same whether the interval is half a second or a quarter or an eighth of a second long. If the estimate of a half-second interval is wrong by a hundredth of a second, it will be wrong by a hundredth of a second for an interval lasting an eighth of a second. When it comes to time segments longer than half a second, however, the brain seems to use a different strategy. In that case, the size of the error does not remain constant as the length of the interval changes. The longer the time interval, the greater is the amount of the inaccuracy. This second strategy is known as scalar unit timing.

Musicians have almost impeccable time judgment because the fine structure of music is in the domain of absolute timing. The intervals involved are very brief, about half a second or less. In the range of musical tempos from adagio and andante to allegro and presto, the time between beats is only between .63 second and .29 second. That is why performers can anticipate the beat with such precision even when improvising a trill, and why listeners know when a player misses a beat, even by a very small fraction of a second. Mothers, as they clap, rock, coo and sing to a child, usually observe a tempo that is seldom slower than adagio or faster than presto. This interval of about half a second between beats Dr. Beebe calls the "magic range." As a result, the baby can anticipate the beat, and thus predict the timing of mother's behavior, in much the same way that a concert audience can predict the beat of the music.

How does one locate oneself in time, when time is a large expanse of hours or weeks or months, rather than a pattern of moments, as in the split-second world of conversation? It seems that people go far beyond mere intuition or instinct in dealing with time on the large scale, outside the restricted range of "timing," whether that timing is absolute or scalar, and call upon other powers of the mind. William J. Friedman, a psychologist at Oberlin College, who has made a special study of the human time sense, finds that in adults a number of psychological faculties are used to create a knowledge of time, working together in a sort of collaboration. What is more, these different faculties are not equally trustworthy for the task. Adults, for example, may use their subjective impressions to estimate time duration, but they also subject these impressions to powers of logical inference, and usually logic has priority. Its judgment carries more weight. Five minutes spent waiting for the dentist may seem like half an hour, but we know from experience and common sense that it could not have been as much as half an hour, and that is the estimate we trust.

The point Dr. Friedman emphasizes is that it is a waste of effort trying to find a psychological master clock that would supposedly keep track of the passage of time, because such a clock does not exist. "I think one of the most interesting results of recent work on time is that time turns out to be a multiplicity rather than a unity," Friedman said. "Even though we have a single word for time, it is many different things, and probably each of these things is mediated by psychological processes. The appealing notion that we have one clock in the head is very simpleminded when you compare it with the variety of abilities we know now to exist.

"I am particularly impressed by the way people go beyond the raw data of experience when they are dealing with time. It is clear that our knowledge of time is very much a cognitive construction. In large measure, it is based on the conscious representation of time we gradually build up in our heads over the years. Of course, there are some innate properties of memory and perception that help to organize our experience of time. We know, for example, that people have a compelling sense of the relative recency of events, quite apart from their ability to date those events precisely. But we often use logical inferences about duration and we are very unlikely to abandon logic in favor of intuitive impressions. We realize that we are often misled in our perceptual life, and we usually accord more importance to conscious knowing. Innateness does not play a very large role here."

Time, said the poet Paul Valéry, "is a construction." Nowhere is this more true than in the case of the psychological sense of the passage of time. An emerging theme in modern psychology is the idea that what may seem to be a single mental faculty in fact represents a "family" of competencies, each one relatively independent of the others, and these ideally work in concert, but may become uncoupled. This means that the faculty itself, seen as a whole, is bound to vary from one person to another and to change in various ways as an individual grows up, since every competence need not have appeared fully formed at birth. Some competencies improve with learning and practice during childhood and youth, and all do not improve at the same rate, or necessarily are perfected during a lifetime. The mind is pluralistic, according to this view, and contains a number of autonomous but loosely related skills and capacities which the Harvard psychologist Howard Gardner calls "frames of mind."

In order to master space, for example, not one but several competencies must be brought into play. These include being able to perceive a form or an object, turn the object this way and that in the mind's eye, appreciating how it would look if seen from a different angle, rearrange spatial relationships mentally and recreate aspects of a visual experience. The possibility of acquiring a given spatial competency may be innate, but practice may be needed to develop it, and a person may be better at some of these skills than at others, depending on age and experience. Young children are able to find their way around a space, even if it is unfamiliar to them, but they are very poor at trying to describe the space in words and at drawing a map of it. Children of school age can negotiate specific sections of a city with ease, but fail when they try to coordinate their separate journeys into a single unified mental view of the city as a whole. Eskimos in the Arctic can easily operate in a nearly featureless landscape in which Caucasians would soon get lost. Wide differences have also been found in the way people use visual imagery. Someone of only average intelligence may be able to summon up vivid images of past experience, while it is perfectly possible for a person to have an IQ approaching the genius level and yet be stymied when asked to visualize the scene at that morning's breakfast table. Aldous Huxley could only conjure up faint mental pictures, and he did so with great effort. That is one reason, Gardner thinks, why Huxley began to take hallucinogenic drugs, which allowed him to perceive a reality "no less tremendous, beautiful and significant" than the world as seen through the eyes of a poet like William Blake.

Roger Shepard notes, however, that fiction writers are often superb constructors of mental images. Joan Didion builds her stories on what she calls "pictures in my mind," vivid and detailed images which dictate the nature of the work, even down to syntax and the arrangements of words in sentences. Joyce Carol Oates has said of her characters, "I do so much daydreaming about them that I sort of let them have their head, like fish swimming, and let them do what they're going to do before I start to write."

Musical ability also seems to be a collaboration, in which a family of competencies are loosely coupled. Being sensitive to pitch and being responsive to rhythm are both necessary for musical intelligence, but these functions sometimes dissociate from one another and develop or break down separately. The same is true of language. Mastery of syntax, the basic structure of sentences, seems to be relatively effortless and does not depend on special talents. Our knowledge of the rules of syntax becomes conscious only when we actually speak sentences, just as the rules of how to swing a baseball bat are brought to the surface of awareness only while swinging one. Yet syntax is only one component of the complex array of skills needed to carry on an ordinary conversation. The meaning of a sentence is usually encoded in many different ways, of which grammatical structure is just one, and how it is interpreted depends not just on the order of the words, but also on intonation, expression, gesture and the context in which the words are used. That is why people differ so widely in their ability to use language successfully in concert with others, even though they may all be equally proficient at syntax.

In the same way, "temporal intelligence" can be regarded as a conglomerate of mental powers. Getting to know about time is a gradual process in childhood, and it involves some very sophisticated concepts. Temporal intelligence, which includes much more than the brain's automatic tuning in to the micropatterns of speech and music, is closely related to a child's IQ. Children of kindergarten age have a good grasp of the daily rhythms of daylight and darkness, sleeping and waking, but not of the less immediately familiar cycles of summer and winter, spring and autumn. They know something about clock time, but are uncertain as to exactly how clock time corresponds to lived time. The following exchange took place when a kindergarten teacher was discussing the question of time with her class:

JENNIFER: Time is always going.
TEACHER: Going? How do you mean?
KENT: It's not going someplace, like if you are going to your grandmother's. It's not going like that, but it's always changing to a different time. Say one o'clock, then one-thirty.
TEACHER: Would you say that it changes slowly or quickly, or how?
KENT: Well, it's about half an hour between them.
TEACHER: Suppose all the electricity went off. And all the clocks stopped moving. Would time stop?
CHORUS: Yes!

For children of this age, clock time is not properly integrated with notions of time based on daily routines, such as having breakfast, going to bed, coming home from school. Connections between personal experience and the abstract number system of the clock have to be made by means of mental processes which develop slowly, as the child grows up, and they are formed little by little rather than all at once. One of the chief tasks a child faces is to locate himself in a stable present and past and to separate subjective time from objective time. For young children, the present seems to be more expansive than it is for teenagers and adults. The present lacks sharp, clear edges, so that it is not easy to tell where the past stops and the future begins.

Early in the growing-up stage, a child has a sense of duration, of events lasting over a span of time. But the present remains ill-defined and variable in extent. It expands and contracts, like Alice in Wonderland. Only later does the present become distinct enough to be used as a reliable marker fixing the boundary between what was and what will be. This is only to be expected, because the present is highly subjective in childhood and is therefore less uniform, less bound by realistic constraints which are the same for everyone. When a young child uses a past verb form, the verb often refers to specific previous events that have some relevance to the present. The child may say "cup broken" or "milk allgone." Verbs are generally learned before adverbs, and the present tense is learned before either the past or the future tenses.

Lorraine Harner, a psychologist at Brooklyn College, believes that children tend to confuse the boundaries between past, present and future because they cannot yet coordinate the dual nature of the present, which is both an objective reference point for past and future and also a category of experience in its own right. The present is public as well as private, and these two properties of the "now" interact in building up a person's full, mature sense of time. It is relatively easy for children to learn that past and future are different domains of time, and that one excludes the other, but it is more difficult for them to grasp the relation of both domains to the present. Children have to separate their experience of the present from their idea of the present. This is called "decentering," and it calls for the ability to understand various time relationships from different points of view in time. One test of the ability to decenter is to show young children a series of pictures of the beginning, middle and end of some simple actions, such as a postman putting a letter in a box or a girl brushing her teeth, and then ask them to point to the picture that best illustrates a statement like "The girl will brush her teeth," or "Show me the girl after she brushes her teeth." To answer the first question a child simply has to point to a picture of the girl putting toothpaste on the brush, and to answer the second one, point to a picture of her washing the toothbrush. But children do not find this easy to do, because it requires them to make a distinction between past and future in a special way. The distinction is made, not with reference to the subjective, personal present, but to a present shown in pictures that the children had already seen, and thus were part of their personal past. They had to shift perspective, so as to view the past or the future from the point of view, not of the actual present, but of an abstract *idea* of the present, one represented by a picture of a girl in the throes of brushing her teeth, or a mailman in the act of putting a letter in a box. Under such circumstances, children find it easier to understand "before" than to understand "after."

The family of competencies that together compose temporal intelligence includes the mental manipulation of space, so that space and time come together in psychology as they do in modern physics. That is only to be expected if one accepts Harry Jerison's speculation that during evolution the brain acquired special mechanisms for coding space in terms of time and time in terms of space. Decentering itself is a highly spatial metaphor, implying as it does a position along a line in space. In fact, children have to learn to decenter in space as well as in time, and this is accomplished gradually, not suddenly. It does not come easily to children even by the age of seven. People draw on their understanding of space as one of a variety of different concepts and mental manipulations which help them to understand time. They "watch" time in their mind, just as they watch scenes in the world around them. The months of the year may be imagined as being laid out on a ruler, with each month taking up an inch, or as points around an imaginary circle. These inner representations are like pictures, in that information from all parts of the image is available simultaneously, rather than being accessed in sequence, one by one. Both space and time challenge human powers of comprehension, and even the inventiveness of the brain, at more than one level. Imagery, as well as logic, plays a role in thinking about time. There is intriguing evidence that people construct mental maps of time, similar to mental maps of space.

One of the curious facts about the human ability to conjure up spatial images, even in a dark room with the eyes shut, is that the brain can picture a remembered scene so that the relations among the various parts of the scene are stable and consistent, even though viewed from a perspective quite different from that taken when the scene was actually experienced, "live." A man may visualize himself running to catch a train as if through the eyes of an observer watching from a distance. He is on the outside of the experience looking in, rather than on the inside looking out, as was the case in real life. The reason such a scene can be watched from many vantage points in the theater of the mind is that the image conjured up by the brain is spatially integrated, and therefore provides reliable information from whichever angle the mind's eye chooses to look at it. Probably the brain's temporal maps of the hours, days and years of the past are also integrated, so that we may go through the daily routine of life from hour to hour and from day to day without consciously linking the hours and moments into a continuum of time, but later we are able to form a mental concept of how they are related. This is an impressive accomplishment, because it provides us with information about time relationships which are not consciously learned while going about our affairs. In this respect, too, adults are different from children. The mental maps of space that young children construct are not integrated. They consist of not one but many independent perspective maps, each taking a different point of view. One may not be coordinated with the rest. Only as the child grows up are these maps integrated at a higher level by more general representations in the brain. The same seems to be true of mental maps of time. Such maps in children may consist of a number of separate perspectives on past time, which are not well coordinated and are more strongly dependent on content, on remembered actions, than the more abstract images of adults, which are personal and idiosyncratic, and partly influenced by the conventional markers of public time, such as clocks and calendars.

A time image, like a space image, can be viewed in broad panorama, with many of the details missing, and then examined selectively in close-up. Asked to summon up the image of a Siamese cat, a person may not be aware that the animal's eyes are blue until he actually focuses on them with his mind's eye. In the same way, we can "zoom in" on certain features of a time map, bringing into view a particular Wednesday in June, or the hour before sunset on the last day of a summer vacation. Such flexible use of time images is rare before a child reaches the teens, but younger children are able to draw pictures of the months of the year, usually realistic ones of April showers and March winds, by the age of nine.

All this suggests that the human concept of time, far from being based on one or more biological clocks, is a construction which calls for the participation of a remarkably diverse array of mental strategies, engaging the whole of the mind and not just an isolated part of it. Intuition and logic, emotion and reason, palpable experience and high abstractions, the power to transform space into time and time into space in the mind, all play a role in the construction of our *everyday* knowledge of time. Some of these strategies are difficult and need to be learned, although the potential for acquiring them is innate, in varying degrees. Certain kinds of people learn the strategies faster or better than others. Musicians are especially good at judging tempo and athletes can predict lap times with exceptional accuracy. Perhaps it is no exaggeration to say that the construction of psychological time is as complex as the construction of a spoken sentence; both linguistic intelligence and temporal intelligence share in common the ability to represent real experience in symbolic terms.

In fact, the richness of psychological time suggests that it is one of those faculties, like language, which mark a clear dividing line between human beings and the rest of the animal kingdom. Scientists have spent much time in acrimonious discussion as to whether nonhumans can be said to possess language. Some attention is now being given to the question of whether they also are able to construct a sense of time that is anything like that of humans. David Premack, at the University of Pennsylvania, has examined the time sense in primates. He has found that chimps are consciously aware of past and future, and are not stuck in the present. Elizabeth, one of the chimps Premack tested, was able to arrange plastic words on a board so as to make a comment on some simple act of play she had just performed, such as cutting an apple. The words would say something like "Elizabeth apple cut." She only attempted this about 25 percent of the time, but each of her attempts was faultlessly accurate. In a few special instances, Premack and his colleagues could not make out what Elizabeth was trying to say, at first. Then they noticed that her statements were not about what she had just done, but about what she was just about to do. That kind of comment was rarer than the others, but it was definite proof that Elizabeth could do more than simply record the past; she could symbolize the future. She was consciously and explicitly thinking ahead. She made predictions about her future behavior, and her predictions were just as accurate as her comments on the past.

"If we had been on our toes, we could have taught Elizabeth right then and there to make a distinction between past and future," Dr. Premack said. "We could have said to her. 'Okay, kid, when you write a string of words that is a prediction, add this special word. And when you write a description of what you have just done, don't use that special word, or use a different one.' Had we tried that and succeeded, then we could say that the animal has the equivalent of a distinction, at the linguistic level, of past versus future."

In spite of Elizabeth's prowess, however, Dr. Premack is sure chimps do not possess anything resembling a human time sense. He regards the two psychological worlds as being profoundly different. All vertebrates can discriminate time in the weak sense of the term. They can tell the difference between short and long intervals of time. They know that "it happened a moment ago" is not the same as "it happened five moments ago." An animal can be trained to turn left when the interval between two sounds is five seconds and right when it is ten seconds. What a chimp like Elizabeth can do is an advance on that weak set of concepts, but it is still a long way from the highly evolved human awareness of time.

"We cannot say that it is just a matter of humans and chimps having a similar awareness of time, except that humans can deal with much longer spans of time, going back months or years," Premack said. "In the case of primates we are not talking about my recollection of what it was like to be a child in my home town, or sitting in a room contemplating the future, or comparing the feelings about romantic love I had as a young man with those I had as a not so young man. Nothing of that sort is remotely at stake in the case of chimps. I don't think it is possible to speak of the human sense of time without including in it the ability to represent yourself to yourself. If we think of the concepts that are presupposed by a sense of our own mortality, one of those concepts would certainly be knowledge of time in the full, human meaning of the word. It is very likely that in all the animal kingdom we are alone in that knowledge. Such concepts are among the richest in consciousness. I believe that the chance of them being found in nonhumans is really very slender."

SEVENTEEN

Darwin and the Return of Martin Guerre

Among the family of mental competencies that go to make up the human time "sense," none is more important than memory. Yet just as temporal intelligence is not one thing, but a diverse collection of mental strategies, memory, too, is manysided and complex in its own right. The more closely psychologists examine memory, the more pluralistic it seems to be.

A critical distinction can be made between short-term memory and long-term memory, one that has important consequences for understanding how the human brain deals with time. Short-term memory is a sort of buffer device which delays the flow of experience until it is stored in the brain in permanent form. The buffer can hold its small hoard of information for only about 15 seconds, unless items in store are consciously rehearsed. After that, the information must pass to long-term memory or else be lost. The limitations of short-term memory need no explaining to anyone who has stood in a pay-phone booth after being told a telephone number by the operator, frantically repeating the number over and over again, perhaps out loud, while fumbling for the coins needed to make the call. Another property of short-term memory is that it is literal-minded. It preserves the time order in which events actually happen. In long-term memory a string of words is usually organized according to meaning, not temporal sequence. If we hear the words "red, dog, rose, bark, perfume," the words are held in short-term memory in that order. Once transferred to long-term memory, however, the words are likely to be recalled in a different order, based on categories of meaning: "red, rose, perfume, dog, bark." Because short-term memory is so literal-minded, so indifferent to sense, no synthesis of information occurs, and its contents are available to consciousness almost immediately, unlike the sometimes laborious searching and reconstructing needed to retrieve material from permanent store. It takes only about 25 thousandths of a second for the brain to retrieve one letter from short-term memory. Such rapid access would have been a great asset in evolution to animals faced with intense competition for food and territory and needing to act quickly on a minimum amount of reliable information; presumably, the best strategy in that case is to hold a few items in immediate memory so as not to be bogged down in a Proustian abundance of recollections. In humans, short-term memory holds about seven items at a time. That is a structural limitation, one that is dictated by the peculiar architecture of the human brain.

Not only humans but many other animals as well possess short-term memory, but the time limit varies from species to species. Scientists are able to measure the length of time information can be held in this type of memory because it is easily affected by disturbance or shock, which is why people hurt in street accidents often can remember nothing about the accident itself and have amnesia concerning the few moments before the accident. If an animal is given a slight shock or chill, the information in its short-term memory store is jolted out and lost, instead of being passed to the safekeeping of long-term memory. Using this procedure, researchers find that anything a bee remembers can be disrupted up to five minutes after the act of remembering, but after five minutes, it is in permanent store and does not need to be processed actively in order to be retained. The world champions in length of short-term memory, as far as we know, are fruit flies, which take 45 minutes to transfer information into long-term form.

Short-term memory is an example of a built-in constraint, imposing clearly defined limits on how the world can be known, and the constraint is nonetheless an advantage, because an animal can be more effective in dealing with the present if it carries light mental luggage, suitable for a mobile existence where action and quick decisions are essential to survival.

Long-term memory, which retains information for up to a lifetime, is not a simple or single psychological entity. It seems to be made up of more than one and perhaps as many as four interrelated systems, each having a unique function and operating according to its own particular rules. The final word has not been spoken on this subject, and the theory of multiple subsystems of memory, some of which appeared in evolution earlier than others, is still open to fierce debate and undoubtedly will be modified as time goes on. Three of these proposed memory systems have been called episodic, semantic, and procedural. A fourth possible system, which has to do with the ability of the mind to complete fragmentary information, is so little understood that it has been referred to as question mark (QM) memory.

One critical difference between these various memory systems is in the extent to which they take time seriously. Although the popular view of memory is that it is linked inextricably to time past, and to particular times in the past, the reality is not as simple as all that.

Semantic memory does not emphasize time. It is a memory for facts, rules, formulas, dates, words, special telephone numbers, historical events and wordly knowledge of the kind that can be represented by symbols. A feature of semantic memory is that items held in it do not usually have a space-time context. We remember the date of the battle of Gettysburg, or the nine-times multiplication table, or words of the English language, without necessarily recalling when or where we learned them. The information is indifferent to time and space. We must have learned such facts as that President Garfield was shot, or that Scarlett O'Hara was the mistress of Tara, on a definite day and in a unique place, but that space-time context has disappeared, leaving only the bare item of knowledge. Semantic memories are constant. The name of a friend, which side won the Civil War, how Socrates died, is knowledge that does not change from one year to the next. The rule that "9 times 9 is 81" is immutable wherever and whenever it is applied, whether to apples or potato chips or to the abstract symbols of mathematics. Similarly, the rules of comprehension hold whether we are reading billboards, newspapers or the back of a packet of corn flakes at breakfast. They are among the invariant properties of the human information environment.

Since semantic memory stores rules and formulas, it can be used to generate information as well as to be a mere repository for it. The mind can make inferences and generalize: "If it's Tuesday, this must be Belgium," or, "The fact that a police car is following me may have something to do with the fact that my speed is 83 miles an hour." Semantic memory underlies the ability to make mental models of the world and manipulate them in the mind's eye. It also makes it possible to recall the same list of items in a different order, by applying different rules. We can know that Friday follows Tuesday in the conventional sequence of days of the week, and also that Friday precedes Tuesday in alphabetical order, so that it is possible to name the days of the week alphabetically, in the correct sequence, without any additional knowledge, even if we have never done so before in our lives.

Procedural memory includes memory for skills. It is a "how-to" memory. Skills are a form of semantic knowledge, but they seem to be stored at a level more distant from consciousness than facts and are even more independent of time and space. Being able to play a musical instrument, ride a bicycle, solve certain kinds of puzzles, are accomplishments acquired after many hours of practice, but it is not necessary, or even desirable, to remember each practice session as a unique experience, when it occurred and where it occurred. Ideally, each session should be remembered as being exactly the same as all the other sessions, since the whole point is to produce a skill that is uniform and always performed in the same way. It is fatal to think about how one practiced riding a bicycle on a particular day in the past while actually riding it. Skill memories are acquired by repetition and they are not easy to unlearn.

Episodic memory, while also long-term, has properties that are very different from the other systems. Whereas time and place are unimportant in semantic memory, they are essential to the very existence of episodic memory, whose contents are not bare items of knowledge but autobiographical episodes of personal experience in which context, especially the context of when and where, has not been removed. Remembering that Venice is in Italy is semantic knowledge, acquired perhaps at school. We probably forget who first informed us of that fact and under what circumstances. But actually going to Venice and walking around the Cathedral of San Marco is an episode. The memory of it is linked to a personal experience at a unique time and in a unique place. It has a calendar date and a place in the chronology of the rememberer's life. Context sticks to the episode when it is recalled. We remember not just the cathedral but the gondolas and the pigeons, the time of day and season of the year, whether it was warm or cold, roughly what age we were then, the storm that suddenly came up over the lagoon during dinner and even the taste of the fettuccine.

If rules, logic, concepts, are the timeless stuff of semantic memory, episodic memory is organized around a person's knowledge of his own identity. Episodes emphasize differences and uniqueness rather than sameness. Being able to read, knowing the name of a friend, is knowledge that is constant and needs no space-time context, but reading *Gone*

With the Wind or actually meeting the friend after a long absence is a unique, unrepeatable, personal event. Reading *Gone With the Wind* for the second time is a different experience from reading it for the first time, and affects our memory of that first reading. Whereas information can be retrieved from semantic memory without affecting the contents of that type of memory, recollecting an episode is an episode in itself, part of our life's experiences. It is not only an output from memory, but an input as well, and the contents of episodic memory are altered as a result.

Endel Tulving, a psychologist at the University of Toronto, was the first to describe clearly the properties of the semantic as opposed to the episodic memory system. He believes that the two systems involve different kinds of consciousness, one a "knowing" consciousness, the other "self-knowing." It is possible for a person to know something about an event that happened in his own experience, even if he has no recollection of it actually happening. That is the mode of consciousness peculiar to semantic memory. But it is not remembering in the full sense. Episodic memory, which Tulving sees as a specialized subsystem of semantic memory, involves a consciousness of how events are related in subjective time. The subjective awareness of time seems to be a necessary complement to episodic memory, which makes it possible to make mental journeys backward in time. Semantic memory alone can give us no feeling of personal identity; for that we need semantic memory's subsystem, episodic memory. It enables us to represent ourselves to ourselves, in David Premack's words. Semantic memory, in turn, is a subsystem of procedural or skill memory, which is the only one of the three that can work on its own without help from another memory system. Procedural memory is used in response to the here and now and does not represent the past to conscious awareness.

The unit of information in episodic memory is not a fact or a rule, but an event, and events are temporal; they occupy a certain span of time. These event units are stored in an organized fashion, so that a pattern of relations exists among them. That is true also of semantic memory, where facts are richly interconnected, so that, for example, knowing that the Italian cruise ship on which one's aunt once sailed for a European vacation had something to do with the Mona Lisa helps us to remember that the ship's name was the *Leonardo da Vinci*. The units in episodic memory, being events, are related, not in a web of word associations, but by temporal links, one event following, preceding or happening at the same time as other events. This temporal organization is looser than the type of organization in semantic memory, where facts and their relationships have probably been learned many times on many different occasions and are reinforced in that way. We have all read or been reminded repeatedly that Leonardo painted the Mona Lisa. That is not the case with the episodic memory system, where an event unit is unique, unrepeatable, restricted to a particular interval of time. What is more, the event is further limited to one that we personally experienced or observed. For the *Leonardo da Vinci* to be part of episodic memory, we must have taken the cruise ourselves, rather than learning about it at second hand, through a postcard from the aunt. The time in episodic memory is subjective time, the time of our own experience, not that of clock and calendar. That is why it is tied up with our personal identity. All this makes the contents of episodic memory less firmly fixed, and hence less reliable, more easily changed or lost, than the units of the semantic system, which do not "wander around freely in the structure." When information is organized tightly and durably, it is easier to summon up than information that is organized loosely. If the mental link between the painting and the ship were to come adrift, we would find it much harder to remember that the ship is called the *Leonardo*. When we make mistakes in recalling information from semantic memory, the error is more likely to be that of leaving something out than of putting something in that was not there in the memory system already. Episodic memory, like semantic memory, is not haphazard. Our experiences of the fourth dimension are as orderly as they are in the other three. But the order is of a different kind.

Tulving has compiled a list of the important differences

between the semantic and episodic memory systems. They include the following:

EPISODIC MEMORY	SEMANTIC MEMORY
Sensation	Comprehension
Events, episodes, the self	Facts, ideas, concepts, the universe
Experience	Symbols
Truth based on personal belief	Truth based on social agreement
Feelings more important	Feelings less important
Limited inference	Rich inference
Cues for retrieval of memories: "When?" "Where?"	Cue: "What?"
Depends more on context	Depends less on context
Experience recollected is the remembered past	Experience recollected is actualized knowledge
Very vulnerable	Not very vulnerable
Reported in the form of "I remember"	Reported in the form of "I know"
Late in development	Early in development
Deliberate access	Automatic access
Less general utility	More general utility
Application to artificial intelligence questionable	Application to artificial intelligence excellent

Semantic memory is more important than episodic memory in the basic, everyday activities of life. Skill memory is a fairly primitive system which appeared early in evolution but was not discarded as the brain became more complex and sophisticated. Semantic memory, which is more explicit than skill memory and may retain some autobiographical content, was built on top of this more archaic system. Episodic memory came later and had to coexist with the other kinds of memory already in place. This speculative account is plausible in view of the fact that the three types of memory develop in a maturing individual in the same sequence as they supposedly did in evolution: first skill, then semantic, then episodic. It often happens that mental faculties which appeared relatively late in evolution also mature at a later stage of an individual's life history. Baby monkeys are born with skill memory, but do not acquire fact memory until they are more than a year old. The same seems to be true of human beings. Episodic memory is especially slow to develop, only emerging in its rich, fully fledged form in adulthood, after semantic memory has been established, which could explain why an adult's memory of early childhood is so sketchy and incomplete.

It is an open question whether episodic memory, which takes subjective time so seriously, is unique to humans. But the autobiographical memory subsystem certainly reaches its zenith in human beings, with their strong sense of personal identity and the highly developed "self-knowing" consciousness. It also marks a dividing line between the human brain and the computer, whose memory for facts is imposingly reliable but lacks the rich context of time and place which is the hallmark of the self-knowing category of consciousness.

"If you go backwards in time in episodic memory, you come to a real event in time and space," Tulving explains. "But if a computer simulates this kind of memory, you travel back in time and come upon a programmer who fed instructions into the machine about an event which did not happen in the computer's subjective time. That is an important limitation of computers, and one which marks a real difference between human and artificial intelligence." It might be possible that a programmer could supply a computer with such an immense amount of contextual information about visiting museums, say, that the machine could keep up an intelligent conversation about its tour of a museum and the important people it met there. "Nevertheless, we know that the computer would only be manipulating symbols according to certain rules," Tulving maintains. "It would be talking about words, rather than about original experiences organized in time in its personal past. In the episodic memory of a human being, such experiences are related to their sense of personal identity, of continuity in subjective time. A computer telling us about meeting Pope John Paul II at the Museum of Modern Art in New York is like a child telling us about Alice on her trip to Wonderland." He concludes: "I rather doubt that computers' memories can ever approximate people's remembering of personal events."

We can see at once, from the list of properties compiled by Dr. Tulving, that memory for episodes is something of a frill, an extra, if life is seen as an unrelenting struggle for survival. Episodic memory is much more inward-looking and subjective than semantic memory. Its contents are heavily impregnated with a sense of self and with the belief that knowledge can be gained from an exploration of the self as well as from the facts, ideas and concepts of the real world. Episodic memory is more fragile than semantic memory. Information is less easily, less automatically retrieved from it, and the information is of more limited "usefulness," in the narrow sense of the word. In a *gedanken* or thought experiment, Dr. Tulving selectively deprived an imaginary group of people of the use of either their semantic or their episodic memory systems, and then told them to go about their ordinary daily affairs. The loss of episodic memory turned out not to be a serious handicap, whereas the loss of semantic memory prevented many of these fictitious individuals from functioning at all.

If autobiographical memory and the self-knowing consciousness are a sort of luxury, not part of the hard-core survival equipment needed to adapt to a Darwinian world, why did they appear at all? A possible reason, Dr. Tulving suggests, is that when we retrieve information from episodic memory, we have a heightened sense of certainty that it is true, that what we remember happening at a specific time and in a specific place really happened, even if it did not. This belief, whose psychological origins are mysterious, is simply "there," and is strong when there is no other evidence to support it. That sense of certainty, not so marked in the case of semantic memory, makes a person more willing to act on his episodic knowledge and therefore more effective in planning for the future. It helps us to be successful in a competitive world and so to be "fit" in the Darwinian sense.

Yet there is no denying the dependence of the episodic memory subsystem on the older and firmly established system of semantic memory, nor the power of the knowing consciousness versus the self-knowing one. In fact, there is a marked tendency in the human brain to abstract away from episodic knowledge, with its unique, personal, space-time context, and generalize across contexts, distilling knowledge that is timeless and placeless. Logical thought depends on such a process, because logic, which is essentially atemporal, integrates and connects facts which may be distantly separated in time and space. A person can manipulate information using the rules of logic easily and efficiently when the information is stored in the brain in a nontemporal fashion.

The Cornell psychologist Ulric Neisser has shown that even someone with a very good memory and a sophisticated legal training, who earnestly tries to remember events as pure episodes, with all their space and time context intact, may still slip into generalization and abstraction. A dramatic case in point is that of John Dean, who testified at the Senate Watergate hearings in June 1973. Dean, the former counsel to President Richard Nixon, described dozens of meetings with John Mitchell, Robert Haldeman, Charles Colson, Gordon Liddy and Nixon himself. Dean's testimony strengthened the suspicion that high officials of the U.S. government had engaged in a coverup of the fact that the White House took part in planning the burglary at Democratic Party headquarters at the Watergate.

By comparing Dean's recollections of conversations at those various meetings with the actual transcripts made from tape recordings arranged by Nixon himself, Neisser found that Dean, whose memory had seemed so impressive in the hearing room that journalists referred to him as "the human tape recorder," was by no means always accurate on specific conversations. Yet, he was correct in his general view that a coverup had existed. Neisser concluded that Dean's testimony was not semantic, in the accepted sense of the word, since he was ostensibly describing unique episodes from his personal experience, but the testimony was not strictly episodic either, since Dean's memory of a particular episode was often incorrect. The testimony was somewhere in between the two, trying to be episodic but showing semantic tendencies.

What appears to be specific in John Dean's recollections is actually based on abstractions made from a number of separate episodes. Dean extracted common themes which remained invariant across many different conversations and incorporated those themes into his testimony. In an account of a September 15, 1972, meeting in the White House Oval Office, for example, Dean ascribes remarks to Nixon that show the president knew of and approved the coverup. Nixon "was pleased that the case had stopped with Liddy," and Dean himself warned the president that he could give no assurances that "the day would not come when this matter would start to unravel." Neither of these statements was actually made. Dean was also wrong about small details, as, for instance, when he said that on entering the Oval Office, he was asked by the president to sit down. Dean was not lying to the senators, however. The White House transcripts make it clear that on September 15 Nixon was fully aware of the coverup, but they do so implicitly, not explicitly. When the coverup was discussed openly in front of him he showed no surprise, although his words were not the words Dean had reported him as speaking.

Episodic memory is of great interest because it is such an important member of that family of mental strategies that together make up the human time sense. With its strong emphasis on the self and the autobiographical past, and with its preference for personal experience over objective facts, episodic memory is closely linked to the subjective awareness of time past. Short-term memory, the most literal-minded of all, retains the time order of events just as they occurred in the real world. Semantic memory, indifferent to time, can rearrange events according to certain abstract rules or restore them to their original order by the use of other rules. For example, semantic memory can list the novels of Charles Dickens in alphabetical order or in the order in which Dickens wrote them. In episodic memory, the novels would be arranged according to when in his own life the rememberer had read each novel. This time order is individual and varies from one person to another. In semantic memory, however, the order is the same for everyone, because the rules are universal and do not change. "The semantic memory system handles temporal concepts as it does others, with reference to the world that exists independently of the individual," Tulving comments. "In that world, events have temporal relations, but these relations have nothing to do with personal time."

The profoundly different psychological properties of semantic as opposed to episodic memory may be one reason for the lack of intellectual rapport between scientists and artists, the first striving to discover laws that are timeless and the same under all circumstances and to manipulate abstract symbols according to the rules of logic, the second dealing in the unique, the distinctive, the unrepeatable and the personal. The tension between the two kinds of knowledge can be seen very clearly in the recent debate among historians on the importance of narrative in the writing of history.

For centuries, an "episodic" approach to the past, one that took time seriously, was regarded by writers of history as the best and most natural one. The unit of such an approach was the event and the person. Historians were to a large degree storytellers, narrating events, or biographers, tracing the lives of individuals caught up in events, or both at once. They dealt with the unique and the particular, with the experiences and actions of people in an unrepeatable context of time and place. That long tradition began to be belittled, however, by influential groups of scholars, the "new historians" who emerged in the 1930s and 1940s, who emphasized the "semantic" elements of their craft. They preferred analysis to description, general laws to single instances, societies to individuals. They took time and place less seriously than their predecessors, the narrative historians, had. Statistics were the grist that fed their intellectual mills, voluminous data about mass behavior, large-scale economic conditions and population movements. Many were influenced by the ideas of Karl Marx and the methods of social science. The new historians were less interested in storytelling than in quantifying and mathematicizing. Some of them aimed to make history into a science rather than an art, and they pressed one of science's own tools, the computer, into the service of this ambition. Whereas the older school of historical writers tried to convey what it was like to live in a real world of the past, almost to "remember" it as if it were an episode in their own subjective time, some of the avant-garde scholars made models of abstract worlds which had never existed, testing them with computer programs and mathematical algorithms.

The semantic school did not win a lasting victory in their revolt against the episodic mode, however, The resurgence of narrative, championed by another generation of "new, new" historians, really does seem to suggest that it may be a permanent part of the nature of historical writing, as it is of human nature itself. In a manifesto that is now famous, published in November 1979, Lawrence Stone, a Princeton historian, launched a fierce attack on the "scientific" school, and announced that he detected "an undercurrent which is sucking many prominent 'new historians' back again into some form of narrative." Stone dealt caustically with those who wrote history by crunching numbers with machines. The computer printouts which are the raw material of that kind of scholarship, he complained, were inaccessible to the public and often unintelligible to other historians. Worse, they had explained none of the outstanding puzzles, the big "why" questions that have bothered historians for so long, such as the causes of the great revolutions. A statistical, massively quantitative study of slavery in America, for example, left scholars as far from agreement as ever.

"It is just those projects that have been most lavishly

funded, the most ambitious in the assembly of vast quantities of data by armies of paid researchers, the most scientifically processed by the very latest computer technology, the most mathematically sophisticated in presentation, which have so far turned out to be the most disappointing," said Stone. What they have produced are "huge piles of greenish print-out gathering dust in scholars' offices; there are many turgid and excruciatingly dull tomes full of tables and figures, abstruse algebraic equations and percentages given to two decimal places."

Clearly, the task of explaining the "why" of historical change in terms of timeless, universal laws and structures has proved to be fraught with unexpected difficulties and obstacles. "A belated recognition of the importance of power, of personal political decisions by individuals, of the chances of battle, have forced historians back to the narrative mode, whether they like it or not," Stone concluded. "More and more of the 'new historians' are now trying to discover what was going on inside people's heads in the past, and what it was like to live in the past, questions which invariably lead back to the use of narrative."

Stone's statement, entitled "The Revival of Narrative," was almost shocking and made a tremendous impact, because he had been one of the founders of the semantic school. Other voices took up the same cry in the 1980s. In a speech to the American Historical Association in 1981, Bernard Bailyn, of Harvard, said the greatest challenge facing historians in the years ahead "is not how to deepen and further sophisticate their technical probes of life in the past, but how to put the story back together again." This was also the main concern of Natalie Davis, another professor of history at Princeton, in her book *The Return of Martin Guerre*, which was originally a film. Professor Davis tells the true story of a sixteenth-century peasant from the French village of Artigat, who, one day in 1548, decided to leave his wife and infant son to go to the Spanish wars. After eight years, during which not a word was heard of the absent warrior, someone claiming to be Martin Guerre came back to Artigat.

He was suspected of being an impostor, and a lawsuit was brought by the real Martin's wife. The evidence in the case was ambiguous, and so was the suit itself. Just as the judge was about to deliver a verdict, a moment of pure melodrama occurred. A man with a wooden leg walked into the courtroom and announced that he was the true Martin Guerre, returned to claim his wife, family and property. Professor Davis uses an "episodic," narrative approach in her book, but also offers an interpretation of the story that provides general, "semantic" knowledge about life and society in sixteenth-century France. One of her main concerns is to emphasize the "pastness" of the past, to make it seem different from the present. She has a story to tell, one that is set in a unique time and place, involving people who are not types but individuals; yet by stopping the narrative or by using various rhetorical devices, Professor Davis manages to show how this singular episode of the true and the false Martin Guerre helps us to frame theories, concepts, rules, which embrace more than the limited context of the tale itself. "My intent was to make the story flow back on the character of peasant life, village and family life," she said. "The narrative tells not only about a weird and fascinating case of identity, but also of the ways in which peasants tried to manipulate institutions for the sake of getting a little bit of maneuverability in marriage or indenture. A clever piece of trickery in this one instance poses general questions about how prevalent deception was in France in the sixteenth century, and why. Lawyers, royal officers and would-be courtiers knew all about manners, the shaping of speech, gesture and conversation—what Montaigne called 'self-fashioning'—that helped them rise to higher positions. But where did self-fashioning stop and lying begin? Long before Montaigne put that general question to his readers, the impostor masquerading as the real Martin Guerre posed it in a particular instance to his judge. The court case was also a study in how upper class people, learned men of letters, look at peasants. I try to make the storytelling and analysis work together in *Martin Guerre*."

For Dr. Davis it is very important to present, not just a macrochronology on the large scale, showing that big causes led in some vague way to a certain big result, as for example, it was thought that the Reformation was caused by a surge in literacy, or by urbanization, or by the abuses in the Catholic church. She also wants to show cause and effect at work in small, microchronological detail, so as to learn how the precise ordering of things in time might make a difference to the final outcome. Even the accidental, some turn of events that could not be predicted because it is not part of any general law, could play a role in producing the end result, as did the surprise return of the real Martin Guerre to France. Microchronology, more episodic and particular than macrochronology, sheds light on how individual people interpret their personal experience of historical change, even though that change may be the result of great, general and perhaps impersonal forces. "For me it is always extremely interesting to look at the exact temporal order of events to see what difference that order makes to the experience itself," Dr. Davis said.

"What is new in this approach, I think, is the very great attention given to how narrative unfolds, what happens first, second, third, and also to the ways in which people interpret that chronology afterward, how they make sense of the ordering of events in time. Some of the things that happen along the way may not be part of your larger theory of cause and effect. If you are investigating the reasons for the Reformation, for example, the time order of events in a particular episode might not fit the general theory that the causes of the Reformation were literacy, urbanization and church abuses. The chronology of the local episode might not tell you all the causes of the Reformation, but it would certainly have implications for the way in which the Reformation was experienced in a specific place at a unique time, and the way in which it was remembered by the people who lived there. The difference time order makes would not necessarily destroy the general theory, but it could enrich it, and suggest new general causes to look for somewhere else."

Thus narration, the description of a unique series of events set in a definite context of time and space, the style of episodic memory, seems to have remarkable staying power as a medium in which to express the historical process. Semantic knowledge alone is not up to the task. The American historian Eugen Weber thinks that "description is narration; narration is explanation; there is no historical explanation in the scientific sense. . . . Nothing is more concrete than history, nothing less interested in theories or abstract ideas. The great historians have fewer ideas about history than amateurs do; they merely have a way of ordering their facts to tell their story." A similar view is taken by Sir Isaiah Berlin, who argued that the search for a science of history, complete with its own unchanging laws, is as bogus as the dream of inventing a perpetual motion machine. Natural scientists, Sir Isaiah said, must concentrate on sameness, not differences, and reduce everything, "so far as possible, to timeless, repetitive patterns." Historians are a different breed altogether. They need to have a sense of the unique, because history is a "rich brew of apparently disparate ingredients." Computers could not do what historians do. No universal, timeless laws of history have been established, not even "moderately reliable maxims." In a science such as physics it is more rational to have confidence in general laws than to trust specific facts, but a rational approach of this kind is not successful in the writing of history. Supposing a historian tried to discredit eyewitness testimony that Napoleon was seen wearing a three-cornered hat at a specific time of day on the battlefield of Austerlitz, by means of a theory or law which states that French generals or heads of state never wore three-cornered hats during battles, that historian's judgment would not be respected. "Addiction to theory—being doctrinaire—is a term of abuse when applied to historians; it is not an insult if applied to a natural scientist," Sir Isaiah points out, adding that a good historian must be an actor rather than a detached observer, and must enter into the time of past events. This he calls the "inside view," and it requires self-consciousness as well as a knowledge of the facts.

Episodic memory does not necessarily enhance the fitness of a species, in the Darwinian sense. It does not carry the same force of biological necessity as semantic memory does. Episodic information is recalled more effortfully than semantic information, more deliberately and slowly. The greater the amount of knowledge stored in semantic memory, the faster it can be retrieved, and the slightest cue, a few words spoken, a face half-glimpsed, will set it into operation, whereas episodic memory depends on the appropriate circumstances, a certain state of mind. Only when Proust dipped his madeleine into a cup of tea did reminiscence of his past life come flooding in.

Yet episodic memory, this latecomer in the evolutionary story, is essential both to the emergence of consciousness and to the subjective sense of time, in their fully developed, human form. Without this type of memory, self-knowledge, literature, history, and the idea of the importance of the individual might never have flowered in the way that they did. The world might have had a Newton and an Einstein, but not a Shakespeare and a Dickens. It is tempting to imagine that episodic memory, this "luxury" in which time is of the utmost importance, evolved through the telling of stories, at a stage of prehistory when early man no longer needed to spend all his waking life in the struggle for bare subsistence, falling asleep as darkness came, but instead stayed up well into the evening, sitting around the fire whose glow extended the span of wakefulness into the nighttime hours. In this leisure time, a person with a gift for narrative, real or invented, would be a highly valued member of the group. Onto mere knowledge of timeless fact, useful for coping with life's harsh realities, would have been superimposed the episode. The emergence of narrative, so closely linked to the unique complexity and richness of human subjective time, with its peculiar strengths and weaknesses, uses and uselessnesses, was perhaps as glorious and momentous a leap forward by the mind as the gift of speech itself.

PART FOUR

ASSEMBLING THE SENSE OF TIME

EIGHTEEN

Taking Time Apart

Often it is possible to learn more about the workings of memory by examining what happens when memory is impaired or breaks down. Paradoxically, the process of forgetting may give new insights into the phenomenon of remembering. Thus, since memory is such an important member of the family of mental operations that together make up temporal intelligence, knowing more about the reasons it fails can lead to a more complete understanding of the human time sense.

One well-known example of the breakdown of a specific type of memory, much studied by psychologists, is a form of amnesia known as Korsakoff's psychosis, an organic brain disease which affects the ability to remember while leaving many other functions of the mind intact. Amnesia, a condition that appears in many different guises and is still something of a puzzle, is an important source of new knowledge about the structure of memory.

Korsakoff's psychosis is found in people suffering from chronic alcoholism and vitamin deficiency. Its victims, prized by some researchers as examples of pure amnesia, lack the ability to transfer the contents of short-term memory into certain kinds of permanent store. They can hold a conversation, but new information imparted to them by others is not retained for longer than a few seconds. Richard Restak, a neurologist, has described an interview at a New York hospital with a Korsakoff sufferer he calls Lester Anderson, "a tall, thin, undernourished man with a flushed face and a mouth half full of yellow, decaying teeth," who had been found wandering in the streets seemingly intoxicated. When the young intern in charge asked Mr. Anderson how he was, the conversation went like this:

"Fine."

"Do you remember who I am?"

Silence.

"I am Dr. Gary Lawrence, who admitted you to the hospital. Please tell me how long you have been in the hospital."

"I came in last night."

"Do you remember the name of this hospital?"

Silence.

"It's St. Vincent's Hospital, and you have been here three days, not since just last night. What's my name again?"

"Dr. Joseph Smith."

In Korsakoff victims, a lack of vitamins in the B complex gradually results in the loss of brain cells, destroying the specific ability to form new long-term memories for episodes. Restak compares such a damaged memory with a fishing net containing some large holes. These patients are, however, still able to acquire new memories for skills, which they learn and remember without being conscious of doing so, and their powers of logical reasoning are not abolished. They can master puzzles, identify geometrical figures, recognize entire words from fragments of those words and read print upside down and reversed in a mirror. We know that Korsakoff patients acquire a deep, long-term understanding of certain puzzles because after learning them they are able to solve the puzzles from different and unorthodox starting points. At riding a bicycle, they are as good as anyone else.

To cover up the embarrassment of memory loss, Korsakoff patients make up stories and invent facts and situations as a substitute for the reality they are unable to recall. This type of behavior is called confabulating. Lester Anderson, when prompted by Dr. Lawrence, the intern, during the hospital interview, concocted a fictional memory of working in a men's store called Barney's and selling Dr. Lawrence some shirts.

"Do you remember how many I bought and how I paid for them?"

"Yes, you bought six shirts and paid with a check."

"Mr. Anderson, what is my name and where are you?"

"You're Mr. Barney and I'm in a clothing store."

A patient may converse with a doctor for 30 minutes, and then, if the doctor leaves the room and returns a few minutes later, behave as if he knows the doctor. If pressed, however, the patient will admit he has forgotten who the fellow is.

It is typical of Korsakoff patients that they are disoriented in time. Robert Hicks, a psychologist at Chapel Hill Medical Center in North Carolina, has studied many people with this disease, especially from the point of view of their time sense. He discovered that one way to detect a malingerer, as opposed to a genuine case, is to test the patient on his awareness of time. Bogus cases of Korsakoff's syndrome have a normal judgment of the passage of an extended span of time, whereas genuine cases are hopelessly at sea. After he has been with a patient for about an hour, Dr. Hicks might ask him: "By the way, how long have we been talking?" A malingerer would reply by giving a roughly correct estimate of elapsed time. A genuine case behaves in quite a different fashion, often making a wild guess, which is obviously wrong. He confabulates and says: "Oh, at least a minute." Then he might veer to the other extreme and blurt out: "All day." Pressed to say if he is really sure of the amount of time that has passed, a typical patient will reply: "Well, not really. But I had to say something."

A series of experiments carried out by Dr. Hicks on these patients at Chapel Hill shows the folly of trying to explain the human sense of the extended passage of time in terms of a simple internal clock or counter. Psychological counting mechanisms do exist, but their role is limited and they are unreliable as tellers of "real" time, the time of clocks in the external world. In one test, Korsakoff patients with normal IQs were asked to count at what they felt to be a rate of one count a second during a prescribed interval of time. The counting was done by having the patient move a tape on which was printed a random sequence of numbers. Each time the tape was moved, a new number appeared in a small window, and the patient had to read each number aloud at what he thought was a rate of one every second. If Dr. Hicks let an interval of 100 seconds pass, and a patient had read 100 numbers, the patient's sense of the passage of time was almost as accurate as a stopwatch. At the end of the interval, however, the patient would be asked to estimate how much time had gone by. He could not rely on reading the numbers to give him a correct answer because they were in random order, and he was therefore compelled to fall back on his own resources.

The performance of the Korsakoff patients on this task is illuminating. The rate at which they count is similar to that of people without the peculiar memory impairment typical of the Korsakoff syndrome. Two control groups of non-Korsakoff subjects were used in the experiment. One control group consisted of perfectly normal, healthy individuals, and the other of alcoholics, matched in age and social class with the Korsakoff-patients. No matter how long the time interval, the speed at which Korsakoff patients moved the numbered tape did not differ from that of people in the control groups. What is more, as long as the interval did not exceed about 20 seconds, their estimate of the entire interval, after it was over, appeared to be perfectly normal.

A critical difference was noticed in the behavior of the Korsakoff patients when they were asked to judge the length of intervals of time that lasted longer than 20 seconds. Then, they simply made a wild guess. They confabulated about time, as Lester Anderson confabulated about selling shirts in a men's clothing store. After an interval of a minute, during which a patient had counted numbers at a completely normal rate, he might say, when asked to estimate the length of the interval: "ten seconds," or "five seconds," or "thirty minutes."

The time interval of about 20 seconds is not far from the 15-second span of short-term memory. It corresponds roughly to what the psychologist William James, borrowing an already existing term, called the "specious present." James made a clear distinction between the psychological experience of time as it is going by and the experience of time as it has gone by. We perceive time from moment to moment, James said, through the medium of the specious present, specious because it gives the false impression of being an instant of time separating the present from the past and the future, as a mathematical point on a line divides the line. In fact, the specious present has a certain breadth, and includes a small amount of the past. It spreads over an interval of time varying in length "from a few seconds to probably not more than a minute."

James held strongly to the opinion that consciousness cannot consist of a succession of sensations and images, each one separate from the others, like beads on a necklace. If consciousness were to have a particle structure of that kind, we would never be able to have knowledge of anything except the "now," the immediate instant. Once a sensation ceased it would be gone forever and we would be incapable of acquiring experience. The present of our actual experience, James said, "is no knife-edge, but a saddle-back, with a certain breadth of its own, on which we sit perched, and from which we look in two directions in time. The unit of composition of our perception of time is a *duration*, with a bow and a stern, as it were—a rearward and a forward looking end." James said, as a joke, that Adam, who is supposed to have been created complete with a navel, might also have been provided with a specious present, so that even in the very first moment of his life, Adam would think he had been in existence for some small amount of time.

In Korsakoff patients, the specious present appears to be completely normal. They experience time as it passes, within that limited span, in the same way as everybody else does. But they cannot embed the contents of the specious present into a time *context*. It is this loss of context that marks the crucial dividing line between Korsakoff patients and healthy individuals. Events experienced during the span of immediate awareness float free in time.

The reason Korsakoff patients are able to measure time accurately while it is going by, even when the interval exceeds 20 seconds, is that timing the present appears to be a context-free activity of the brain, like riding a bicycle or reading print upside down. It belongs to that aspect of semantic memory, perhaps the more archaic and deeply buried aspect, which stores knowledge of skills. When pedaling down the street on a bike, there is no need to have a conscious memory of the circumstances under which we learned to ride. The skill was acquired by practicing on various bicycles on many different occasions, and it would only lead to confusion if each day were to be recalled as a unique episode. We might fall off. The skill needs to be uniform, the same on all occasions, and differences must be smoothed out. Thus riding a bicycle, like other skills, is most successful when the knowledge needed to ride it is free of a space-time context. The same appears to be true of the "skill" of estimating time as it is going by.

Most of the damage to brain cells caused by Korsakoff's psychosis is in the diencephalic areas, the "between brain," located above the brainstem and below the cerebral cortex. Here the mammillary bodies, named for their bulging, breastlike shape, and part of a large cluster of nerve cells known collectively as the thalamus, are impaired as a result of vitamin deficiency. The thalamus is a major integrator of information arriving from the eyes, ears and other sense organs, on its way to the cerebral cortex, the most highly evolved of all brain structures. All sensory information passes by this route. Specialized circuits in the thalamus analyze these incoming signals, and the processed message is sent on to the cortex.

These between-brain areas, which are impaired in Korsakoff patients, are richly interconnected with the frontal lobes, the largest and least understood region of the cortex. The frontal lobes, located just above the eyes, make up about a quarter of the whole mass of the cerebral hemispheres. They are late in evolution and do not grow to their full size in a human individual until about the age of seven. They have been called the seat of personality. Endel Tulving suggests that the "self-knowing" type of consciousness, associated with episodic memory, has an anatomical basis in the frontal lobes. It is in this forward part of the brain that vigilance is regulated and the most complex kinds of purposeful activity are controlled. The frontal lobes are not responsible for sense perception as such. They play a very important role in forming conscious plans and intentions, and they modify the state of activity of the brain cortex, energizing it so that it is properly fired up and ready to carry out those plans. We are not talking here about a simple, instantaneous response to an event in the present, but about forming and executing a program, a sustained, intricate chain of actions making up an intellectual or behavioral task, and verifying its results. The frontal lobes mediate purpose, and purpose implies the future; it implies time. They also regulate stable, selective attention to relevant events and inhibit responses to irrelevant events, enabling the brain to concentrate on completing a plan or intention. Among such stable intentions is the intention to memorize information arriving from other parts of the brain.

Of great interest for our purposes is the fact that there are clocklike cells in the frontal lobes which deal with moment-by-moment time, time as it is going by. In monkeys, which have frontal lobes that are much bigger than those of most animals, though smaller than our own, the timing cells literally count down. When a light goes on for an interval of time, certain cells in the frontal lobes begin to fire and then stop. Other cells start to fire just before the interval is over, and a third class of cells in the same part of the brain simply fire for as long as the interval lasts. Timing is in the frontal lobes for the very good reason that they mediate plans and intentions, and these processes, by their very nature, take time seriously.

In Korsakoff victims, the timing cells in the frontal lobes are left intact. That is why their judgment of time as it is going by is as good as that of healthy people. The main damage is in that part of the brain where messages arriving from the senses en route to the higher centers of thinking, planning and remembering are interpreted and marked for significance. Dr. Hicks speculates that an event or an episode which for a normal person is important and full of meaning is marked as such by the thalamus, and the marked message is passed on to the higher brain centers, labeled in some way for storage in episodic memory. In Korsakoff victims, however, messages of personal significance are not marked in this way and so are not laid down in permanent form, with a space and time context. "If the information is not interpreted by the thalamus, the frontal lobes are not going to be told, 'Hey, this is important, tell the rest of the brain to store it,' " Dr. Hicks explains.

It seems that time as it is going by and time as it has gone by are not dealt with by one and the same psychological strategy. At least two members of the family of temporal intelligence are at work. The tick-tock of the timing cells enables us to keep track of the present. Korsakoff patients can count seconds for as long as you like. But in order to have a normal sense of time as it has gone by, the brain must be able to retain in episodic memory the events that happened in time, and this is something that Korsakoff patients cannot do. Their time sense is like a clock that marks each second as it passes but shows no record of the number of seconds that have passed beyond a narrow, 20-second window. It is a clock with no numbers and no dial. Once a time interval exceeds the span of immediate awareness, once a context is needed, Korsakoff patients are wildly erratic in their time judgment. They are able to learn new, complex sequences of actions, such as puzzles or skills, and that is one of the special functions of the frontal lobes. They can also use rules of reason and logic. But they do not know where or when they learned the skills, nor do they retain in conscious memory the order of the consecutive actions that make up a skill. Sometimes they use reason and logic to get around this deficit in the knowledge of the time order of a chain of actions. Asked if he has eaten lunch before he caught a taxi to the hospital, a Korsakoff patient might answer "yes," not because he remembers which came first in the sequence, but because it is reasonable to suppose that he had eaten lunch before being driven to see the doctor, since no lunch is provided at the hospital.

Korsakoff patients can drive a car normally, and that involves intricate, carefully timed actions which they know without knowing that they know them. But they are unlikely to remember what happened on the ride to the hospital, because that is an episode. They might invent a story about what happened, if pressed, but when the story is checked with another person who rode in the car, it turns out to be fiction.

"Timing by itself, counting numbers during an interval, is more like semantic memory," Dr. Hicks said. "It is a context-free activity. But being conscious of a particular stretch of time which is beyond the span of immediate awareness is like an episodic memory. If you cannot remember the context of an interval, then you have no feeling that time has gone by. Korsakoff patients really have no extended awareness of time as it has gone by. Their days are simply disjointed in the sense that what they are doing right now, this minute, totally occupies their consciousness. There is no continuity between what they are doing now and what has been going on during the rest of the day. Nor is there any integration of the immediate moment with planning for the near future. They are locked into the present." Endel Tulving has described an amnesic patient who, asked to say what it is like when he tries to think about tomorrow, replied: "It's like swimming in the middle of a lake. There's nothing there to hold you up or do anything with."

In order to have a normal sense of the passage of time, there must be a stream of consciousness in which are stored events that demarcate time. Such events may originate externally, in the environment, or internally, in the brain. We may not be consciously aware that our overall feeling of extended time is based on them, but when the ability to store events breaks down, we become prisoners of the present. "If you want to experience extended time, you must be able to summon up a context," Dr. Hicks said. "I am using the word 'context' in the sense of a unique matrix of events that defines any one particular episode."

The sense of moment-by-moment time, based as it seems to be on specific mechanisms in the frontal lobes, can be manipulated and influenced quite easily, by drugs or by altering a person's state of mind; simply scaring him is often enough to produce a change in his sense of time as it is going by. This is shown by a set of studies Dr. Hicks carried out on marijuana smokers at Chapel Hill.

For many years, it was thought that acute marijuana intoxication produced a psychological state which closely resembles amnesia. The reason for this belief was that the first cognitive function to be disturbed under the influence of marijuana is memory. A relatively small dose of the drug, smaller than the dose needed to interfere with movements of the body, can produce a memory deficit similar to that seen in Korsakoff patients.

In the recent work of Dr. Hicks, however, it has become clear that there is another function that is even more sensitive to marijuana intoxication. That function is the ability to estimate the passage of time. One way of testing the accuracy of a person's time sense is to ask her to behave like a clock, signaling the start and finish of a time interval she judges to be of a certain length. She "produces" the interval. In people who have smoked marijuana cigarettes, a drastic distortion of the time sense can be observed. When tested by this method, the smokers produced intervals that are much briefer than those timed by the stopwatches of the scientists supervising them. If a smoker is asked to produce an interval lasting one minute, her interval might last only 30 seconds as measured by a stopwatch. The smoker underestimates the length of the interval. That means her own private time is running twice as fast as public time. One of the smoker's own, subjective, psychological seconds is equal to two stopwatch seconds. The more marijuana she has smoked, the shorter is the interval she produces with respect to "real," or stopwatch, time.

This underestimate of public time is a little unexpected, because it is quite different from what happens in amnesia, a state that marijuana intoxication was thought to induce. In fact, the effect of marijuana on the time sense is exactly the opposite of amnesia's. When a Korsakoff patient is asked to produce a specific time interval, he overestimates stopwatch time. For a marijuana smoker real time seems to race along at high speed, but for a Korsakoff patient it seems to go very, very slowly, almost coming to a complete stop.

There is another, highly significant difference between changes in the human time sense that occur in amnesia and those that are seen in marijuana intoxication. When Korsakoff patients count numbers on a tape, they do so at a rate that corresponds quite closely with real, or stopwatch, time, no matter how long the interval may last. The situation is entirely changed, however, in marijuana smokers. When a smoker counts numbers as an interval of time passing, using the same procedure as for Korsakoff patients, the rate of the smoker's counting speeds up, racing ahead of real time. His moment-by-moment sense of time is accelerated, and his estimate of the whole interval is much too short. For him, real time seems to be stretched out, slowed down.

Why does marijuana have this accelerating effect on the time sense? The answer seems to be that it exaggerates an entirely normal psychological process. When we are waiting for some important event we expect to occur in the near future, real time seems to drag, to run slowly. Clocks dawdle and minutes creep by. By comparison with external clock time, subjective time races along: five minutes seem like twenty. A watched pot or a tea kettle never comes to the boil. There is an intense, exclusive focus of attention on the passage of time itself. Dr. Hicks believes that this state of mind is carried to extremes during marijuana intoxication.

Science has a means of measuring changes in the state of the brain when it is waiting for an anticipated event. This is done by detecting and charting the wave patterns created by electrical activity in the brain, the same procedure that brought such a wealth of new insights into the internal structure of sleep. Watching at this window into the mind, scientists are able to observe the changes in the wave patterns that occur when some specific event in the outside world is registered by the brain, as distinct from the background noise of spontaneous, internal electrical activity. Brain responses of this kind are called evoked potentials. Evoked potentials are thoughts made visible as wavelike marks on paper. They represent "what's in the brain that ink might character." One such wave, known as the P300, appears on the charts whenever a person is surprised by sequences of events that are unpredictable, and represents one's inner state of mind while making decisions under conditions of uncertainty.

Small electrical events occur in the cortex of the brain as a result of hearing a sound or seeing a flash of light. Changes in the size, or amplitude, of the waves on the chart correspond to the amount of attention a person is paying to the sound or to the light. One important type of evoked response is called the "expectancy wave." Someone is told that two signals will be given one after the other, perhaps a light flash followed within two seconds by a sound. The light is a warning signal, preparing the watcher for the second, "imperative" signal, to which he is told to respond in some specific way. During this waiting interval between the two signals, a wave of increasing amplitude shows up on the charts, and the size of the wave is directly related to the intensity of certain features of the individual's psychological state, including his motivation, interest and ability to focus attention. This expectancy wave has a technical name: contingent negative variation or CNV. It is measured by means of electrodes pasted onto the scalp, over the frontal lobes of the brain.

The expectancy wave can be reduced in size by giving depressant drugs, but under the influence of marijuana intoxication, the increase in its amplitude is huge. What seems to be happening is that the brain damps down all activity elsewhere in order to get ready for the imperative signal, because it knows that signal will be an important event, requiring a response. The brain mobilizes its mental resources in preparation for the response and at the same time screens out distractions. The mechanism prevents interference by anything not relevant to the imperative signal. In many respects, marijuana is like a sedative-hypnotic drug in its effects, reducing alertness and inhibiting arousal, as barbiturates do. In the case of the expectancy wave, however, marijuana acts exactly like a stimulant such as amphetamine. It increases the amplitude of the wave. It makes the brain concentrate on one thing and one thing only, quieting other circuits.

One striking property of this time-obsessed mental state is its autonomy. After producing an inaccurate interval, a marijuana smoker is remarkably immune to corrective feedback in the form of a report on his performance. The smoker can be told again and again that his judgment of time is distorted, that external time is not passing as slowly as he thinks, yet he will not correct the error. Informed that the time interval he produces is too brief, he will still make it too brief on the next occasion, and the next. This is quite different from what we find in normal experience. Young children can be taught to estimate short intervals of time correctly as long as they are given feedback from a teacher. With the marijuana smoker it is different. Only if he is distracted while the interval is in progress, with nontemporal information, will his judgment of time match stopwatch time. Given a list of words to memorize during the interval, the smoker no longer experiences the racing of subjective time. The more difficult the words, the less distortion occurs in his sense of time. The reason seems to be that as long as the brain is mobilizing some mental resources for the task of memorizing words, a person's whole attention cannot be given to time. By introducing nontemporal information, a normal experience of time is restored, one that has a more stable correspondence with the time of the external world.

Generally, the CNV is a highly reliable indicator of the state of a person's time judgment. The larger the expectancy wave, the more inner time speeds up, making external time seem to run slower and slower. It is clear that the CNV represents the activity of the timing cells in the frontal lobes of the brain.

A brain cell generates electrical impulses and fires them down the outside of the long tail of the cell, called an axon, which terminates very close to the surface of another, "target" cell without quite touching it. That is one element of the process by which cells communicate. But the electrical pulses do not actually cross the tiny gap between the end of an axon and the surface of the target cell. They are signals to release a neurotransmitter, a chemical substance which, as we have seen, comes in hundreds of different types. The chemical is pumped along the hollow interior of the axon and sprayed against the surface of the target cell, changing its rate of firing.

The idling speed of a brain cell is about one electrical impulse a second, but the cell can be made to increase its rate up to about sixty impulses a second in the twinkling of an eye when large amounts of the right type of neurotransmitter are sprayed against receptors on its surface which respond to transmitter molecules of a specific shape, as a lock turns only to a key that fits.

To Dr. Hicks, it now looks very much as if the expectancy wave, the CNV, is generated by cells producing and responding to the neurotransmitter dopamine. The CNV, which is associated with the acceleration of subjective time, can be made to disappear using drugs that block cell receptors sensitive to the shape of the dopamine molecule. One of the best known of these drugs has the trade name Haldol. The effect of Haldol on these cell surface receptors is like that of putting an Impala key into a Chevy Nova car. It happens that the Impala key slides into the Chevy ignition lock all right, but it will not start the car. While it is in, it keeps all other keys out. Similarly, Haldol prevents the dopamine key from binding to the appropriately sensitive receptors by occupying them without altering the cell's speed of firing. As a result, the drug wipes out the expectancy wave completely and slows down subjective time, making objective time seem to go faster. Scientists can also work this trick in reverse, speeding up instead of slowing down subjective time, by administering drugs that increase the production of dopamine. Such drugs include the amphetamines, to which lay users have given the entirely suitable name "speed." In the 1960s, doctors in the emergency rooms of hospitals noticed that people on amphetamines often had hallucinations closely resembling those of schizophrenics. That is hardly surprising, since schizophrenia involves the dopamine systems of the brain. Drugs that help to allay the symptoms of schizophrenia do so by blocking the action of dopamine.

The speeding up of subjective time in conditions of emergency, intense anticipation or danger seems to be more an accident due to the nature of the brain than some evolutionary strategy having to do with time alone. Dr. Hicks believes it is a secondary effect. What seems to be happening is that when an individual faces a threat or waits for an important event, the brain is hypervigilant, and cells of the frontal lobes involved in planning and coordinating behavior are working in overdrive, firing at a terrific rate so as to meet the threat or respond to the event. This toning up of brain circuits in anticipation of something to come is a primary function of the frontal lobes. It is what produces the expectancy wave.

But because the electrical-chemical communications network of the brain is very poorly insulated, increased activity in one set of cells is likely to spill over into an adjacent set, and make them fire at a faster rate too. It so happens that the cells in the frontal lobes that do timing are near the cells that are responsible for planning and coordinating behavior, both systems apparently being dopamine-driven. In a state of intense vigilance, when planning circuits are on red alert, the hyperactivity of those circuits causes the timing cells to speed up. Instead of firing rather slowly, at ten impulses a second, the timing cells might double the rate to twenty a second, so that for every ten units of stopwatch time, there are twenty units of cell time. That would explain why stopwatch time seems to creep by under such circumstances. Time distortion occurs because of the proximity of cells that plan and cells that time and because of the leakiness of the brain.

"When you are waiting for a tea kettle to boil, you are in a state somewhat like that of an animal watching for a predator," Dr. Hicks said. "You are totally focused on one event in the near future to the exclusion of everything else. You're not watching televison or having a conversation. You're just focusing on time. The animal is completely focused on one future event, a specific threat. It's not going to look around for berries or wonder what's over the next hill. Both the person and the animal are in a state of hypervigilance, which is one where subjective time is accelerated. You can put a person in that state just by scaring him. Or you can produce the same effect with drugs that produce a large expectancy wave. Of all human cognitive functions, subjective time is one of the easiest to manipulate with drugs."

All this shows that there are many aspects of the human "sense of time." Some are clocklike, but variable because of an accidental feature of the geography of the brain, while others form a context for the experience of time as it is passing. Some are more amenable to alteration by drugs than others. Certain systems operate on time as it is going by, while others are responsible for time after it has gone by. Different kinds of memory mediate different kinds of time experience. A context of nontemporal information seems to be as indispensable to a normal sense of time as the temporal information generated by the clocklike cells of the frontal lobes. Awareness of pure time, time on its own, is not enough. Nontemporal information also acts as a corrective to the speeding up of subjective time that results from marijuana intoxication. Given lists of words to learn, the marijuana smoker is distracted from attending to moment-by-moment time; then, his estimate of an interval of time becomes normal and corresponds quite well to public time. Reason and logic play a role, enabling people who lack episodic memory to construct a temporal sequence of their past actions even though they have no recollection of that sequence. Logic also comes to the aid of healthy people, by reminding them that even though subjective time ran fast due to brain activity (reflected in the amplitude of the CNV), that does not mean objective time really ran slow.

William James remarked upon the radical difference that exists between the sense of moment-by-moment time and the sense of time extending into the past. A time filled with varied and interesting experiences seems short while it is in progress but long as we look back on it. In retrospect, the time seems to collapse. How does this curious anomaly come about? The reason, James thought, is that whereas we "intuit" the specious present directly and almost automatically, we conceptualize an interval that has already passed by, remembering events that occurred in it, and use those events to symbolize a span of time. That is why a tedious day, in which nothing significant happens, seems to stretch out interminably while it is going on, because the mind attends to the passage of time itself, and is in the state of a person waiting for a tea kettle to boil. There is little nontemporal information to distract the attention from focusing on time. Once the day is over, however, and the mind looks back, the time shrinks and shortens, because the number of memorable events that symbolize its length are sparse. So a distorting effect is always at work in our sense of time that has gone by, James said, because we reproduce the events and do not perceive them directly. They are refracted by the lens of memory, so that there is a sort of "perspective projection" of the past onto present consciousness, like the projection of a wide landscape onto a film in a camera.

"Time" is not many things. But the psychological bag of

tricks by means of which the human mind deals with time is many things, and the properties of one may not be the same as the properties of another. The concept of an absolute time is as hard to defend in psychology as it is in physics. The Oxford philosopher John Lucas has said that "Time is puzzling not only because it is unlike anything else in our conceptual universe, but just because it is part of our conceptual structure and is connected to a number of other parts. These different connections give us different perspectives on the concept of time, indeed almost different concepts . . . no single approach can reveal the whole nature of time, and each needs to be supplemented by others."

In the 1980s, psychologists and others are beginning to identify the mental structures that make up the totality of subjective time, the connections that link those structures, and what happens when the links are broken. The approach is not single, but pluralistic and wide-ranging, and the results are leading beyond the brilliant guesswork of William James, building a foundation for a new and more complete theory of the nature of the human mind.

NINETEEN

The Codes of Time

It is clear that among the most important elements of a normal, human sense of time are memory, attention and context. In psychology, these three are intimately connected. The particular way in which they are combined, the part each one plays in the mind's construction of time beyond the specious present, helps to determine temporal experience.

One member of this trio of mental operations, whose significance may have been overlooked, is attention. The role played by attention is critical, because it determines the breadth or narrowness of context. If attention is restricted, context is reduced, and reduced context plays havoc with the sense of time. Attention that is focused on time and time alone, ignoring nontemporal information, whether external or internal, screens out events that can be used for the symbolic construction of time that has gone by. Without such event landmarks, a person is adrift in time, as a tourist wandering around a strange city is adrift in space, unless the tourist pays attention to distinctive features of the landscape and the routes connecting them. During the Apollo 14 moonwalk in 1970 astronauts Alan Shepard and Edgar Mitchell came within 75 feet of their intended destination, the rim of Cone Crater, but turned back to the spaceship without reaching the rim. They had become lost and confused because of the absence of recognizable landmarks on the lunar surface, and had no idea they were so close to their objective. The terrain came close to being pure space, because of its unrelieved sameness. The temporal equivalent of the two astronauts' dilemma is to concentrate one's attention on pure time.

In his novel *The Magic Mountain,* which is called a "time romance," Thomas Mann ruminated on the curious way in which subjective time swells and shrinks depending on the nature of the events that occupy it, and the attention we pay to them. The young, impressionable hero of the novel, the engineer Hans Castorp, languishing at a sanatorium in Switzerland, grows accustomed to a routine in which one day is very much like the next, and many hours are spent reclining on a chair under blankets so that he can "see time spacious before him," and enjoy the sense of its passing without having to attend to anything nontemporal, even a book. Such periods of eventless sameness, Mann maintained, weaken the human perception of time. "Great spaces of time passed in unbroken uniformity tend to shrink together in a way to make the heart stop beating with fear," he wrote. "Habituation is a falling asleep or fatiguing of the sense of time, which explains why young years pass slowly, while later life flings itself faster and faster upon its course."

The brain is made in such a way that attention drifts away from the familiar and the humdrum, the predictable and the unsurprising. It tends to latch onto the unusual and the interesting, preferring contrast and change to sameness. Attention discriminates. It is an organizing principle which imposes form on the formlessness of raw impressions. "Experience is what I agree to attend to," William James said. "Only those items which I *notice* shape my mind—without selective interest, experience is an utter chaos." The opposite of selective attention is distraction, a confused, scatterbrained state in which "the eyes are fixed on vacancy, the sounds of the world melt into confused unity, the attention is dispersed so that the whole body is felt, as it were, at once, and the foreground of consciousness is filled, if by anything, by a solemn sense of the empty passing of time."

The importance of the shaping power of attention is nowhere more apparent than when it ceases to function or is at a low ebb. This happens every night, during REM stages of sleep, and it helps to explain why time is distorted in dreams. Dreams are an example of a mental process in which attention seems to become uncoupled from consciousness, or at least is greatly diminished. In waking life, attention is a source of control, giving order and form to experience, sifting what is important from what is irrelevant, separating the meaningful from the meaningless.

In the dream state, our freedom to select the wheat from the chaff of experience is lost. Scenes, characters, plot unfold as in a movie theater, where the script is already written, the camera angles preordained, the closeups unalterable. The freedom to choose what to notice and what not to notice, to linger over or ignore, is almost completely absent. In a movie, the breadth or narrowness of context is determined by the director's attention, expressed in his camerawork; wherever he looks, the audience must look also and accept the limits imposed. The only other option is to leave the theater. In dreams even that option is usually closed.

Such freedom to choose what we attend to begins to be taken away from us even before sleep begins, during certain states of fatigue. It has been suggested that slow-wave sleep promotes recovery from simple physical exhaustion, while REM periods repair the ravages wrought by the stress of the day on specific psychological functions. One of these psychological functions is attention. Not all mental processes show equal wear and tear during the waking day. The ones that deteriorate the most tend to be those that are comparatively recent in evolution and are less automatic, closer to consciousness. Those that require the least effort of attention and are often the most easily performed are less affected. When we are mentally tired, semantic memory works better than episodic memory, which is highly dependent on context. As this type of weariness increases, the psyche seems to relinquish gradually its attempts to deal with the more demanding aspects of reality, and slowly adapts to a simpler mode of functioning. Alertness declines and irritability rises. The brain is unwilling to venture into new and unfamiliar territory or to make long-term plans. The mentally exhausted person tends to be disoriented in time and in space. Apparently, what is happening is that context is being reduced. Attention is narrowing, shrinking the breadth of information that is registered and stored in long-term memory. Ernest Hartmann, a professor of psychiatry at Tufts University, suggests that specific pathways in the brain enable us to focus our attention in a selective way. The pathways are subtle feedback guidance systems which help a person to navigate through space and time. During tiredness, these systems do not work as well as when the brain is fresh, and during REM sleep they are selectively shut down for repairs.

"Time and space seem less well organized in the dream," Hartmann says. "The edges of the visual dream space are imprecise, and one is half unaware of time."

The sense of time certainly is not missing entirely from dream experiences. The dreamer knows roughly how long he has been asleep and how long a dream has lasted in real time. Also, there is a thread of temporal continuity as the various stages of the sleep cycle follow one another in their clocked sequence. REM dreams often probe progressively farther and farther back into the past of the dreamer's life as the night proceeds. But the sense of time in dreams is definitely unlike that of waking consciousness. Often a dreamer will find himself hurtling around a strange city, lost in space as well as in time, rushing to catch a train for which he is scandalously late, while all kinds of bizarre distractions prevent him from arriving at the station before the train is due to depart. There was ample opportunity to make an early start, but time was simply ignored until the stage of panic was reached, and even then there is no clear sense of the clock. In waking life the dreamer may be a person of impeccably punctual habits, making a fetish of getting to the station with half an hour to spare.

The clinical reason for the weakening of the time sense in REM dreams, Hartmann believes, is that a specific class of brain chemicals, the catecholamines, are depleted after a day of emotional stress or hard intellectual work. These substances are neurotransmitters. The most important of them is norepinephrine, which has a close chemical relationship to the hormone adrenaline, another catecholamine, which mobilizes the body in stressful circumstances. Because such chemicals give off an intense green fluorescence when treated with a specially prepared formaldehyde vapor, a scientist can watch them under a microscope and follow their paths as they travel through the communication network of the brain. Norepinephrine has two main pathways, which researchers have identified. One pathway lies mainly in the back of the brain, while the other is closer to the front. All the cell bodies of the first pathway are found in one specific part of the brainstem called the locus ceruleus, or "blue place." From the blue place, the route traveled may turn downward to the cerebellum, the part of the brain that deals with very rapid and finely coordinated motor movements, or upward to the cerebral cortex, where the so-called higher mental processes go on.

In Hartmann's opinion, catecholamines are involved in the brain's ability to keep attention flexible during wakefulness. When the mind is tired and even more when we are in the REM stage of sleep, this flexibility of focus is lost. In dreams, the power to shift attention rapidly and to be aware of a number of different people and things all at the same time or one after the other is noticeably diminished. Much reduced, too, is the capacity to recognize and test patterns in the environment, including time and space patterns, a mental competency Hartmann believes is mediated by catecholamine chemistry. A dramatic reversal of this effect is seen when a person is given an overdose of amphetamines. Then, exactly the opposite type of behavior occurs. Attention is exaggerated, and so too is the recognition of patterns, so that the brain insists on finding pattern or meaning in the environment, even if there is none, and this is carried to the extreme in paranoia. Hartmann points out that the great chess masters, born imposers of pattern, are often at least a little paranoid. He suggests, as a plausible speculation, that waking functions which keep a person securely located in time and space are under the influence of the catecholamine systems, which are shunted out for servicing during REM sleep. Consequently, the dream state can show us how the brain operates in the absence of the catecholamine influence.

A dreamer has little or no insight into the delusional nature of the dream experience, accepting wildly improbable and even impossible scenes as being real, On waking, one realizes at once that they were a fabrication. A dreamer is naive, gullible. One reason for this curious failure of insight, this absence of sophistication, may be a lack of context. Events in dreams are taken at their face value, as they happen, and are not related to the rich store of worldly knowledge which forms a mental context for perceived events during the waking day. A naive person may accept the trumped-up lies of a confidence trickster as plain truth because he lacks the information born of experience which would tell him instantly that he is listening to a piece of fiction. The story he is being told stands on its own, stripped bare of the context that a more mature consciousness would provide. Similarly events in a dream stand alone. Dreams are full of symbolism and they speak in metaphors, but a dream cannot be interpreted while it is going on. Only after waking can a context of knowledge be applied which will unravel and elucidate the meanings of the cryptic messages a dream communicates.

The critical roles played by attention, memory and context in human subjective time were evaluated in a series of experiments carried out by Robert Ornstein. His findings are especially interesting, because he rejected the idea of a single, psychological clock, or any one mental faculty that could be thought of as an "organ" of time, and instead tried to examine time perception within the mainstream of modern research into how the brain knows, learns and remembers. Ornstein investigated the human experience of what he calls "duration," periods of time that are longer than the specious present but not usually as expansive as days or weeks. A duration in this sense has a minimum length of ten seconds and no clearly defined upper bound.

Ornstein concluded that our experience of duration, our sense of whether an interval of time seems short or long, depends on information that is attended to while the interval is actually in progress, and the manner in which that information is stored in memory. We construct the experience of time from such stored data, so that the sense of duration can be regarded as a concept, similar to our concepts of order and chaos. The mind does not necessarily intuit order directly, without performing mental operations on it, but it may construct a subjective order out of what is, objectively speaking, random.

The more information the mind attends to during a given interval, the more will be stored in memory, and that makes the interval seem to last for a longer time than it would if less information had been attended to and stored. A decrease in the amount of information makes subjective time shrink. An increase makes it expand. But that is not the whole story. The experience of time is also affected by the way in which the information is coded for storage in memory, and here the question of context enters.

In the modern theory of communication, a message is said to convey a large amount of information if it is unexpected, if it comes as a surprise and is difficult to predict in advance. The person to whom a message is sent must be in a state of some uncertainty as to what it will say; otherwise there is no point in sending it. When a message is actually received, this uncertainty disappears. The greater the prior uncertainty, the greater the information content of the message. Novelty, differences, change, all increase the uncertainty, and therefore the amount of information, whereas familiarity, sameness, uniformity, all reduce it.

Messages are usually coded in such a way that they are partly unpredictable, but not completely so. Aspects of the message that are new are set in a context of aspects that are familiar. Change is embedded in sameness, and the result is to reduce the uncertainty in the mind of the person to whom the message is being sent. The letters that form words and the words that form sentences on a page of printed text do not all come as a surprise, because the rules of spelling, syntax and meaning, agreed upon in advance, make certain sequences of letters and words more likely to occur than other sequences. Looking at letters and words, the reader can guess ahead at the next letter or words in the sequence, using his knowledge of the rules. The first sentence of Jane Austen's novel *Emma*—"Emma Woodhouse, handsome, clever, and rich, with a comfortable home and happy disposition, seemed to unite some of the blessings of existence, and had lived nearly twenty-one years in the world with very little to distress or vex her"—gives the reader a good deal of new, unpredictable information. It reveals the heroine's age and circumstances and her outlook on life, which could not have been guessed in advance. On the other hand, not every letter or word in the sentence conveys novel information. Some can be predicted on the basis of what has gone before. The rules of spelling, for example, tell the reader that the string of letters *unit* will be followed by either the letter "e" or the letter "y" and that *distres* will be followed by another "s." Rules of syntax make it likely that a noun or adjective will come after the article "a" and that the word "seemed" precedes the infinitive form of a verb. Rules of meaning lead us to expect that the word after "twenty-one" will be "years" and anticipate that the phrase "distress or . . ." will be completed by a synonym for distress, or at least a compatible word, so that the reader, allowed a number of chances at guessing, might be able to predict the word "vex." If someone wanted to send this sentence by telegram, compressing it as much as possible so as to save money on cable charges, he might reencode it in the following form: "E. Wdhse, handsme clvr rch, cmftbl hm, hppy, 21."

This telegrammatic version of Jane Austen is coded in such a way as to eliminate those elements that are not essential for comprehending the *information* content of her message. Letters can be omitted because knowledge of the rules of spelling enable the reader to predict the letters that are missing, knowledge of rules of syntax make it possible to guess at how the words should be related, and the bare meaning of the sentence is preserved in spite of deletions because "disposition" adds nothing to the information that Emma is happy, while "seemed to unite some of the best blessings of existence" and "with very little to distress or vex her" can be inferred from the fact that she possesses almost every imaginable advantage and privilege, and the phrase "in the world" is superfluous as far as sense is concerned, since it is unlikely that Emma Woodhouse had lived all or part of her life on the planet Mars.

Those elements of the Jane Austen sentence that make the sentence more predictable than it would have been otherwise provide redundancy, to use the technical term of communications theory. Rules of spelling, syntax and meaning are redundant according to this definition, because they are familiar, established in advance, before the message was sent. Such rules are built into the codes of speech and writing. They are redundant because they are constant. If they change, they do so very slowly, over hundreds of years. They represent information we already know, used as a context for information that we did not know. A reader opening *Emma* for the first time can comprehend easily the new knowledge it contains because of the context of familiar knowledge in which that knowledge is embedded. Whereas the information in a given sentence may be unique, different from any other sentence ever written, the redundant rules are the same in all sentences. Redundancy helps to make a message intelligible by making it more predictable. Unless redundancy is in the code, the message is unreliable, because each symbol is quite unpredictable, and so there would be no way of knowing whether errors or misprints had occurred. The message cannot be trusted.

In music the "rules" of meter and rhythm and in conversation the tempo of speech are redundant elements, and they increase the ability of a listener or speaker to predict what is ahead by adding a context of sameness and familiarity to the changes and surprises that any form of communication must produce if it is to communicate at all. In a conversation, new information tends to occur in a predictable time slot, making it easier to comprehend and respond to.

In printed text, redundancy provides connecting links between one part of a message and other parts, because of the rules governing the organization of letters and words in a sentence. That is what makes it possible to guess ahead and spot mistakes and typographical errors which violate the rules of organization. Robert Ornstein has shown that the lack of such connecting links in the codes that the brain uses to process information has an effect on the experience of time. An interval of time seems longer when many different things are happening than when not much is happening. But beyond that, the interval seems longer when happenings are stored in memory in a code that contains little redundancy than when the code is highly redundant. Reading the telegram version of the sentence from *Emma*, for instance, is experienced as taking up more subjective time than reading a sentence containing the same number of letters, written in Jane Austen's usual redundant prose style.

The more random an experience is, the less redundancy it contains, because event follows event unpredictably, without rhyme, reason or rules. Events are unconnected. They lack a context of familiar elements. A random experience is more "dreamlike" than an experience that has a familiar context. In one study, Ornstein played ten different sounds, such as a typewriter key being struck or a sheet of paper being torn, on a tape recorder, either in an orderly sequence, each sound being played and repeated twenty times, or in a disorganized, random fashion. Both tapes lasted for five minutes. He found that the random tape, where the sequences of sounds lacked redundancy, was judged on the average to be 1.33 times as long as the duration of the tape where the sequences were organized and contained a good deal of redundancy.

The simple statement that time is an effect of change, therefore, needs to be amended as far as the human brain is concerned, so that the important roles played by sameness and difference are taken into account. In a random sequence, change is paramount, and there is no uniform, unchanging set of rules to connect one event with the next. In an organized sequence containing some redundancy, sameness is there as a context for change. Subjective time is an effect of both change and sameness, and each influences the experience of time in a different way.

The new or the strange, the unruly and the disconnected, seem to occupy more psychological time than a familiar experience or one that is easily understood. If a well-known musical melody is played backward, listeners think it lasts for a longer time than when they hear it played forward. A melody with many notes also seems to take up more time than a melody with fewer notes. Redundancy may be the reason why a successful day seems shorter than a day full of failure, because success tends to be better organized than the more random structure of failure. Adding redundancy, embedding the unknown in a context of the known, shortens the time as it is subjectively experienced. The ambiguous line drawing that follows was shown to some students in two equal intervals of time, once on its own and the second time with a caption which provided extra information, providing a familiar context and reducing the strangeness of the drawing.

Students who looked at the picture with and without its caption considered that the time seemed longer when the caption was omitted.

The experience of duration, Ornstein concludes, is constructed out of information stored in long-term memory. Messages stored in memory do not correspond in a simple, direct way with messages that arrive at the brain from the outside world, because the brain selects and filters incoming infor-

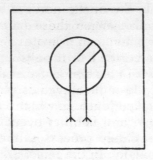

*Redundant caption: "An early bird who caught a
very strong worm."*

mation by means of the flexible focus of attention, admitting
some aspects of it to long-term memory and excluding other
aspects, and encoding with a greater or a lesser amount of
redundancy. That means one person's estimate of how long
an interval lasted may differ from someone else's, because
the deployment of attention and the way in which information is coded in memory is not uniform but idiosyncratic and
depends on what is interesting or meaningful to the individual at a particular time. A humdrum experience, one that is
so familiar as to be boring, may be screened out by the brain's
attention filter, thus reducing the amount of information that
reaches long-term memory. For instance, a commuter remembers little about the train journey to work in the morning
and home in the evening, and in retrospect the time col-
lapses almost to nothing. An experience in which the same
event is repeated many times is likely to seem short when
looking back on it because repetition is redundancy in its
purest form, sameness rather than change. The experience of
duration Ornstein calls "malleable, a personal construction
formed from the stored portion of normal experience."

Karl Pribram has identified specific parts of the brain—
the amygdala of the limbic system and the related frontal

cortex—as critically involved in redundant codes and impli-
cated also in the ability of the nervous system to become
accustomed to novelty. When these parts of the brain are
intact, humans and nonhuman primates get used to an event
if it is repeated and attend to it only when the repetition
varies, when it contains an element that is new. Damage to
the brain in the region of the amygdala may result in a "déjà
vu" or "jamais vu" experience, in which an event that is new
seems to be familiar, or a familiar event seems to be new.
The brain inappropriately processes novelty as repetition
and repetition as novelty. In the intact brain, a variation in a
pattern is felt as something new and worth paying attention
to, independently of actually recognizing the variation. This
explains, Pribram thinks, why music, which contains a tremendous amount of repetition and variations on repetitions,
arouses deep-seated feelings in the listener, quite apart from
the aesthetic pleasure of noticing patterns of sameness and
change. In music, variation, or change, cannot exist without
the previously assumed idea of repetition, or sameness. It
would follow then that meaning in music is derived from the
structure of the musical code, which is more redundant than
the codes of most other art forms.

The frontolimbic parts of the forebrain encode signals
arriving from the outside world in such a way as to enhance
redundancy. They are involved in the generation and control
of feelings produced by repetition and also in the processing
of variations on repetition, especially temporal variations. By
contrast, the back parts of the brain reduce the amount of
redundancy. They act, in Pribram's words, "much like an
editor searching for novelty," coding in such a way as to be
independent of context. The front parts code so as to be sensitive to context, organizing information in a network of associations. Ornstein suggests that the front brain codes
information in such a way as to speed up the rate of subjective time, while the back brain slows it down.

Thus the experience of duration is not simple and not
fundamental, but is influenced by a complex process of selection and organization, and is an effect, not just of "change,"

but of the structure of codes in memory. Without episodic memory, there can be no experience of time that has gone by. Subjective time would cease to exist, not because it is a substance or force that comes to be or passes away in its own right, but because it is a secondary phenomenon which no longer has any meaning when the psychological events from which it is derived do not occur, rather as the time of physics ceases to exist in the absence of physical events. Attention and memory are more primordial, more basic psychological entities than the sense of duration.

Subjective time, beyond the range of the specious present, is an "emergent" psychological property. In physics, there are also emergent properties, such as the temperature and the pressure of a gas. Temperature and pressure are not fundamental, but are large-scale effects of the motion of millions of microscopic particles. A single particle by itself cannot be said to have either temperature or pressure, because both these "emerge" as properties of the whole and not of the individual parts. In the same way, a newspaper photograph "emerges" from the hundreds of small dots that compose it. The mind itself may be regarded as an emergent property of brain circuits. Sir Alan Cottrell, the English physicist, points out that most of the ordinary properties of bulk matter are emergent, since the only fundamental properties of physical systems are the motions of elementary particles, and our knowledge of those motions is imperfect. Large-scale properties of matter are a product of our ignorance of what is happening on the small scale, what the individual particles are doing. The natural inclination of the human mind is to "crave understanding and seek only such general knowledge as is necessary for that understanding," Sir Alan says. The flow of time "appears to be an emergent property of ourselves, a subjective embellishment of the more austere concept of physical time as no more than a coordinate along which events exist timelessly. . . . When we touch a hot surface, we detect a molecular motion, but our experience of this sensation is nothing like kinematics. The feeling of *hotness* is a different kind of sensation altogether. Again, when we look at a newspaper picture we see, not an array of dots, but the image of a well-known face. In both cases, we subjectively dredge collective properties out of the complex of our sensations, and these collective properties belong to entirely different categories of experience than their physical causes."

In a sense, therefore, subjective time is a construct, a whole made up of psychological parts. And because that construct is so intimately associated with the mental processes of attending, knowing and remembering, the quality of subjective time must reflect the quality of those fundamental processes. The experience of time, Thomas Mann said, is "so closely bound up with the consciousness of life that the one may not be weakened without the other suffering a sensible impairment." Such a correspondence between being and time has been demonstrated in modern experimental psychology, confirming in the laboratory the earlier speculation of philosophers that upon the success or failure of the mind in its work of constructing subjective time depends the integrity of the self.

TWENTY

Putting Time Together

The intimate connection between the human experience of time and episodic memory means that consciousness of time and consciousness of self are tightly interwoven. Episodic memory is autobiographical, and the information stored in it represents events in the rememberer's own past. It is a basis for defining an individual's personal identity. The same cannot be said of semantic memory, with its store of words, facts, rules, formulas, symbols, ideas, concepts and relations. Semantic memory, as well as being atemporal, is essentially impersonal and refers more to the objective world. It is the personal belief of the rememberer that determines whether an event is true in episodic memory, whereas the truth or falsehood of the contents of semantic memory is decided by a consensus of other people. Memory for an episode is memory for time and place, and for the personal meaning of that particular time and place in the life of the rememberer. Semantic knowledge is not associated with personal identity in the same way, but always has a connection, no matter how oblique, to an aspect of the real world, and that aspect may or may not include the knower.

Consciousness of the self, in the fullest sense of th..t term, may have been a late arrival in evolution, much later than is generally supposed. If so, the subjective consciousness of time also must be a comparatively recent innovation in the history of the human species. Julian Jaynes, a Princeton psychologist, thinks that human beings possessed well-developed semantic memories at a very early stage of their history, but were deficient in episodic memory until as late as the latter part of the second millennium B.C., a period when civilization and culture were already well advanced in certain respects. In Homer's *Iliad,* written down in about 900 B.C. from an oral tradition that goes back to about 1230 B.C., there is no evidence that subjective consciousness, as we understand it today, existed. The *Iliad,* in Jaynes's view, stands at the turning point between a human mind that was incapable of reminiscing and one in which this faculty was awakened and began to flower. For the most part, the characters in the *Iliad* have no introspections. They do not "sit down and think what to do," but let the voices of the gods direct their actions. There are no words in the poem for consciousness or mental acts. And there is little sense of personal time, of someone's past life being an internal narrative set in a space-time context. In the *Iliad,* Jaynes maintains, time is treated in a careless, inaccurate fashion where it is treated at all. The poem is a window onto an era in the early history of civilization when people did not imagine but only experienced. They were individuals who "did not live in a frame of past happenings, who did not have 'lifetimes' in our sense, and who could not reminisce because they were not fully conscious."

The gods take the place of subjective consciousness. Jaynes writes: "It is impossible for us with our subjectivity to appreciate what it was like. When Agamemnon, king of men, robs Achilles of his mistress, it is a god that grasps Achilles by his yellow hair and warns him not to strike Agamemnon. It is a god who then rises out of the gray sea and consoles him in his tears of wrath on the beach by his black ships, a god who whispers low to Helen to sweep her heart with homesick longing, a god who hides Paris in a mist in front of the attacking Menelaus. . . . The beginnings of action are not in conscious plans, reasons, and motives; they are in the actions and speeches of gods."

Only late in the *Iliad* do any of the characters talk to themselves, and they are rather surprised to find themselves doing so. On two occasions Agenor and Hector address themselves in soliloquy, but these passages appear to be later interpolations, since the meaning is at odds with elements of what has gone before. Afterward, both heroes exclaim, "But wherefore does my life say this to me?" as if taken aback by the subjective nature of the experience. Quite a different impression is given by the language of a later work attributed to Homer, the *Odyssey*, which shows a tendency to spatialize time, an essential step toward constructing a mental map of it. The *Odyssey* contains many time words and makes increasing reference to the future. It may be no coincidence that this later epic is itself a story of personal identity, a voyage of self-discovery through many wanderings and adventures, a myth of the beginnings of subjective thought.

We invent or assume a space in consciousness where we can narrate aspects of our lives, past or future. Such a space does not exist in reality. In fact, consciousness has no particular location, as we know from "out-of-body" experiences, those in which a person sees himself as if from a remote vantage point, perhaps high up in the ceiling. Mind space is a metaphor or analog of real space, and the ability to make metaphors is an essential element of consciousness. Aspects of the actual world which are not spatial are made spacelike so that they can be "looked at" with the mind's eye. Spatialization, Jaynes says, is characteristic of all conscious thought, so that something quite abstract can be laid out in the mind as if on a table, all its parts put side by side for our mental inspection. Spaceless things can be made spacelike in this fashion. If they were not, we could not be conscious of them at all. We certainly could not be conscious of time. Jaynes comments:

"If I ask you to think of the last hundred years, you may have a tendency to excerpt the matter in such a way that the succession of years is spread out, probably from left to right. But of course there is no left or right in time. There is only before and after, and these do not have any spatial properties whatever—except by analog. You cannot, absolutely cannot think of time except by spatializing it."

In the metaphoric space and the spatialized time of consciousness, we narrate episodes in which an analog of the self acts out the stories of our lives. The analog "I" moves in this constructed space-time of the mind, having adventures, and it is only in this way that we can be truly conscious of ourselves, reminiscing about the past, imagining future actions and the likely outcome of those actions, deciding, choosing, rehearsing, solving personal problems. We short-cut behavior in the real space and time of the world by watching the analog self behave in the make-believe space-time of consciousness. We can "see" not only where our life is leading, but also who we are. Without the ability to create a spatial metaphor of time, there could be none of that sense of past, present and future which is the basis for this kind of internal narrative; and the ability, in Jaynes's view, is not unique to the human species in general but unique to humans of historical times, because it arose from the power of language to create metaphor. Here the themes of subjective time, narration, evolution and consciousness of the self all come together. Man cannot explain himself until he can narrate himself, and for this he needs inner space and inner time. For ancient man, it was quite different. Then it was the gods who did the explaining.

Before the period in which the *Iliad* was written, according to this hypothesis, much of human knowledge was of the kind that is the special province of semantic memory or of skill memory for such activities as dancing or playing a musical instrument, which cannot be contemplated introspectively. Knowledge of this sort is not set in a space-time context and is not the basis for defining personal identity. A great deal of it, more than is generally realized, is unconscious. Much of what we call "thought" is not really conscious at all. Often, decisions are made below the level of awareness, and then are made to seem rational when the conscious mind explains them to itself. The mind works faster than consciousness can keep up with. When we speak, we are not aware of searching for words or of stringing them together to form phrases and sentences; we are conscious only of meaning. "So we arrive at the position that the actual process of thinking, usually thought to be the very life of consciousness is not conscious at all and that only its preparation, its materials, and its end result are consciously perceived," Jaynes says. Reminiscence, on the other hand, based on episodic memory, is a highly conscious process which is beyond the scope of nonhumans, he maintains, and came into existence only in post-*Iliad* times.

The development of time consciousness in its highest form parallels the development of a sense of self. It may be no coincidence that in REM dreams, where the normal framework of time and space is weakened, a loss of the sense of self occurs. The normal waking feeling of having a definite, continuing identity which remains the same through time, is greatly diminished in this kind of dream as is also the feeling of free will. This may be due to the selective shutting down of the catecholamine systems, which are involved in attention and memory. Such a loss of the sense of self also occurs in patients who have had prefrontal lobotomy surgery. The prefrontal regions of the brain, at the very front of the frontal lobes, play an important role in purpose, intention and long-term goals, and when they are impaired, the patient prefers sameness in the environment and is completely absorbed in the events of the moment, making no plans or preparations. The horizons of the intellect shrink, and so do the horizons of time. In this condition there is no way to reminisce about the past or project into the future, so that conversation becomes stuck in the here and now, lacking all direction and continuity. In the end, communication virtually breaks down.

The diminished time horizons that result from smoking marijuana also are accompanied by a lack of a continuing personal identity, so that the self is experienced as strange and unreal. The drug tends to depersonalize the smoker, as it distorts subjective time. Frederick Melges, a Stanford University psychiatrist, who with Karl Pribram has investigated the brain's methods of processing temporal information, has shed some intriguing light on the connection between the sense of self and the sense of time by questioning people who undergo altered states of consciousness. Melges set out to learn more about the extent to which the breakdown, under the influence of marijuana, of the ability to organize subjective time is associated with the feeling of losing a sense of personal identity. He devised what he called a "temporal integration inventory," a list of statements that express the degree to which the structure of psychological time is disturbed under abnormal conditions. The list contained such judgments as these:

"My past, present and future seem like separate islands of experience, with little relation to each other."

"My past and future seem to have collapsed into the present, and it is difficult for me to tell them apart."

"My past, present and future seem all muddled up and mixed together."

"Things seem to be happening to me rather than my making things happen."

"My sense of direction seems to be impaired."

The people who took the test were shown the list and asked to say how well each statement described their own experience under marijuana, on a six-point scale, ranging from "not at all" to "extremely." They were also asked to respond in the same way to a second list, a "depersonalization inventory," containing such statements as "My body seems detached, as if my body and my self are separate," and "I feel like a stranger to myself."

The results showed that the disorganization of subjective time and the depersonalization of the self are closely linked. As each smoker's internal mental construct of a past, present and future became less coherent, he or she simultaneously became more depersonalized. There seems to be a uniform relationship between the two states. THC (tetrahydrocannabinol), the main ingredient of marijuana, enhances the feeling of living from moment to moment, making the experience of a time line running from past to present seem broken up, so that the past is unrelated to the present. The future appears remote and far off. This disruption of future perspective is most drastic at the time of peak drug effect, about one and a half hours after taking it, and is intensified by stronger doses.

In the Melges study, most people had a strong impression of being in the here and now and of giving little or no thought to the future. One person said, "I don't understand it. I don't. There seems to be no transition, no changes in time. It's just *it*. I'm just here now. I have no sense of time. I'm just floating along." Another could not think of the future, because he was unable to "keep an idea long enough—I can't remember my future." When pressed to give an example of a future event, he looked at a chocolate bar in his hand, and said, "I'm going to take a bite of candy." Before taking the drug, this same student had talked about plans to obtain a doctoral degree and his career goals for the next six years. A third person said he was "just savoring the present."

"When time becomes disorganized, so does the person," Melges and his colleagues conclude. "The personal past, present and future constitute a fundamental subjective framework through which an individual views and identifies himself."

The idea that subjective time is a construct of consciousness in which different mental systems play a collaborative role, and is closely linked to the very existence of a sense of the self, is relatively new in science, but is no stranger to philosophy. Such a doctrine forms a central thread of speculation in the writings of the existentialist philosophers. In fact, the work of the existentialists complements to a remarkable extent the experimental work carried out in the 1970s and 1980s by psychologists and medical researchers, who are beginning to understand the complex ways in which the components of body time and mind time are linked to form a single system, and how these components can uncouple and break apart.

Existentialism, an umbrella term for the work of a number of thinkers whose ideas are often in conflict, is remarkable for the importance it attaches to time, and in particular to subjective time. In fact, the writers of this school of philosophy have been criticized for giving too exalted a status to inner time, while relegating objective, public time to a minor role in human life. The viewpoint of existentialism also sits comfortably with the modern concept of biological clocks, since existentialists tend to define time in terms of human concerns, of when it is appropriate for an event to happen in the life of an individual. In the writings of Martin Heidegger and Jean-Paul Sartre, especially, subjective time is inseparable from the discovery of meaning, purpose and value in life. But discovery of this kind, which enriches existence, does not occur in a programmed way, by biological necessity. It is a highly conscious process which may be successful or may be unsuccessful. Unless the components of temporal intelligence are put together to form an integrated structure, subjective time will disintegrate into a meaningless succession of moments. There is no "time integrator" based on a mechanism built into the brain at birth, similar to the mechanisms that regulate breathing or seeing. Putting subjective time together requires a deliberate, resolute effort, a choice, an exercise of freedom on the part of each individual.

Existentialists insist on the finiteness of human life as a central fact of their philosophy and on the uniqueness of each person, who is born, lives and dies on a definite date in history and in a definite place in the world. Time and place, therefore, are of fundamental importance in the making of human individuality. What is more, episodic knowledge is of greater worth than semantic knowledge to this way of thinking. That is one reason why existentialism is opposed to the kind of universal humanism that seeks to identify with all mankind, instead of concerning itself with the experience of actual individuals. We cannot understand human beings, these thinkers insist, by considering them in the abstract, "semantically," drawing general conclusions about them as a species, stripped of a context of time and place. The radically temporal character of human existence makes it impossible to escape into a world of timeless, eternal ideas. Existentialists would agree with Bertrand Russell when he said, "I do not even wish to live with eternal things, though I often give them lip service; but in my heart I believe that the best things are those that are fragile and temporary," except that they would go farther and say that lip service is about as far as we can possibly go in our relationship with the eternal. There is no ejection seat that can fling us clear of history, out into timelessness. Part of the anguish of the human condition arises from a realization that we are stuck in what Søren Kierkegaard called "the straitjacket of existence."

One of the defining features of human beings, as distinct from other animals or lifeless objects, in the existentialist view, is the fact that we find various possibilities open to us, and it is only because we are finite creatures, radically temporal in our natures, that we are able to do so. In the work of Sartre, human consciousness, or what he calls the *pour-soi*, the for-itself, is always in the process of separating from the *en-soi*, the in-itself, which represents pure being. The *en-soi* is timeless and unchanging. It is the domain of objects that are simply there, self-sufficient and without meaning or value. The *pour-soi* is quite different. It is temporal through and through and, far from being always the same, is unstable and restless. Human consciousness is continually in rebellion against the timelessness of pure being, and that is an inescapable fact of human life. It is largely through the construction of subjective time by the *pour-soi* that existence acquires meaning and purpose.

The *pour-soi*, consciousness, is an organizer of time, a synthesizer. It creates a unified whole out of past, present and future, dominates these subsystems and gives meaning to them. Such a synthesis does not come readymade. It is not a given. But it happens, because human consciousness is able to rise above the present and the past and escape from clock time. Consciousness is a licensed rover, independent, free in time and free in space, a sort of psychological time machine dispersing itself in past, present and future simultaneously. The one-wayness of time which prevails in the physical world is no impediment to the *pour-soi*. It goes back and forth, unifying our tomorrows with our todays and yesterdays. Whereas the *en-soi* is stuck in the world of things as they are, consciousness can traffic in things that are not, but perhaps could be. Its spontaneity is not even shackled by the laws of logic and deterministic cause and effect. It has the power to doubt and imagine, and in order to imagine it must be able to escape from the world. Its nature is to conjure up possibilities. The *pour-soi* is master of the past and present, and more powerful than both, because it is more free. Thus the past is not the ruling influence in shaping an individual's life. By unifying past and future, consciousness modifies the past by giving it a new context. The content of our lives up to the present moment is a fact, and must stand as it is. But we can interpret and reinterpret our past from the standpoint of what we are and of our future possibilities. As a result, the meaning of the past is always being altered, much as in a novel later chapters make the reader see earlier ones in a different light.

Consciousness, in Sartre's scheme of things, is temporal, not because it dwells in a temporal world, but because its nature is to move, to thrust toward the future, where possibilities are. Since movement presupposes time, consciousness becomes time as a result of this very mobility. The *pour-soi* is not conscious first and then temporal afterward. It is temporal just by being what it is, and its temporality is not the same as universal time, the time of planets and galaxies, clocks and calendars, all of which impinge on us from outside. It is closer to what Einstein called "the free constructions of the mind." Temporality is not a frame or container for events. It is just the special moving, disengaging, unifying and synthesizing activity of consciousness, and such activity is as natural to the *pour-soi* as breathing is to the lungs or beating is to the heart.

Temporality, the synthetic whole, gives meaning to its

parts, which include the past and present. I sit at a desk with a piece of paper in the typewriter and an empty teacup at my elbow. A lamp shines on the desk. By itself, this situation has no meaning. It is simply a collection of objects, if the mind focuses solely on the present moment, on the now. Seen in this way, such a state of affairs might even seem absurd. Only if consciousness moves beyond the moment, out toward the possibility of writing a page, a chapter, a whole book, does my situation acquire meaning, because then the future, what is not, is brought in and placed alongside the present, what is. We can understand what is, Sartre emphasized, only through what it can be. Man is not the sum of what he has, but the totality of what he does not yet have.

Many novelists of his day, Sartre pointed out, distorted time in their works, so as to capture the pure present stripped of context. Time is desynthesized, taken apart. As a result, the worlds these writers portray are curiously lacking in meaning. Sartre said William Faulkner, for example "decapitated" time by robbing it of a future. Faulkner's stories show us events only after they have happened; we are always looking backward to what was and not forward to what will be. His characters are the sum of what they have been, not of what they might become. The future simply does not exist in Faulkner's novels, which may explain why he found time irrational and human life absurd. Without the unifying power of the *pour-soi*, its future-seeking mobility, meaning disappears, so that the absurdity Faulkner found is one he put there himself. Not that human life is not absurd, Sartre is quick to add. But it does not have that particular kind of absurdity.

It is an aspect of the freedom unique to human beings that subjective time does not have a uniform structure, but is put together in different ways by different individuals. Of course, if subjective time were to consist merely of a succession of instants, coming out of the future and vanishing into the past, and if only the span of immediate awareness, the specious present, had any psychological reality, it would no longer be strictly subjective but would be the same for everyone. The more closely a person approaches this state of living effortlessly in the present without any attempt to construct an integrated mental time structure, just letting moment succeed moment and day succeed day, the less authentic is that person's mode of being. It cannot be called "existence" at all in the sense of the word as the existentialist philosophers use it.

The secret of authentic existence is to integrate past, present and future into a unity. Each one of these three dimensions is different from the others and has its own special properties, but all can be fitted together to form a single temporal structure in consciousness. In the work of Martin Heidegger, who had a strong influence on Sartre, the three dimensions—or "ekstasies," as Heidegger called them—are seen as being interdependent. As long as the individual holds them together, time is not experienced as a mere succession of moments. Far from being a domain of nonexistence separate from the rest of a person's temporality, the future guides present choices which in turn enable the past to be reinterpreted, so that both the future and the past are part of a person's unfinished life. This unity of past, present and future Heidegger calls "timeishness."

Heidegger gives the highest priority to the future, because the future is the direction in which a person's possibilities lie, and human beings are above all creatures of possibility. Held in this way, past and future have a definite existence in a person's life, but they are held only with effort and resolution. Unless such an effort is made, the components become uncoupled and break apart; subjective time loses its authentic structure and disintegrates into a sequence of moments. Then the life of an individual might just as well be that of a stone or an insect, because it is the structure of subjective time that determines whether or not a person is fully human.

An important property of timeishness, however, is that it is finite and closed. Endless time, in Heidegger's view, is empty of meaning. He regards the public time of clocks as empty, for the very reason that it is public, is not "owned" by a single, unique individual, whose life-span has a beginning and an end. It is an endless succession of nows, belonging to anybody and nobody. Human consciousness cannot throw itself forward into the future indefinitely. At some point it comes up against the boundary of the individual's life, the limit of his life-span, and rebounds back. The limit returns the throw of consciousness as a tennis player returns a serve. Subjective time, and therefore being itself, has a circular structure. Man brings his possibilities back from the future toward himself, and these possibilities are constrained by present circumstances and past choices; by the unique, specific life as it is lived by a particular individual in his place and in his time. The individual is both ahead of himself and thrown back on himself, defined both by his possibilities and by the limits on his possibilities. In this respect man is very different from an inanimate object, whose possibilities are general, applying equally well to a whole class of objects, in any place and at any time. A rock, if pushed over a cliff, will fall, because that is the universal way of all rocks, no matter the history and circumstances of the particular rock in question. It is quite otherwise with a human being, whose possibilities are uniquely his own and not those of the human species in general, because they arise from his own timeishness, the special way in which he unifies his future, present and past. Heidegger turns upside down the notion that because a human existence is bounded by time it must be without meaning or point. Instead, he argues in a highly original fashion that endless being, like endless time, would eliminate meaning and purpose because it would lack the essential circular structure.

By its very nature, the sort of temporal unity created by human consciousness varies from one person to another and from one circumstance to another. Sometimes the future will be given greater emphasis at the expense of the past and present, leading perhaps to unrealistic strivings which take no account of the limits imposed by what is already given. There will be a preoccupation with the past, with its traditions and routines, or a dwelling in the present, the plight in which man most often finds himself. As John Mcquarrie says: "This dwelling in the past has, like the other cases, various ways in which it manifests itself. Always, however, it is typical of the man who, in common parlance, has no will of his own—the irresolute man, the man of bad faith, the scattered man."

Modern psychology has extended Heidegger's thesis, by showing that in order to have a *normal* human experience of time, the moment-by-moment span of the present must be fitted not only into a context of the subjective past, as provided by episodic memory, but also into a context of the subjective future. Unless both contexts are available, the sense of time is impoverished or distorted, and when that happens, the psychological balance of the individual is impaired. The experience of time, and personality itself, break down. In a set of experiments conducted by Bernard Aaronson, the contexts of past and future were systematically removed from consciousness by hypnotic suggestion. A hypnotized student was told: "When I wake you up, the past (or the future) will be gone. There will be no past (or future)." Aaronson found that when the past was removed, time estimates became distorted, and the student seemed to lose both inhibition and the ability to discriminate meanings. When the future was removed, a euphoric, semimystical state was experienced on waking in which everything seemed to occur in a boundless, immanent present. All anxiety vanished. The student spent his time savoring the present. When he was told that past and future had been taken away, however, a very different state resulted. The student became immobile, made no response when his name was spoken, and behaved as if in a catatonic stupor. On waking, he expressed great fear and relief, saying that he felt as if he had been reduced to a machine, being aware of everything that had happened under hypnosis, but simply registering it passively, as if he were a tape recorder. The memory of the experience "struck him with horror."

What is interesting about this result is that the student's condition during the time when he was deprived of both past and future was almost identical with the condition when only the present was removed. The elimination of the present also produced a catatonic state in which the student felt reduced to the level of a machine. All meaning was obliterated. The only difference was that in the absence of a present, the student's body became very rigid and when his name was called there was a mild fluttering of the eyelids. So severe were his symptoms that the experiment was cut short soon after it had begun. Evidently, when the contexts of both past and future are removed, the psychological effect is the same as when the present alone has ceased to exist.

In fact, some members of the family of mental competencies that collaborate in providing us with the experience of time past also play a role in forming a sense of time future. The contents of episodic memory, and therefore the structure of subjective time, are not recalled automatically and with a high degree of accuracy, as are the contents of semantic or skill memory, but are constructed consciously and sometimes erroneously. A fictional event invented in the present may be mistaken for a factual one belonging to the autobiographical past. Sometimes a memory of the past, like an imagined episode of the future, can be pure invention, and yet be accepted as true. It is the very firmness of our belief in an episodic memory, even if it is untrue, that suggests a reason why this particular type of memory arose in evolution, to provide us with a subjective certainty that is a spur to action. The famous Swiss psychologist Jean Piaget had a very precise and detailed memory of being kidnapped as a small child while in his baby carriage. He was able to recall the exact place where the kidnapping took place, the struggle between his nurse and the thief, the intervention of passersby and the arrival of the police. There was only one thing wrong with this recollection. None of it ever happened. The story, in its entirety, was made up by Piaget's nurse. When Piaget was fifteen, the nurse wrote a letter to his parents admitting that she had invented the entire incident, fabricating signs of a struggle, even inflicting scratches on her own face. Yet Piaget experienced the imaginary episode as a memory. He heard his nurse tell the fictitious story to his parents, who believed it to be true at the time, and out of it narrated a make-believe visual "memory" which never left him.

Imagination, belief, probability, logical inference, narrative power, all contribute to the construction of the past, as they contribute also to the construction of the future. The past is partly and the future wholly an invention of the mind. In both cases we use information that is true to generate information that may or may not be true, but is possible and plausible.

Until the 1970s, a behaviorist view of temporal intelligence prevailed, one which held that the mind is swept along on the flow of external, public time. For a behaviorist, the environment is paramount. Today, the individual is seen more as a subject than as an object of time. One is, or can be, an "author" of time, to use a word in vogue among psychoanalysts. That is to say, one must decide how moment-by-moment time is fitted into contexts of the past and of the future. The object of psychoanalysis has become, in Jerome Bruner's words, not so much to reconstruct the patient's life archeologically as to help him construct a more contradiction-free narrative of it. In his prizewinning novel, *JR*, William Gaddis depicts characters who fail to be authors of their subjective time and pay a price for it in the poor quality of their lives. In the novel, clocks are omnipresent. They represent the public time on which Heidegger placed so little value, because it belongs to everybody and therefore to nobody. Clocks symbolize time as moment-by-moment succession, free of context. Living in a disconnected now, the antiheroes and antiheroines of *JR* do not construct a context of the past and future, in which to narrate the stories of their lives in mental spacetime. For that reason they are existentialist failures. They are on the way to being in the condition of Dr. Aaronson's hypnotized student whose past and future were removed by hypnosis. The characters suppress, forget and lie about their own past lives and are unable to invent a future. Gibbs, the main character of the novel, can be the author neither of his own subjective time nor of a literary manuscript he sits over pointlessly, unable to make headway. He cannot go forward into fictional time any more than into personal time.

The Freudian theory of human behavior held that our present psychological state is determined by the past. In the light of what has been learned in the past few years about the structure of subjective time, that doctrine must be regarded as inadequate. In the present, the mind constructs and organizes the past and imagines a future, using the many different cognitive strategies that together make up temporal intelligence. The clocks of the body are coupled to form a system, and only when these couplings hold can the "meaning" of biological time be expressed. When they fall apart, the value and usefulness of temporal organization for the organism is diminished. The time of the body is irrelevant to the time of the world. In a similar way, the various components of psychological time must be integrated into a system if subjective time is to have meaning, and this requires some of the "resoluteness" mentioned by existentialist thinkers. It demands of the individual that he be an active participant in the formation of that internal timescape which plays no small part in the authorship of his own life.

TWENTY-ONE

"There's an Idea Missing"

In the existential philosophy of Sartre, "human nature," as we commonly understand the term, has no place. The reason for that omission is that Sartre does not emphasize structure in his theory of the psychological makeup of man. He argues that man is what he is precisely because of an absence of structure, because of a primordial emptiness at the core of his mental being. We must fill this inner vacancy in our consciousness, this gap, with acts, thoughts and perceptions which we are almost boundlessly free to choose. We do not have fixed natures which determine choices; rather our choices, freely made, constitute our nature. An interior space, when filled, is a human being's essence, but essence can be constructed only after the individual has come into existence, not in advance of that event. Essence is not handed down to us as a birthright, coded in the genes. Man is free, but freedom is not a property of his nature; it has no essence.

A mechanical device, such as a watch, is created from a preestablished concept, a blueprint, which guides the watchmaker in his work. The essence of a watch, the idea of its design, either on paper or in the mind of the craftsman, precedes its existence. The opposite is true of humans, Sartre maintains. For our species, existence comes first, and essence must be created as a individual's life proceeds. God was regarded by some philosophers as a sort of craftsman, creating human beings as a watchmaker builds a watch, according to a concept which existed prior to that creation. An idea of man was in the mind of God before man himself came into being, so that God knew exactly what he was doing when he made Adam. "Human nature," according to this doctrine, is exactly that blueprint residing in the divine mind. Even certain nonreligious philosophers in the eighteenth century retained the notion of human nature as a universal idea of which each individual person is a unique expression. Voltaire, though not a Christian, argued the case for theism by saying that the existence of a man implies the existence of a supreme intelligence, in the same way that a watch implies the existence of a watchmaker.

In a famous lecture, given in Paris in 1946, Sartre proclaimed that "there is no human nature, because there is no God to have a conception of it." He went on to say, "if indeed existence precedes essence, one will never be able to explain one's actions by reference to a given and specific human nature; in other words, there is no determinism—man is free, man *is* freedom." Mary Warnock comments: "In the nothingness which lies at the heart of human beings there is an endless number of possibilities. Since there is no Human Nature as such, there is no necessity for a man to determine himself in one direction rather than another. His possibilities include the possibility of answering 'No' to every suggestion, not only of what he should do, but also of what he should think, or even how he should describe and categorize what he perceives in the world. When a man really sees for the first time that this nothingness exists within himself (in other words, that he is free to do and to think whatever he chooses), he suffers Anguish. He is unable to bear the thought of his boundless freedom, and in order to escape from this anguish, he often adopts the cover of Bad Faith. This takes the form of pretending to himself that he is not as free as he actually is."

If existentialism rules out the notion of human nature, and deemphasizes structure, today's life sciences tend to move counter to that view. We have seen that modern biology, in its search for an understanding of the nature of life, has revealed the existence of more and more structure in the living system. The body is heterogeneous, richly organized in time as well as in space. It possesses an elaborate architecture, extending downward in scale into the invisibly small microcosm of the cell. A similar, "architectural" approach to the brain is taken by certain influential schools of modern psychology, especially those that study the brain as a processor of information. Cognitive science, as one such school is called, investigates the ways in which people think, learn, perceive and remember. It assumes the existence of various specific mental structures which are common to all human beings, and which encode, select, retain and transform information, of internal as well as of external origin, in different ways. There is a cognitive system, and it consists of various elements and components that play specific roles in the processing of information and have a certain amount of autonomy. One important feature of the system is that there are built-in limits on the amount of information certain components can handle, a notion highly unpopular with other schools of psychology which consider that the structure of the mind is determined by the structure of the information reaching it from the outside world. Cognitive scientists emphasize such things as the restricted capacity of attention and the fixed constraints on short-term memory, which in general cannot hold more than about seven items for longer than about 15 seconds. A seminal influence for this school was an essay published in 1956 by George Miller, who argued that the "magical number seven" limits many of our mental operations, such as the ability to distinguish phonemes from one another and to estimate numbers accurately, as well as the workings of short-term memory. The seven-item constraint seems to be incorporated into the design of our nervous system. A distinction must also be made between highly constrained components and systems at a higher level, the so-called central processor, which coordinates and directs the activities of perceiving, remembering and thinking. There is a definite architecture of the mind, universal to all humans, and the architecture is an aspect of what we refer to as human nature. It is implicit in the blueprint of DNA, the genetic message, as the blue-print of man was once supposed to be implicit in the mind of God.

Such an account of the structures of knowing is especially congenial to computer scientists. People who design machines that mimic the operations of the human brain have found that only by building in a large amount of highly specific structure in the form of components, circuits and programs is there any hope of making an artificial intelligence system that resembles human intelligence even remotely.

Psychologists who try to understand how it is that human beings know what they know do not all give equal emphasis to structure. Some talk more about function, stressing the "why" rather than the "how" of knowing. These include the behaviorists, who largely ignore inner structure and concentrate instead on the external conditions that modify behavior and on mental activity as responses to those conditions. A mental act occurs because it is reinforced or not reinforced by a stimulus from outside the organism.

For those who take structure seriously, however, it is clear that an important distinction must be made between elements of the cognitive system that are relatively free in their operations and those that are more fixed. It is the particular mix of free and fixed elements that determines the nature of the mental architecture. Certain structures process information in a highly specific way, and the freer parts of the brain are not able to interfere with those processes, but must accept the finished product on a take-it-or-leave-it basis. Among these fixed structures are the perceptual systems of seeing, hearing and language comprehension, and, of more interest for our purposes, a number of timekeeping mechanisms. The mental structures that keep track of time as it is going by, that monitor the present, appear to be more fixed than those that govern our sense of time as it has gone by, the sense of time's extended passage. Unfree elements say to the freer ones, in effect, "This is the way I represent the world to you, and if you don't like my representation, you can lump it, because I cannot represent it in any other way."

The theory of various types of psychological structures, some highly constrained and specialized, others much less constrained and more general in their operations, sheds fresh light on much of what has been said in preceding chapters about time and the human brain. The moment-by-moment sense of time is quite different from the ability to map and measure time past. Mental strategies by which the mind deals with the specious present and with split-second slices of time are not the same as those by which it is able to be aware of time's extended passage. An infant tunes in to the rapid time sequences of its mother's gestures and movements automatically, without thinking of what it is doing. At quite an early age, children are able to discriminate the time order of sounds in the stream of speech and the beat and rhythms of music. In fact, not only is the brain able to perceive the time order of the constituent sounds in a spoken sentence, it is unable *not* to perceive them in that way. If somebody coughs while another person is speaking, the cough floats free in time, because the nature of the perceptual structures involved is such that we are obliged to discriminate the time order of rapid speech sounds, and are obliged not to discriminate the time order of equally rapid nonspeech sounds. This is true for every human being, except a few who are disabled in special ways.

Quite different are the mental strategies used in comprehending extended time, having a perspective of past and future, and integrating both into the present. These are not automatic and fixed, but are fallible, idiosyncratic, voluntary and depend on learning and experience. A baby cannot manage such a task. In fact, moment-by-moment time discrimination and a sense of extended time are so distinct in the brain that it is possible for one system to fail while the other operates as usual. The Korsakoff patients who could not store information in episodic memory, and were therefore unable to estimate how many minutes, hours or days had gone by, still measured time as it was going by just as easily and as accurately as healthy people. The mental structures underlying the sense of moment-by-moment time, and those mediating the sense of extended time, seem to be genuinely distinct, and the difference is that the first are obligatory and specialized, while the second are free and general.

Jerry Fodor, a philosopher and psychologist at MIT, has introduced a theory that certain psychological structures which manipulate information in the brain are "modules," in the sense that they are to some extent self-contained and do not share information freely with other systems of the brain. Modules are like little autonomous computers. They may be capable of very elaborate and extremely rapid feats of computation, but their range of knowledge is restricted. Modules do not know as much as the brain as a whole knows. Other, "higher" centers of the brain, such as those involved in thinking, judging and imagining, are not restricted in this way and share information without hindrance. Modules are confined to certain specific domains of knowing, which means that there is a limited number of questions about the world that they can answer. The higher, nonmodular mental processes, "central systems," are able to range freely across domains. Central systems are denied access to the internal operations of the modules, and therefore must be content to work with the special representations of the world that modules provide. They cannot say "no" to what the modules give them.

Good examples of modules are the perceptual systems of seeing, hearing and language comprehension. The visual system is not like a camera, simply accepting an image of the world as it falls upon the retina of the eye. Unprocessed, an image of this kind is flat, wrong way up and highly unstable. The brain center of vision, which is located at the back of the head, reconstructs and interprets this dance of light so that we "see" a three-dimensional scene, where objects are stable, and on which the constancies of color, brightness and size have been imposed. The visual system ensures, for example, that people see rooms as rectangular and human faces as convex, even if the room is specially altered by psychologists so as to be crazy—one wall much higher than the others, the floor and ceiling sloping away at a mad angle, and the "faces" of the people in the room covered with concave masks. We are all of us compelled to "see something *as* something," in Roger Shepard's words. Looking at this drawing, which Shepard calls his "martini-bikini" figure

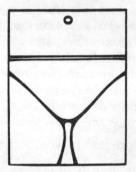

we see either a cocktail glass with an olive being dropped into it, or a not very prepossessing person of indeterminate gender wearing a swimsuit. The figure is ambiguous and sketchy, but the visual system of the brain is forced to make something of it, to recognize it as familiar, by means of its special kind of computations.

The language perception system has similar properties. If someone speaks and we attend, we are unable *not* to hear the sounds the speaker makes as anything but words, and words in a particular order. The overwhelming majority of people cannot hear speech as pure sound, even if they try very hard to do so. Our conscious, thinking minds cannot say "no" to the language comprehension modules of the brain, just as they must accept what the visual modules provide. There is no way that consciousness can get inside the modules and experience the raw data of the senses, unaltered by the computations of the perceptual systems. In general, the only way to avoid hearing speech sounds as words is to switch off attention or stop the ears with cotton wool, and the only way to see the world unconstructed is to shut the eyes.

The "higher" mental processes of thinking and imagining, learning and judging, are altogether more liberated than the modules of perception. Whereas we can perceive the world only in certain special modes, we can think about the finished products of perception, embellish them and manipulate them in many different ways. Dr. Fodor has suggested some games to play that illustrate the difference between the constrained, modular processes of perception and the free processes of thought. We are all bound, he says, to hear Shakespeare's *Hamlet* as sounds encoded as words, and words in a specific order. But now try *thinking* about *Hamlet*

> As a revenge play
> As a typical product of Mannerist sensibility
> As a potboiler
> As an unlikely vehicle for Greta Garbo

The possibilities are endless. We could imagine the play as if directed by Alfred Hitchcock, rewritten by Harold Robbins, or converted into an afternoon soap opera for television. In the same way, we see a brick as a unique object in space with a constant shape, size and color, manifestly a brick and nothing else, even if viewed from an unusual angle or in a bad light. But now think of sixteen different ways to use the brick. Again, there is no limit to the number of imaginary uses. (To raise the level of water in a bath during a drought? As a plain surface on which to write a check payable to the IRS, just to show that taxpayers can also be a nuisance?) And notice the contrast between trying to hear "To be or not to be" as an acoustic object and thinking about it as an acoustic object. The first task is almost impossibly difficult, while the second is easy and fun to do.

Thought is free to make the world different, new or strange, while perception is constrained to represent it always in the same fashion, presumably because in evolution it was important for survival that animals should have, in all places and at all times, reliable information about what is actually going on in the environment, not what they think is going on. "We have only the narrowest of options about how the objects of perception shall be represented, but all the leeway in the world as to how we shall represent the objects of *thought*," Fodor says. He goes on:

"No doubt there are *some* limits to the freedom that one enjoys in rationally manipulating the representational capacities of thought. If, indeed, the Freudians are right, more of the direction of thought is mandatory—not to say obsessional —than the uninitiated might suppose. But the quantitative difference surely seems to be there. . . . By contrast, perceptual processes apparently apply willy-nilly in disregard of one's immediate concerns. 'I couldn't help hearing what you said,' is one of those cliches which, often enough, expresses a literal truth; and it is what is *said* that one can't help hearing, not just what is *uttered*."

Modules are not as stupid as reflexes, because they can make inferences and perform dazzling feats of computation, operating actively on sense impressions rather than just accepting them passively. In fact, modules could be called intelligent, in a limited sort of way. But they are like reflexes in that they are set off automatically by certain kinds of stimuli in the environment. Light patterns falling on the retina of the eye start vision working, whether we want it or not, and likewise speech and music sounds trigger hearing. Modules are obligatory, whereas thought is voluntary. Also, modules are encapsulated; in other words, what we know may have absolutely no effect on what we perceive. For instance, we can look at an optical illusion in which two lines of equal

length are made to look unequal

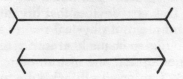

and even after being told that the lines are equal, having the trick explained to us, we *still* perceive them as being unequal. Visual perception is modular and is insensitive to information that may be in the possession of the thinking brain.

What central systems are good at most of all is reasoning by analogy. That is because they are not restricted to one particular mental domain, as the modular perceptual systems are. They share information freely. Central systems are global, whereas perception is local. The wiring of their neural circuits is thought to be diffuse rather than specific, unlike the circuits of modules, and might even be random. Information flows "every whichway" in central systems, with connections changing from moment to moment. Central systems bring the totality of what the mind knows to bear on the special representations of the world that the modules of perception furnish.

Modules do not reason by analogy. They cannot, because of the way they are built, being specific to one particular psychological domain. A module sets up boundaries. Analogies and metaphors, products of the central systems of the brain, break down boundaries, cross domains, talk about one thing in terms of something else. Metaphor (a word that means, literally, a "carrying over") carries a thought over from one psychological domain to another—say, from the sense of touch to the sense of hearing, or from smell to taste. A violin may sound silky, and a rose smell sweet. Space words may be used to talk about time, and concrete objects be used to represent abstract concepts: "A mighty fortress is our God." Metaphor is not limited to poetry but pervades everyday talk, and seems to be necessary, fundamental to human thinking at the higher levels. When the language faculty is impaired by disease, metaphors are the first to go. Analogies and metaphors are not mere embellishments to adorn speech. They are devices that bridge the gap between the outer environment of physical reality and the inner environment of thought. Hannah Arendt asserts that there are not two separate worlds, one of true being and the other of mere appearances, as some philosophers have supposed, because metaphor unites the two worlds. Metaphors "are the threads by which the mind holds on to the world even when, absent-mindedly, it has lost direct contact with it, and they guarantee the unity of human experience. Moreover, in the thinking process itself, they serve as models to give us our bearings lest we stagger blindly among experiences that our bodily senses with their relative certainty of knowledge cannot guide us through."

Metaphors have helped scientists to make important discoveries through their power to link the domain of the world we can see and touch with the domain of ideas and theories. Einstein, who gave us a new way of thinking about space and time, drew extensively on vivid, mind's-eye images of direct physical experience before expressing his discoveries in words and mathematical symbols. At the age of sixteen, Einstein thought about what he might see if he were able to ride alongside a beam of light at the same speed as the light as it moved through space. On this imaginary journey, he decided, he would see a "spatially oscillatory electromagnetic field at rest," a paradox which suggested that the laws of electromagnetics would not be the same for an observer at rest as for an observer moving at the speed of light. Such freedom of the human mind to think about the world in unusual ways is remarkable in comparison with the much more limited scope of perception.

The ordinary human consciousness of time as past, present and future is a mental process which moves across psychological domains, and in that respect is not unlike scientific discovery. That is what makes it radically unlike the consciousness of time as it is passing. The automatic, immediate detection of rapid temporal patterns in the environment is a "perceptual" act rather than an act of thought, and as such is modular, obligatory and local. This timekeeping faculty of the brain operates in its own right, without interference from the central systems, which are voluntary and global. The chief role of perception is to keep the individual aware of local space and time, not what is hundreds of miles away or months and years in the past. Central systems are able to deal with the faraway in space and with the distant in time. Since modules are insensitive to what the thinking brain knows, the well-known effect of time distortion, which makes an hour by the clock seem an eternity in the dentist's waiting room, but rapidly fleeting if spent in the theater watching an enthralling play, might be somewhat akin to the optical illusion shown above, where two lines exactly equal in length appear to be unequal because their pictorial contexts differ. The central systems of the brain know that a tedious hour and an exciting hour both contain sixty minutes, just as they know that the first line in the optical illusion is not shorter than the second line, though perception continues to represent it as shorter.

The mental structures that mediate awareness of the specious present are presumably less free than those that underlie awareness of the past and future. Short-term memory, which holds information about the "now," appears to be a classic instance of an encapsulated system limited to a specific psychological domain. There is evidence that short-term memory and long-term memory are distinct structures in the architecture of knowing, the first kind strictly limited in its capacity, less influenced by context and organizing its contents in time order rather than according to meaning, as long-term memory does. Short-term memory, in a word, looks more modular than long-term memory, where context is important and central systems would therefore be expected to play a role.

In order to have what we could properly call a consciousness of time, the brain goes beyond the raw, "perceptual" impressions of time and crosses psychological domains. We make use of analogy, imagery and general knowledge. We think and reason about time, make metaphors of it, in order to be at home in a temporal world. That is an everyday necessity, not a luxury or a frill; since we cannot depend on mere impressions, we must construct a past and future using any mental strategy that works. Tuning in to the tempo of a conversation, discriminating the time order of music or speech sounds, is akin to hearing *Hamlet* as words and sentences, whereas constructing a personal past and future is a mental process more like thinking of *Hamlet* as a vehicle for Greta Garbo. These two types of "temporal intelligence" are profoundly different.

The psychologist William Friedman believes that as many as three distinct cognitive systems underlie even something as seemingly simple as a person's knowledge of the months of the year. The first system is based on a list of the names of the month, arranged in order from December to January, which is how a computer typically stores information. In this form, the list is difficult to recite backward from December to January. The second system works with images. The months are actually spread out in mental space, in the form of an oval or circle, in lines, rows or columns, so that the months can be read off backward or forward, as if displayed on a calendar. Images of time tend to be personal and idiosyncratic. Reggie Daker, a character in John Buchan's time-travel novel, *The Gap in the Curtain*, who from childhood had been in the habit of seeing abstract things in concrete form, visualized each day of the week as a different shape. "Monday was a square and Saturday an oval, and Sunday a circle with a segment bitten out." (For Reggie each of the Ten Commandments had a special color. The third was dark blue and the tenth pale green with spots.) The third type of mental strategy underlying awareness of the months is an associative network in which the names of the months are linked to many other concepts and propositions, such as the fact that March is usually windy and April is the start of the baseball season.

Images of time are among the many kinds of analogies or metaphors that the central systems of the brain are free to construct. Thinking about time as if it were space has often been regarded as a vice, a weakness, impeding progress toward a correct theory of time. But if the making of analogies is somehow central to the way human beings understand things, then it is only to be expected that we should think about time in terms of space. The philosopher Henri Bergson vehemently deplored such a mixing of domains. Yet it has not escaped notice that Bergson himself wrote about time in spatial metaphors, describing duration variously as flowing, growing, unfolding, surging, undulating and winding up like a thread into a ball. Not surprisingly, this led him into paradox. But Bergson's lapse was almost unavoidable, since language is riddled with examples of space words used to refer to time. The analogies are built into the way we speak and write. Children seem not to be able to handle temporal order until they have mastered spatial order, and space words tend to be used spontaneously before time words. Does language corrupt our thinking about time, or is language replete with spatial metaphors for time because that is how the central systems naturally proceed? Julian Jaynes answered "yes" to the latter question in the most unflinching terms when he said: "You cannot, absolutely cannot think of time except by spatializing it."

The fact that time cognition is under the control of the central systems, whereas time "perception"—which is confined to the present, often to split-second slices of the present may help to explain why the human consciousness of time has remained a profound enigma for so long. St. Augustine's celebrated and overquoted remark in Book Eleven of the *Confessions*, "What, then, is time? If no one asks me, I know what it is. If I wish to explain it to him who asks me, I do not know," admits to a perplexity which can still be seen in the writings of philosophers and psychologists today, fifteen centuries after St. Augustine.

One reason for the lack of a satisfactory theory to explain psychological time is that the operations of the central systems themselves are deeply mysterious. Dr. Fodor states categorically that even today nobody knows anything about how analogical reasoning works. The more global a psychological process is, the less it is understood, and a very global process, such as the making of metaphors, including metaphors of time, is not understood at all. Local, modular, perceptual structures may be just as complex as central systems, perhaps more complex, but they are less mysterious. Our ignorance of them is not so glaring. That is why scientists have been more successful in building and programming machines that mimic human acts of perception than in designing ones that can think in ways that people think. In fact, the outstanding example of computers imitating thought is the success of expert systems, which can diagnose an illness or give a legal opinion. Expert systems look very clever and sophisticated, but they operate in domains of knowledge that are carefully restricted to specialized fields such as medicine and law. They store a vast amount of information, but the information is not global. It is confined to a particular subject and is manipulated by the computer according to the rules of standard deductive logic.

We are unable to build truly intelligent computers, Dr. Fodor thinks, because we do not understand yet how global as opposed to local systems of the brain work. "You could have predicted several years ago what we are actually getting in artificial intelligence today," he said. "Namely, machines which can reason intelligently in very limited domains. As soon as you ask for a computer system capable of the sort of reasoning where the domains of application are not restricted in advance, it just can't be done. We are not talking about some insoluble metaphysical mystery here. I just think there's an idea missing, and until someone comes up with an idea we are not going to build machines that think in a truly human way."

The failure to mimic analogical thought may also be the reason no machine has been built which has a subjective sense of extended time.

"The main problem in psychology is exactly the same as the main problem in the philosophy of science," Dr. Fodor said in an interview. "That is, it is very hard to understand how you can solve problems where there is an unrestricted amount of data that can apply to the solution of the problem. In the perceptual systems of the brain there are definite, built-in constraints on the kinds of information that can be brought to bear on the problem of knowing the world. The visual system, for example, might make a hypothesis about what it sees, and confirm or refute that hypothesis by searching the available data, and the class of data to which the system has access is restricted ahead of time. You can imagine, roughly, how that search is carried out, and mimic it with a computer program. But in the case of central systems, which seem to have free access to an unlimited amount of data, and depend on general considerations of economy and simplicity and symmetry and so forth in the solving of problems, there just aren't any computer models. The information is too diffuse. We don't have a general notion of relevance, a notion that tells us when we should consider a hypothesis or a piece of data relevant to the problem at hand."

In Dr. Fodor's opinion, this is exactly the reason we have trouble understanding how confirmation works in science, where the background data is unlimited. "Speaking crudely," he says, "everything that a scientist knows is relevant to determining what else he ought to believe. In principle, our botany constrains our astronomy, if only we could think of ways to make them connect."

An important feature of central systems is that they are voluntary, whereas the modular perceptual structures start to crank out their special computations willy-nilly, whenever they are exposed to appropriate messages from the environment. Being voluntary, central systems need not be used indiscriminately under any and all circumstances, or they may be partially used, so that their rich powers of generating knowledge, including knowledge of time, are not exploited to the full. Time "cognition" appears late in evolution, is late developing in the life of the individual, and is located at a higher level of the nervous system than time "perception." It alters as a person grows up; it is influenced by culture and education. The spontaneous ability of the brain to lock on to the real time of speech and music does not change a great deal as a person matures. Toscanini, for example, hardly varied his tempo throughout his long career as a conductor. The stability of his beat over the years is almost uncanny. Studies of tape recordings of two performances of the Brahms-Haydn Variations, ten years apart, which Toscanini made with the NBC Symphony Orchestra, show that the performances differ in their total length by less than half a second, a discrepancy of one part in two thousand. Toscanini wrote metronome marks into the score of the music, but his own timing was vastly more accurate than the metronome marks. Only in case of one of the Variations, where Toscanini had changed his conception of how the music should be conducted, was a difference in timing significant. It is unlikely that the great maestro had a conscious "memory" of his tempo of ten years earlier. He must have produced the tempo unconsciously, from an inner pulse.

By contrast with this innate timekeeping ability, time cognition develops gradually in a person. It evolves and changes as the years go by. Spatial images of time order are understood by children at about the age of four, but images of the relative lengths of different time intervals are usually acquired later, at about seven. Such images tend to become more abstract and flexible, less concrete and literal, as the child grows up. Until the age of about eight, children use mental lists to represent the order of the months of the year. Lists are typical of the way digital computers store and retrieve information. Between eight and ten, however, children shift from lists to images such as lines and circles. After age ten, a child is able to construct an image of time for a particular purpose, as opposed to simply storing a set of fixed images. It is at this stage that a profound advance in time perspective occurs, enabling the child to make mental models of long stretches of historical time, to "see" temporal landmarks over the span of centuries in the mind's eye, and thus to date events in a serial order. The ability to make a metaphor of historical time as a stream flowing onward does not usually appear until about twelve or thirteen, when a feeling for narrative may emerge in its full form. Before that age, history is represented in the mind as a series of still pictures, like a slide show as opposed to a movie.

But such nonmodular ways of thinking about time, ranging freely across psychological domains, do not develop automatically, as if by an innate program, which invariably unfolds in the same way and at the same rate in every individual. They depend on cleverness and attention and effort, and are linked to IQ. Even after years of being taught history in high school, a first-year university student may still have little better than a child's conception of the past, seeing it as a jumble of chaotic facts and a smattering of dates. Knowledge of time, in the rich sense, is an important element of what we commonly refer to as intelligence. Paradoxically, even though memory is an essential element of time cognition, much of what we are able to remember of the past is determined by what we know about time.

If the world of "true being" and the world of appearances are not separate, because metaphor connects them and makes them one, can the same be said of the two worlds of body and mind? The answer seems to be yes. The body, like the mind, makes metaphors of the temporal aspects of the external world. It internalizes predictable time schedules, such as that of daylight and darkness. Circadian clocks are a kind of metaphor, because they "carry over" the day-night rhythm, caused by the spinning of our massive and solid planet on its axis, into another domain, the fragile domain of the living system. Actual, geophysical rotations are represented in the body in terms of cells and tissues and nerves. Biological analogies of this kind are fairly complex. As we have seen, the circadian system contains models of the 24-hour cycle which distinguish between subjective day and subjective night, and the distinction is built into the structure of the clocks. Time cues in the form of light signals from the outside world have different effects on the clocks, depending on what time of subjective day or night it happens to be when a light signal arrives.

Biological metaphors of time and psychological metaphors of time are different in many respects but similar in others. In one special case, that of the week, it appears that the mind has constructed a metaphor that carries over experience, not from the world of physical objects and events, but from the domain of physiology to the domain of conscious thought. Instead of internalizing a cosmic cycle, the brain externalizes an innate body rhythm. The week, once regarded as a purely arbitrary creation, a cultural artifact which rests on convention alone, seems to be rather a mental representation of a biological structure.

Body clocks are "modular," in the sense that they are highly specific, innate, involuntary and restricted as to the amount and kind of information that is available to them. The clocks do not respond indiscriminately to external cycles of any and every sort, but only to those with a particular frequency. The most prominent of these frequencies are 24 hours, a week, a month and a year. Clocks sensitive to other frequencies may yet be discovered as research into the structure of biological time goes forward at a rapid pace, but there is no reason to suppose that for every imaginable frequency in the environment there is a body clock designed to synchronize with it. Thus the clocks, though complex, are highly constrained. In this respect they resemble the locked-in mental structures that mediate time "perception."

Yet the time metaphors built into the body and the time metaphors made by the mind have this in common: they both help the individual to function successfully in the context of his environment. The body has a "close and wise" relationship with the world, Claude Bernard said more than a century ago. Today, we are finding out that the relationship is wiser than Bernard himself could have surmised, because he did not know the extent to which the body's timetable is organized so as to be appropriate to the timetable of the environment, and appropriate not to the present moment alone, but also to the future. A circadian rhythm, which is a metaphor of the day-night cycle, enables the body to anticipate what the world will do next, as well as to keep track of what the world is doing now. The body is put into a chemical state which makes it ready for the waking day before the day has actually broken.

The brain, too, is wise in this respect. It analogizes external time in different ways, so that the individual can tune in to the rapidly fleeting temporal patterns of human communication which are so important a part of the human environment, and have an automatic, effortless awareness of the present. Yet the brain also has the power to construct knowledge of the extended past and ideas of the future. Without the ability to tune in to the present, we could not communicate, and without the power to model the past and future, as opposed to simply memorizing the facts of the past, we could have no subjective sense of time, and therefore no art, no history and no self-consciousness except of the most limited and impoverished sort.

The body, like the mind, is not a complete instrument for manipulating time. Structures have evolved which respond to specific temporal patterns in highly constrained ways, while relatively unspecialized systems of the brain construct knowledge of time in a more general, but less reliable and uniform fashion. The "time sense" is incomplete, and in every species it is incomplete in a different way. To define the incompleteness in a particular species is to define the temporal identity of that species, and human beings are no exception to the rule. The mind cannot know everything. Man is the supreme generalist of the animal kingdom, but he still specializes in particular modes of cognition. Built-in psychological constraints limit his scope. They put a crimp in his freedom. On the other hand, the constraints enable him to enter into a wise relationship with the world, as a birthright, and without needing to be exceptionally lucky or bright or driven. In that sense, the obligatory limits on freedom, dictated by man's evolutionary past, can be said to increase his freedom. We can hardly call a person free who cannot understand a spoken sentence or comprehend music, or synchronize the rhythms of his body with the rhythms of his environment. Yet all those abilities are made possible by specific constraints which are architectural, part of the temporal identity, the special kind of incompleteness, of the human individual.

Sartre saw man's freedom as potentially limitless, not

because man is unconstrained, but because he can, if he chooses, say "no" to the constraints. Man's biological and psychological givens, the circumstances in which he finds himself, need not determine what he is or what he makes of his life, because the individual decides what meaning the givens will have. What we know today of the structure of biological and psychological time compels us to question that doctrine. The time schedules of the body are specified by the genes, not by individual choice, and their "meaning" is to be found in the evolutionary strategies that produced them millions of years ago. Biological clocks do not leave the body free to be in any state at any time. Our biology is not the same at noon as it is at midnight, and that rule applies whether we like it or not. Some body rhythms constrain more tightly than others, and some are more easily overridden or ignored than others by the higher systems of the brain. But there comes a point at which the conscious mind is not free to say "no" to the body's timetable. The sleep-wake rhythm may be a forgiving tyrant, but it is a tyrant just the same, and the cortisol rhythm is perhaps more tyrannical than forgiving. Similarly, the modular systems of the brain constrain the manner in which we perceive the world. They perform their computations autonomously and always in the same way, and the thinking brain has no choice but to work with what they provide.

It is hard to avoid the conclusion that the essence of an individual is partly given and partly constructed, fixed as well as free, and that includes his temporal essence. Human nature is heterogeneous, because the architecture of body and mind is heterogeneous, a mix of the obligatory and the voluntary, the innate and the learned, the specific and the general, the modular and the nonmodular, clock systems and mental constructs. In each of those antitheses, the first term is less mysterious than the second because it is less free. We can explain the fixed features of the time sense, but still grope in mists of uncertainty to understand the free ones; when we do, we shall arrive at a theory, not of the time sense alone, but of human nature in its highest manifestations.

NOTES

INTRODUCTION

18
Time like a mermaid. Jean-Paul Sartre, 1943. *Being and Nothingness*. New York, Pocket Books, p. 208.

19
Einstein wondered aloud. Jean Piaget, 1969. *The Child's Conception of Time*. Translated by A. J. Pomerans. London, Routledge and Kegan Paul, p. ix.

ONE: THE CLOAKROOM TICKET AND THE OVERCOAT

23
Einstein on beef tea and overcoats. Quoted in Hilaire Cuny, 1963. *Albert Einstein, The Man and His Theories*. Translated by Mervyn Savill. Jacques Ahrweiler, ed. London, Souvenir Press, p. 129.

24
"Out yonder there was this huge world." In Paul Arthur Schlipp, ed., 1949. *Albert Einstein: Philosopher Scientist*. La Salle, Illinois, Open Court Press, p. 95.

24
"the flow of time has been abolished." Thomas Gold, 1977. "Relativity

and Time." In Ronald Duncan and Miranda Weston-Smith, eds. *The Encyclopaedia of Ignorance*. New York, Pocket Books, p. 100.
24–25
"an unbridgeable gap." Sir Alan Cottrell, 1977. "Emergent Properties of Complex Systems." *Ibid*. pp. 134–135.
25
"to 'physicalise' space and time." Alexander Moszkowski, 1972. *Conversations With Einstein*. Translated by Henry L. Brose. London, Sidgwick and Jackson, p. xvi.
25
"time and space are modes by which we think and not conditions by which we live." Quoted by John A. Wheeler, 1982. "Bohr, Einstein and the Quantum." In Richard Q. Elvee, ed. *Mind in Nature*. San Francisco, Harper and Row, p. 22.
27
"time came into being with the heavens." Plato, *Timaeus*. Translated with an Introduction by H.D.P. Lee, 1965. Baltimore, Maryland, Penguin Books p. 51.
27
Primordial chaos timeless but not spaceless. Francis M. Cornford, 1939. *Plato and Parmenides*. Indianapolis and New York, Bobbs-Merrill, p. 18.
31
Bruno "imprisoned on account of his wicked words." Vatican manuscript Urbane 1068 (Doc. Rom. xxxi and xxxii). Quoted in Dorothea Waley Singer, 1950. *Giordano Bruno, His Life and Thought*. New York, Henry Schuman.
31
"separation of time from its physical content." Milič Čapek, 1973. "Time Measurement." In Philip P. Wiener, ed. *Dictionary of the History of Ideas*. New York, Charles Scribner's Sons, Vol. IV, p. 392.
32
"this grand book, the universe." Galileo Galilei, 1623. *The Assayer*. Rome. Quoted by Stillman Drake, 1957. *Discoveries and Opinions of Galileo*. New York, Doubleday and Co., pp. 237–238.
32
"sprightliness" of judgment. Galileo Galilei, 1632. *Dialogue Concerning the Two Chief World Systems*. Translated by Stillman Drake, 1953. Berkeley and Los Angeles, University of California Press, p. 301.
32
Galileo forced to revise his approach. Alexander Koyré, 1978. *Galileo Studies*. Translated by John Mepham. New Jersey, Humanities Press, pp. 67–78.
32
"rid himself of the old circular obsession." Arthur Koestler, 1968. *The Sleepwalkers*. New York, Penguin Books, p. 483.
33
They are "quacks." Isaac Barrow, 1735. *Lectiones Geometricae*, Lecture 1. Translated by E. Stone. London, p. 4.
33
"for time has length alone." Isaac Barrow, 1735. *Ibid*. p. 35.

34

"Absolute, true and mathematical time." Sir Isaac Newton, 1686. *Principia*. Translated by Andrew Motte, 1971. Berkeley, University of California Press, p. 6.

34

Newton "better aware of the weaknesses." Albert Einstein, 1934. *Essays in Science*. New York, Philosophical Library.

35

"By refusing to accept an axiom." In Dimitri Marianoff and Pamela Wayne, 1944. *Einstein, An Intimate Study of a Great Man*. Garden City, New York, Doubleday Doran and Company, Inc.

38

"Time does not today stand in splendid isolation." John A. Wheeler, 1979. In G. Toraldo di Francia, ed. *Problems in the Foundations of Physics*. *Proceedings of the Course LXII*. Amsterdam, New York, Oxford, North-Holland Publishing Company, p. 469.

38

Spinoza's influence on Einstein. John A. Wheeler, 1980. "Beyond the Black Hole." In Harry Woolf, ed. *Some Strangeness in the Proportion*. Reading, Massachusetts, Addison-Wesley Publishing Co., Inc., p. 354.

39

"the biggest blunder of my life." Quoted in George Gamow, 1970. *My World Line*. New York, Viking Press.

39

"If you don't take my words too seriously." In Philipp Frank, 1957. *Philosophy of Science*. New York, Prentice-Hall, p. 123.

TWO: THE UNIMPORTANCE OF TIME

44

"*Milieu intérieur*." In 1853 the French histologist Charles Philippe Robin had written the phrase "*Milieu de l'Intérieur*," but he did not develop the concept. Bernard's first recorded use of the term appears in a lecture he gave as the occupant of a newly created chair of physiology at the University of Paris on December 9, 1857, the first in a series entitled "Lectures on the Physiological and the Pathological Alterations of the Liquids of the Organism." In this lecture Bernard said that in living beings, by contrast with inanimate bodies, there is a spontaneous organic evolution which, although it needs the external environment to manifest itself, is nevertheless independent of that environment in its course because "in the living being, the tissues are, in reality, removed from direct external influences and protected by a true internal environment (*milieu intérieur*), mostly constituted by fluids circulating in the body." In L. L. Langley, ed., 1973. *Homeostasis, Origins of the Concept*. Stroudsburg, Pennsylvania, Hutchinson and Ross, Inc., p. 85. Some details of Bernard's life are drawn from Eugene Debs Robin, ed., 1979. *Claude Bernard and the Internal Environment, A Memorial Symposium*. New York, Marcel Dekker, Inc.

44

"a confusion of dispersed limbs." J.M.D. Olmstead and E. Harris Olmstead, 1961. *Claude Bernard and the Experimental Method in Medicine.* New York, Collier Books.

45

"seemed to have eyes all around his head." Paul Bert, quoted by J. Tarsis, 1968. *Claude Bernard: Father of Experimental Medicine.* New York, Dial Press.

45

a shorthand expression of contempt. Fyodor Dostoevsky, 1881. *The Brothers Karamazov.* Translated by Constance Garnett, edited and with a foreword by Manuel Komroff, 1957. New York, New American Library, p. 605. "Enraged by the tone in which Rakitin had referred to Grushenka, he [Dmitri] suddenly shouted 'Bernard!' And when, after Rakitin's cross-examination, the President asked the prisoner if he had anything to say, Dmitri cried loudly: 'Since I've been arrested, he has borrowed money from me! He is a contemptible Bernard and opportunist, and he doesn't believe in God: he fooled the Bishop!' "

46

in through the door and out by the chimney. J.M.D. Olmstead and E. Harris Olmstead, 1961. *Claude Bernard and the Experimental Method in Medicine. Ibid.* p. 68.

46

Henri Bergson maintained. Henri Bergson, 1975. *The Creative Mind.* Totowa, New Jersey, Littlefield, Adams and Co., pp. 201–208.

47

"in a kind of hothouse." Claude Bernard, 1878–79. "Lectures on the Phenomena of Life Common to Animals and Vegetables." Second Lecture. In L.L. Langley, ed., 1973. *Homeostasis, Origins of the Concept. Ibid.* p. 146.

47

"participates in the universal concert of things." Claude Bernard, *Ibid.* p. 130.

47

Three categories of life. Claude Bernard, *Ibid.* p. 129–147.

49

"Walter, be good to the world," Jean Mayer, 1965. "Walter Bradford Cannon—An Autobiographical Sketch." *Journal of Nutrition* 87:1–8.

49

"Homeostasis." First used by Walter Cannon, 1926. "Physiological Regulation of Normal States: Some Tentative Postulations Concerning Biological Homeostatics." In Jubilee Volume for Charles Richet, Paris, pp. 91–93.

49

"economy is secondary to stability." Walter Cannon, 1939. *The Wisdom of the Body.* New York, W. W. Norton and Co., Inc., p. 317.

53

free to seek "unessentials." Walter Cannon, 1939. *Ibid.* p. 323.

54

man needs a stimulus from the environment. Arnold Toynbee, 1946. *A*

Study of History. D.C. Somervell, Abridgement of Vols. 1–6. New York, Oxford University Press, pp. 69–70.

55

loosening of the calendar constraint on mating behavior. Ernest Becker, 1971. *The Birth and Death of Meaning.* New York, The Free Press, p. 11.

56

An entire theory of color vision. Edwin M. Land, 1959. "Experiments in Color Vision." *Scientific American* 200(5):84–99.

56

If the world were to turn strange on us. P.W. Bridgman, 1950. "The Operational Aspect of Meaning." *Synthese* 8:255–257.

57

Bertrand Russell, writing in prison. Bertrand Russell, 1971. *Introduction to Mathematical Philosophy.* New York, Simon and Schuster, p. 205. Thanks to the kindly intervention of Arthur Balfour, Russell was placed in the First Division of the prison system, which meant he could read and write as much as he liked, as long as he did not produce any pacifist propaganda. Once, while reading Lytton Strachey's *Eminent Victorians,* Russell laughed so loudly that a warder came to remind him that prison was a place of punishment. Bertrand Russell, 1951. *Autobiography.* Vol. 1. New York, Bantam Books, p. 29.

60

"stream of consciousness." William James, 1890. *The Principles of Psychology.* Vol. 1. New York, Dover Publications, p. 239.

60

"We do not care." William James, 1890. *Ibid.* p. 460.

60

"not all psychic life need be assumed." William James. *Ibid.* p. 460.

THREE: THE IMPORTANCE OF TIME

64

the first definitive version of the "cell theory." Theodor Ambrose Hubert Schwann, 1839. In R. Wagner, *Lehrbuch der Physiologie.* Leipzig, 1:139–142. Schwann defined a cell as a "layer around a nucleus," which could differentiate itself. He said that a living being is formed, not by some preexisting idea, but by laws of necessity which are intrinsic to biological matter. Life is different from nonlife only in the way in which the parts are combined. Also Jacob Mathias Schleiden, 1838. "Beitrage zur Phytogenesis." *Müller's Archiv* 137–176.

69

One version of the Chanticleer story. Edmond Rostand, 1913. *The Story of Chanticleer.* Adapted from the French by Florence Yates Hann. London, William Heinemann.

70

Chaucer on Chanticleer. Chaucer, *Canterbury Tales.* "The Nun's Priest's Tale."

> Wel sikerer was his crowing in his logge
> Than is a clokke, or an abbey or logge

By nature knew he ech ascensioun
Of equinoxial in thilke toun;
For what degrees fiftene were ascended
Thanne crew he, that it mighte nat be amended.

70

Franz Halberg coined the term "circadian." Franz Halberg, E. Halberg, C.P. Barnum and J. J. Bittner. In R. B. Withrow, ed. 1959. *Photoperiodism in Plants and Animals.* Washington, D.C., American Association of the Advancement of Science, p. 803.

72

A certain type of diatom. John D. Palmer, 1975. "Biological Clocks of the Tidal Zone." *Scientific American* 232(2):70–79.

72

the mind internalizes laws of motion. Roger Shepard, 1984. "Ecological Constraints on Internal Representation: Resonant Kinematics of Perceiving, Imagining, Thinking and Dreaming." *Psychological Review* 91(4):417.

73

"atoms dancing about in space." Roger Shepard, 1978. "The Mental Image." *The American Psychologist* 33:125–137.

75

"seemingly mad" independence from feedback. Irving Zucker, 1983. "Motivation, Biological Clocks and Temporal Organisation of Behaviour." In Evelyn Satinoff and Philip Teitelbaum, eds. *Handbook of Behavioural Neurobiology.* Vol. 6. New York, Plenum Press, p. 16.

75

About weekly rhythms in mermaid's wineglass. Franz Halberg, 1983. "Quo Vadis Basic and Clinical Chronobiology: Promise for Health Maintenance." *American Journal of Anatomy* 168:543–594.

76

About weekly rhythms in humans. Franz Halberg, Erna Halberg, Francine Halberg, Julia Halberg, Charles Leacha and Edapallath Radha, 1984. "Circaseptan (About 7-Day) and Circasemiseptan (About 3.5-Day) Rhythms and Contributions by Ladislav Derer." Unpublished ms., summary of lectures given in Bratislava, October 8, 1984.

78

"One of the most distinctive features of the week." Eviatar Zerubavel, 1985. *The Seven Day Circle.* New York, The Free Press, p. 9.

78

"The fact that the Israelites." Willy Rordorf, 1968. *Sunday.* Translated by A. A. K. Graham. Philadelphia, Westminster Press, p. 18.

79

the social week may entrain the biological week. Franz Halberg, interviews with the author.

FOUR: A TEMPORAL IDENTITY

84

"the system does not simply follow the environment's behavior." Gerald Weinberg and Daniel Weinberg, 1979. *On the Design of Stable Systems.* New York, Wiley, p. 178.

85

"Larks" and "owls." Charles M. Winget, Charles W. DeRoshia and Daniel C. Holley, 1985. "Circadian Rhythms and Athletic Performance." *Medicine and Science in Sports and Exercise* 17(5):498–516.

86

In Russia, researchers. S. I. Stepanova, 1977. *Problems of Space Biology: Current Problems in Space Biorhythmology.* Moscow, Izdatel'stvo Nauka.

86

This relativity principle of biological time. Hugh Simpson, 1965. "Daily Adrenal Rhythm in Equatorial Amerindians." *Journal of Endocrinology* 32:179–185.

88

Cisplatin and adriamycin. Judy Foreman, 1983. "Daily Cycles Affect Many Body Processes." *Boston Globe,* October 24, pp. 41 and 44.

90

Rhythm of cancer cells. Lawrence Scheving, interview with the author.

92

"A calorie is not the same calorie." Jane Brody, "Body's Many Rhythms Send Messages on When to Work and When to Play." *New York Times,* August 11, 1981, pp. C1, C2.

92

we cannot know whether the sun will rise. David Hume, 1955. *Enquiry Concerning Human Understanding.* New York, C.W. Hendel.

FIVE: THE MASTER CLOCK REVISITED

94

An important pacemaker found in the brain. F. K. Stephan and I. Zucker, 1972. "Circadian Rhythms in Drinking Behavior and Locomotor Activity of Rats are Eliminated by Hypothalamic Lesions." *National Academy of Science USA Proceedings* 69:1538–1586. R. Y. Moore and V. B. Eichler, 1972. "Loss of a Circadian Adrenal Corticosterone Rhythm Following Suprachiasmatic Lesions in the Rat." *Brain Research* 42:201–206.

96

"All of a sudden, it became dogma overnight." Lawrence Scheving, interview with the author.

97

"I could not accept that one single group." Ernest Powell, interview with the author.

98

Eppur si muove. This story of Galileo is told by Stillman Drake, 1978. *Galileo At Work.* Chicago, University of Chicago Press, p. 357. The story first appeared in print in English in *The Italian Library* (London, 1757), published by Giuseppe Baretti, an Italian man of letters. The phrase was applied to biological rhythms by Franz Halberg et al. in "Circaseptan (About 7-Day) and Circasemiseptan (About 3.5 Day) Rhythms and Contributions by Ladislav Derer." Unpublished ms., summary of lectures given in Bratislava, October 8, 1984.

98
"They were altered there too." Ernest Powell, interview with the author.

99
The X and Y pacemakers. Martin Moore-Ede, 1983. "The Circadian Timing System in Mammals: Two Pacemakers Preside Over Many Secondary Oscillators." *Federation Proceedings* 42(11):2802–2808.

99
A mathematical model of the two pacemakers. Samuel F. Pilato, Martin Moore-Ede and Elliot D. Weitzman, 1982. "Mathematical Model of the Human Circadian System with Two Interacting Oscillators." *American Journal of Physiology* 242:R3–R17.

100
Experiment in Midnight Cave. Michel Siffre, 1975. "Six Months Alone in a Cave." *National Geographic* 147:426–435.

102
"a scaffolding, not a building." Franz Halberg, interview with the author.

102
"I think our bodies operate with a host of clocks." John Pauly, interview with the author.

103
A study of people living near Three Mile Island. Marc A. Schaeffer and Andrew Baum, 1984. "Adrenal Cortical Response to Stress at Three Mile Island." *Psychosomatic Medicine* 46(3):227–237.

103
Adrenals given higher priority than sexual functions. Franz Halberg, 1953. "Some Physiological and Clinical Aspects of 24-Hour Periodicity." *The Journal- Lancet (USA)* 73:20–32.

104
A "feed-sidewards" effect. Franz Halberg, 1983. "Quo Vadis Basic and Clinical Chronobiology: Promise for Health Maintenance." *American Journal of Anatomy* 168:543–594.

106
A survey taken in California. Joan Sweeney, 1984. "DST Can Be Cruel to Body Clocks." *The Miami Herald*, April 29, pp. 1A and 13A.

108
Cockroach clocks. Terry Page, Vanderbilt University, interview with the author.

108
Meal schedules as a time cue. D. T. Krieger, H. Hauser and L. C. Key, 1977. "Suprachiasmatic Nuclear Lesions Do Not Abolish Food-Shifted Circadian Adrenal and Temperature Rhythmicity." *Science* 197:398–399.

109
Separate food pacemaker in evolution. Martin Moore-Ede, 1982. *The*

Clocks That Time Us. Cambridge, Massachusetts, Harvard University Press, p. 227.

109

Manipulating biological rhythms by changing meal schedules. Franz Halberg, 1977. "Implications of Biological Rhythms for Clinical Practice." *Hospital Practice,* January, pp. 139–149.

110

Effects on biological rhythms of people living alone. R. Wever, 1975. "Autonomous Circadian Rhythms in Man. Single Versus Collectively Isolated Subjects." *Naturwissenschaften* 62:443–444.

110

Lack of time cues and the symptoms of bereavement. Myron Hofer, 1984. "Psychobiological Perspective on Bereavement. *Psychosomatic Medicine* 46(3).

111

Caveman with a wristwatch. J. N. Mills, 1964. "Circadian Rhythms During and After Three Months in Solitude Underground." *Journal of Physiology* 174:217–231.

112

Tight and loose pacemakers. Gene D. Block, Dept. of Biological Sciences, University of Virginia, Charlottesville. Interview with the author.

SIX: A BIOLOGICAL RAINBOW

116

Search for biological rhythms and for extraterrestrial intelligence. Kent Cullers and Jim Stevenson, interview with the author. And see William J. Broad. "Eavesdroppers Listen for Cosmic Hello." *New York Times,* April 10, 1985, pp. C1 and C7.

118

"The physical universe is an aggregate of frequencies." R. Buckminster Fuller in collaboration with E. J. Applewhite, 1982. *Synergetics: Explorations in the Geometry of Thinking.* Preface and contribution by Arthur L. Loeb. New York, Macmillan, p. 249.

120

Rhythm detection and the 17th century revolution in optics. Franz Halberg, 1966. "Resolving Power of Electronic Computers in Chronopathology. An Analogy to Microscopy." *Scientia.* Revue Internationale de Synthese Scientifique. September-October, pp. 1–8.

120

Spectrum of biological rhythms. Franz Halberg, Erna Halberg and Julia Halberg, 1979. "Collateral-Interacting Hierarchy of Rhythm Coordination at Different Organisation Levels, Changing Schedules and Aging." In M. Suda, ed. *Biological Rhythms and Their Central Mechanism.* Amsterdam-New York-Oxford, Elsevier/North-Holland Biomedical Press.

122

"It is superfluous to keep on discussing." Franz Halberg, interview with the author.

124

mistaken "discovery" of biorhythms. In Jeffrey Moussaieff Masson, ed., 1985. *The Complete Letters of Sigmund Freud to Wilhelm Fliess 1887–1904*. Cambridge, Massachusetts, and London, England, The Belknap Press of Harvard University Press.

126

Listening to the harmonics of the wind. Richard Buell "Composer Davies Casts Quite a Spell." *Boston Globe*, March 19, 1985, p. 26.

127

Playing the chord of hydrogen on a piano. Described in K. C. Cole, 1985. *Sympathetic Vibrations*. New York, William Morrow and Company, Inc., p. 112.

128

An underwater air bubble will "ring." The Sound of Air Bubbles." *Science News*, May 4, 1985, p. 281.

128

Resonances "hold the body together." William Hrushevsky, Dept. of Medicine and Laboratory of Medicine and Pathology, University of Minnesota Medical School and Masonic Cancer Center, Minneapolis, interview with the author.

130

"It is surprising to me that about seven-day rhythms were there." Lawrence Scheving, University of Arkansas Medical School, interview with the author.

131

"resonance can make things seem to appear out of nowhere." K. C. Cole, 1985. *Sympathetic Vibrations. Ibid.* p. 275.

131

"The center of the puzzle is that endotoxin is really not much of a toxin." Lewis Thomas, 1983. *The Youngest Science*. New York, The Viking Press, p. 151.

131

A circaseptan doubles into a harmonic of itself. Franz Halberg, 1983. "Quo Vadis Basic and Clinical Chronobiology: Promise for Health Maintenance." *American Journal of Anatomy* 168:543–594.

133–134

"the big revolution in biologic time structure." Franz Halberg, interview with the author.

SEVEN: LIGHT AND TIME: TOGETHER AGAIN

136

a temperature "eye." T. H. Benzinger, 1961. "The Human Thermostat." *Scientific American*, January, pp. 134–147.

138

Modern hospital as a constant environment. Martin Moore-Ede, Charles A. Czeisler and Gary S. Richardson, 1983. "Circadian Time-Keeping in Health and Disease. Part 2. Clinical Implications of Circadian Rhythmicity." *New England Journal of Medicine* 309(9):530–536.

140

Light can tease the circadian system apart. Jurgen Aschoff and Rutger Wever, 1981. "The Circadian system of Man." In Jurgen Aschoff, ed. *Handbook of Behavioural Neurobiology*, Vol. 4, Biological Rhythms. New York and London, Plenum Press, p. 319.

141

Passengers arriving at Heathrow airport. Martin Moore-Ede, interview with the author.

141

Light of high intensity widens the range of entrainment. Thomas Wehr, Clinical Psychobiology Branch National Institute of Mental Health, Bethesda, Md. Interview with the author.

144

"Maybe we're all living in a melatonin haze." Richard Liebmann-Smith, 1985. "The Man Who Patented Sunlight." *American Health* 4(10):32–35.

145

Studying the pineal "like seeing evolution in progress." S. D. Wainwright, 1982. In Russel J. Reiter, ed. *The Pineal Gland. Vol. III Extra-Reproductive Effects*. Boca Raton, Florida, CRC Press, Inc., p. 71.

147

Study of church records in Labrador. Joel R. L. Ehrenkrantz, 1983. "A Gland for All Seasons." *Natural History*, June 1983, pp. 18–22. And interview with the author.

148

Darkness and time. Thomas Wehr, interview with the author.

149

"The most powerful handle we have on biological rhythms." Thomas Wehr, *Ibid*.

151

"Now that says several things to us." *Ibid*.

152

Monday morning cups of coffee. Martin Moore-Ede, Charles A. Czeisler and Gary S. Richardson, 1983. "Circadian Timekeeping in Health and Disease. Part 1. Basic Properties of Circadian Pacemakers." *New England Journal of Medicine* 309(8):475.

153

"It is going to lead to rational pharmacological intervention." Thomas Wehr, interview with the author.

EIGHT: TIMETABLE FOR SIMPLE MINDS

156

An unborn mammal "knows" what time it is. Steven M. Reppert, Laboratory of Developmental Chronobiology, Children's Service, Massachusetts General Hospital and Harvard Medical School, interview with the author. And see Steven M. Reppert, William J. Schwartz, 1983. "Maternal Coordination of the Fetal Biological Clock in Utero." *Science* 220:969–971.

158

Honeybee learning. James Gould, interview with the author. And see James L. Gould, 1979. "Do Honeybees Know What They Are Doing?" *Natural History* 88:66–75. James L. Gould and Carol G. Gould. "Can a Bee Behave Intelligently?" *New Scientist*, April, 14, 1983, pp. 84–87.

163

"in the dark dungeon of the ego." Malcolm Muggeridge, 1974. *The Infernal Grove, Chronicles of Wasted Time Number 2*. New York, William Morrow and Company, Inc., p. 185.

163

Time as a sort of lens. C. S. Lewis, 1946. *The Great Divorce*. New York, Macmillan, p. 125. "Time is the very lens through which ye see—small and clear, as men see through the wrong end of a telescope—something that would otherwise be too big for ye to see at all. That thing is Freedom: the gift whereby ye most resemble your Maker and are yourselves parts of eternal reality. But ye can see it only through the lens of Time, in a little clear picture, through the inverted telescope. It is a picture of moments following one another and yourself in each moment making some choice that might have been otherwise."

163

Hamlet and *Macbeth* as time plays. Wylie Sypher, 1976. *The Ethic of Time*. New York, The Seabury Press.

164

"We are blind to our own blindnesses." James Gould, interview with the author.

165

"Every other reality in human experience." Rollo May, 1981. *Freedom and Destiny*. New York, Dell Publishing Co., Inc., p. 5.

NINE: THE CLOCKS OF SLEEP AND DREAMS

168

"the gentle tyrant." Daniel Goleman, 1982. "Staying Up. The Rebellion Against Sleep's Gentle Tyranny." *Psychology Today* 16(3):24.

169

Wilse Webb's experiments. Daniel Goleman. *Ibid.* pp. 27–28. And interview with the author.

171

sleep breaks into temporal fragments. Scott S. Campbell, 1984. "Duration and Placement of Sleep in a 'Disentrained' Environment." *Psychophysiology* 21(1):106–113.

172

"There is a time to sleep and a time to wake." Daniel Goleman, 1982. "Staying Up." *Ibid.* p. 32.

173

"As against the times when an unremarkable profile." Kingsley Amis, 1970. *The Green Man*. New York, Harcourt Brace and World, Inc., p. 24. Of this experience Amis's hero remarks: "They are not dreams. They might be described as visions of no obvious meaning seen under poor conditions."

175

a sleep-dream clock. J. Allan Hobson, Laboratory of Neurophysiology, Department of Psychiatry, Harvard Medical School, interviews with the author. And see J. Allan Hobson, 1983. *Sleep: Order and Disorder.* South Norwalk, Connecticut, Meducation, Inc. J. Allan Hobson, 1977. "Body Rhythms" in *McGraw-Hill Yearbook of Science and Technology.* New York, McGraw-Hill Book Company, Inc. J. Allan Hobson, 1975. "The Sleep-Dream Cycle: A Neurobiological Rhythm." *Pathobiology Annual.* J. Allan Hobson and Robert W. McCarley, 1975. "Sleep Cycle Oscillation: Reciprocal Discharge By Two Brainstem Neuronal Groups." *Science* 189:55–58.

178

time and space are less well organized in the dream. Ernest L. Hartmann, 1973. *The Functions of Sleep.* New Haven and London, Yale University Press, p. 137.

181

the stream of consciousness. William James, 1890. *The Principles of Psychology*, Vol. 1, p. 239. "Consciousness, then, does not appear to itself chopped up in bits. Such words as 'chain' or 'train' do not describe it fitly as it presents itself in the first instance. It is nothing jointed; it flows. A 'river' or 'stream' are the metaphors by which it is most naturally described. *On talking of it hereafter, let us call it the stream of thought, of consciousness, or of subjective life.*"

181

the "law of constancy in our meanings." William James. *Ibid.* p. 460.

181

"Every definite image in the mind." *Ibid.* p. 255.

181

"We all of us have this permanent consciousness." *Ibid.* pp. 255–256.

181–182

"just going putt-putt." J. Allan Hobson, interview with the author.

182

REM dreams a "return to the basics." David B. Cohen, 1979. *Sleep and Dreaming: Origins, Nature and Functions.* New York, Pergamon Press, p. 4.

184

"It's quite clear that the brain is not a single system." Elliot D. Weitzman, 1983. In Michael Chase and Elliot D. Weitzman, eds. *Sleep Disorders and Basic and Clinical Research.* New York and London, SP Medical and Scientific Books, p. 253.

TEN: THE GATES OF DAY AND NIGHT

187

The existence of REM periods firmly established. E. Aserinsky and N. Kleitman, 1953. "Regularly Occurring Periods of Eye Motility and Concomitant Phenomena During Sleep." *Science* 118:273–274.

187
a paper was given at a sleep congress in London. In G. E. W. Wolsenholm and M. O'Connor, eds., 1961. *Ciba Foundation Symposium on the Nature of Sleep*. London, J. and A. Churchill.

187
Kleitman's book. Nathaniel Kleitman, 1963. *Sleep and Wakefulness*. Chicago, University of Chicago Press.

188
Kleitman brought the enthusiasm of an amateur to his work. Marjorie Dent, ed., October 1957. *Current Biography*. New York, The H.W. Wilson Co., pp. 307–308.

188
"the one related to the other as a trough is to its crest." Nathaniel Kleitman, interview with the author.

189
mental experiences during non-REM sleep. David Foulkes, 1966. *The Psychology of Sleep*. New York, Charles Scribner's Sons, pp. 106–108.

190
"thinking that does not labor under nearly so many constraints." David Foulkes. *Ibid.* p. 110.

190
Discovery of a cycle of daydreaming. Daniel F. Kripke and David Sonnenschein, 1973. "A 90-Minute Daydream Cycle." (Abstract). *Sleep Research* 2:187. Daniel F. Kripke and David Sonnenschein, 1978. "A Biologic Rhythm in Waking Fantasy." In K. S. Pope and J. L. Singer, eds. *The Stream of Consciousness*. New York, Plenum Press.

192
John Caughey's collection of daydreams. John Caughey, 1984. *Imaginary Social Worlds. A Cultural Approach*. Lincoln and London, University of Nebraska Press.

193
"I see the ultradian system as a very heterogeneous one." Peretz Lavie, interview with the author. And see Peretz Lavie and Daniel F. Kripke, 1981. "Ultradian 1½ Hour Rhythms: A Multioscillatory System." *Life Science* 29:2445–2450. Peretz Lavie, 1977. "Ultradian Rhythms in Alertness, a Pupillometric Study." *Biological Psychology* 3:214–218. Peretz Lavie, 1982. "Ultradian Rhythms in Human Sleep and Wakefulness: A Multioscillatory Conclusion." In W. B. Webb, ed. *Sleep and Biological Rhythms*. New York, Wiley, pp. 239–272. Peretz Lavie and Andreas Scherson, 1981. "Ultrashort Sleep-Waking Schedule.1. Evidence of Ultradian Rhythmicity in 'Sleepability.'" *Electroencephalography and Clinical Neurophysiology* 52:163–174. Peretz Lavie is at the Sleep Laboratory, Faculty of Medicine, Technion-Israel Institute of Technology, Haifa Israel.

193
"Perhaps it is too strong a statement to say." Peretz Lavie, interview with the author.

194
The evolutionary origin of the siesta gate. Marcia J. Thompson and David

W. Harsha, 1984. "Our Rhythms Still Follow the African Sun." *Psychology Today* 18(1):50–54.
195
"In evolution, REM seems to have been selected." Peretz Lavie, interview with the author.
199
"the 90-minute enigma." Peretz Lavie, 1982. In W. B. Webb, ed. *Sleep and Biological Rhythms. Ibid.*
199
"Possibly the newest aspect of our knowledge." John Gertz, interview with the author.

ELEVEN: TIME AND THE BIOLOGICAL BRAIN

204
Alfred Adler's trap. David B. Cohen, 1979. *Sleep and Dreaming: Origins, Nature and Functions.* New York, Pergamon Press, p. 110.
204
Proposal that body temperature and mental efficiency rise and fall in synchrony. Nathaniel Kleitman, 1963. *Sleep and Wakefulness.* Chicago, University of Chicago Press.
205
Different rhythms of simple and complex mental tasks. Simon Folkard, Rutger A. Wever and Christina M. Wildgruber, 1983. "Multioscillatory Control of Circadian Rhythms in Human Performance." *Nature* 305:223–226.
207
Circadian rhythms of short-term as opposed to long-term memory. Keith Miller, Brian C. Syles and David G. Wastell, 1980. "Time of Day and Retrieval from Long-Term Memory." *British Journal of Psychology* 71:407–414.
207
Footballers should learn new plays at 3 P.M. Charles M. Winget, Charles W. DeRoshia and Daniel C. Holley, 1985. "Circadian Rhythms and Athletic Performance." *Medicine and Science in Sports and Exercise* 17(5):498–516.
207
Schoolchildren listening to a story at different times of day. Simon Folkard, Timothy H. Monk, Rosamund Bradbury and Joanna Rosenthal, 1977. "Time of Day Effects in School Children's Immediate and Delayed Recall of Meaningful Material." *British Journal of Psychology* 68:45–50.
208
Caffeine a help or a hindrance. William Revelle, Michael S. Humphreys, Lisa Simon and Kirby Galliland, 1980. "The Interactive Effect of Personality, Time of Day and Caffeine: A Test of the Arousal Model." *Journal of Experimental Psychology: General* 109(1): 1.31. Michael W. Eysenck and Simon Folkard, 1985. "Personality, Time of Day and Caffeine. Some Theoretical and Conceptual Problems in Revelle et al." *Journal of Experimental Psychology: General* 109(1):32–41. Michael S. Humphreys, Wil-

liam Revelle, Lisa Simon and Kirby Galliland, 1980. "Individual Differences in Diurnal Rhythms and Multiple Activation States: A Reply to M. W. Eysenck and Folkard." *Journal of Experimental Psychology: General* 109(1):42–48.
208
Owls and larks in close visual contact. W. P. Colquhoun, 1982. "Biological and Performance." In W. B. Webb, ed. *Biological Rhythms, Sleep, and Performance.* New York, John Wiley and Sons, pp. 59–86.
209
"background primers." In Daniel Goleman, 1982. "Staying Up. The Rebellion Against Sleep's Gentle Tyranny." *Psychology Today* 16(3):32.
209
Evolution like a tinkerer. François Jacob, 1982. *The Possible and the Actual.* New York, Pantheon Books, pp. 34–37.
209
"This evolutionary procedure." François Jacob. *Ibid.* p. 37.
210
Sir Winston Churchill's naps. "You must sleep some time between lunch and dinner, and no half-way measures. Take off your clothes and get into bed. That's what I always do. Don't think you will be doing less work because you sleep during the day. That's a foolish notion held by people who have no imagination. You will be able to accomplish more. You get two days in one—well, at least one and a half, I'm sure. When the war started, I *had* to sleep during the day because that was the only way I could cope with my responsibilities." Sir Winston, quoted in Walter Graebner, 1965. *My Dear Mister Churchill.* London, Michael Joseph, p. 55.
210
An ultradian rhythm of efficiency in mental focus. Peretz Lavie, interview with the author.
211
"You can phase-shift people along the 90-minute cycle." Peretz Lavie, interview with the author.
211
System dominance. T. Downing Bowler, 1980. *General Systems Thinking. Its Scope and Applicability. Series Vol. 4.* New York and Oxford, North Holland, p. 42.
211–212
Allan Hobson reduces the effects of jet-lag. J. Allan Hobson, 1984. "Hobson's Choice or Why You Have Jet Lag on Eastbound International Flights. Unpublished manuscript.
213
"could we be paying homage to an archaic force?" J. Allan Hobson, 1977. "Body Rhythms" in *McGraw-Hill Yearbook of Science and Technology.* New York, McGraw-Hill Book Company, Inc.
213
Boeing 707 overshoots Los Angeles airport. Martin Moore-Ede, Frank M. Sulzman and Charles A. Fuller, 1982. *The Clocks That Time Us.* Cambridge, Massachusetts, and London, England, Harvard University Press, p. 333.

214

a language "organ." Noam Chomsky, 1977. *Language and Responsibilty.*
New York, Pantheon Books, p. 83.

215

"universal by biological necessity." Noam Chomsky, 1975. *Reflections on
Language.* New York, Pantheon Books, p. 4.

215

A being from another planet. Roger Shepard, 1978. "The Mental Image."
American Psychologist 33:125–137.

TWELVE: CHEMICAL CLOCKS AND A BIOLOGICAL MIRACLE

218

The discovery of the chemical clock. W. C. Bray, 1921. *Journal of the
American Chemical Society* 43:1262.

219

the curious activity of a solution of citric and sulfuric acids. B. P. Belousov,
1958. *Sb. Ref. Radiats. Med.* Moscow, Medgiz, p. 145.

219

a spectacular variation of it. A. M. Zhabotinksy, 1964. *Biofizika* 9:306.

219–220

a chemical radio. Joseph Higgins, 1967. *Industrial and Engineering Chemistry* 59(5):19–62.

220

"truly miraculous." Barry Bunow, National Institutes of Health, interview
with the author.

221

Cyclic AMP. Ira Pastan, 1972. "Cyclic AMP." *Scientific American,* August,
pp. 97–105.

225

a famous pioneer and genius of modern communications. Lawrence Less-
ing, 1956. *Man of High Fidelity.* Philadelphia, Lippincott.

227

The first biochemical oscillation. A. T. Wilson and M. Calvin, 1955. *Journal of the American Chemical Society* 77:5948.

227

Oscillations in baker's yeast. B. Chance, R. W. Easterbrook and A. Ghosh,
1964. *Proceedings of the National Academy of Sciences of the United
States* 51:1244.

227

Another reason to suspect cyclic AMP. Martin Zatz, National Institutes of
Health, interview with the author.

THIRTEEN: THE CONVERSATIONAL WALTZ

229

an imaginative new approach. Daniel Stern, 1977. *The First Relationship.*
Cambridge, Massachusetts, Harvard University Press, pp. 86–88.

229

"Do you think our Karl has a chance?" "How about that whozis? Clay-Mildenberger fight." *Time*, September 16, 1966.

231

"We are forced to think." Daniel Stern, 1977. *The First Relationship. Ibid.* p. 85.

231

"the serial order of behavior." K. S. Lashley, 1951. "The Problem of Serial Order in Behavior." In K. H. Pribram, ed. *Perception and Action.* Baltimore, Maryland, Penguin Books, pp. 515–540.

231

Lashley "operated an automobile with supreme disdain." Frank A. Beach, 1961. "Karl Spencer Lashley." *Biographical Memoirs.* National Academy of Sciences 35:186–187.

202

"waves of excitation." K.S. Lashley. *Ibid.* p. 531.

233

They are "especially characteristic of human behavior." *Ibid.* p. 517.

233

"The rhythms tend to spread." *Ibid.* p. 531.

233

a landmark article. Fred Delcomyn, 1980. "Neural Basis of Rhythmic Behavior in Animals." *Science* 210:492–498.

235

Flexing a finger and tapping it. A. M. Wing, 1980. "Timing of Movement Phases of a Repeated Response." *Journal of Motor Behaviour* 12:113–124.

236

"Ensemble is what is real about real time." Ron Scollon, 1981. "The Rhythmic Integration of Ordinary Talk." In Deborah Tannen, ed. *Georgetown University Round Table on Languages and Linguistics.* Washington, D.C., Georgetown University Press, p. 343.

236

"bond of immediate temporal predictability." *Ibid.* p. 343.

237

Signal system in conversation. Frederick Erickson, 1982. "Classroom Discourse as Improvisation: Relationships Between Academic Task Structure and Social Participation Structure in Lessons." In L. C. Wilkinson, ed. *Communicating in the Classroom.* New York, Academic Press.

238

a striking example of lost meaning. Frederick Erickson, 1981. "Timing and Context in Everyday Discourse: Inplications for the Study of Referential and Social Meaning." In W. P. Dickson, ed. *Children's Oral Communication Skills.* New York, Academic Press, pp. 241–269.

239–240

a dinner-table conversation. Frederick Erickson, 1981. "Money Tree, Lasagna Bush, Salt and Pepper: Social Construction of Topical Cohesion in a Conversation Among Italian-Americans." In Deborah Tannen, ed. *Georgetown University Round Table on Languages and Linguistics.* Washington, D.C., Georgetown University Press, pp. 43–70.

241

"the fundamental glue." Frederick Erickson. *Ibid.* p. 65.

242

Duple time underlies many kinds of ordinary talk. Ron Scollon, 1981. *Tempo, Density and Silence: Rhythms in Ordinary Talk.* Unpublished manuscript.

243

someone will clear his throat. *Ibid.* p. 14.

245

"The ways in which each person talks and listens." Frederick Erickson and Jeffrey Shultz, 1981. *The Counselor as Gatekeeper.* New York, Academic Press, pp. 7 and 9.

FOURTEEN: EVOLUTION AND THE RETREAT INTO TIME

247–248

"We ourselves are so deeply entrapped." François Jacob, 1982. *The Possible and the Actual.* New York, Pantheon Books, p. 56.

248

"is, in a way, a possible world." *Ibid.* p.56.

249

Critics who condemn music for "telling a story." Constant Lambert, 1948. *Music Ho!* London, Penguin Books, pp. 81–82.

251

"wager that very few people would have detected the trick by ear." Thomas Mann, 1971. *Dr. Faustus.* Translated by H. T. Lowe-Porter. New York Vintage Books, p. 61.

251

the ear cannot perceive temporal symmetry. Bela Julesz, 1971. *Foundations of Cyclopean Perception.* Chicago and London, University of Chicago Press.

252

The auditory system as timekeeper. Bela Julesz and Ira J. Hirsh, 1972. "Visual and Auditory Perception—An Essay of Comparison." In Edward E. David and Peter B. Denes, eds. *Human Communication a Unified View.* New York, McGraw-Hill Book Company.

253

happening to walk one day around an orchestra. John Cage, 1981. *For the Birds. John Cage in Conversation with Daniel Charles.* Boston and London, Marion Boyars, p. 131.

253

"a sudden joy." John Cage. *Ibid.* p. 131.

255

"Some very bright people." Alan Loft. "2008: A Sonic Odyssey. A Survey of What Lies Ahead for Audiophiles." *Stereo Review,* February 1983, p. 65.

256

cerebral cortex the "ganglion" of the distance receptors. Sir Charles Sherrington, 1947. *The Integrative Action of the Nervous System.* Cambridge, England, Cambridge University Press, p. 325.

259–260

Mammals first creatures to construct an internal model of the world fully incorporating the time dimension. Harry Jerison, 1973. *Evolution of the Brain and Intelligence.* New York, Academic Press.

260–261

"The first expansion of the vertebrate brain." Harry Jerison, 1976. "Paleoneurology and the Evolution of Mind." *Scientific American,* January, pp. 90–101.

261

a preparation for a glorious future. Alfred S. Romer, 1954. *Man and the Vertebrates.* Baltimore, Maryland, Penguin Books, Vol. 1, p. 128.

263

A jellyfish would be a conceptual thinker. William James, 1890. *The Principles of Psychology.* Vol. 1. New York, Dover Publications, p. 463.

264

"These have become singularly human hallmarks." Harry Jerison, 1973. *Evolution of the Brain and Intelligence.* New York, Academic Press, p. 429.

FIFTEEN: THE COUGH THAT FLOATED

267

Galileo used his own singing voice. Stillman Drake, 1975. *Scientific American,* June, pp. 93–104.

268

When meaning is lost, temporal order is lost. Richard M. Warren and Charles J. Obusek, 1972. "Identification of Temporal Order Within Auditory Sequences." *Perception and Psychophysics* 12(1B):86–90.   Richard M. Warren, Charles J. Obusek, Richard M. Farmer and Roslyn P. Warren, 1969. "Auditory Sequence: Confusion of Patterns Other Than Speech or Music." *Science* 164:586–587.   Richard M. Warren and Roslyn P. Warren, 1970. "Auditory Illusions and Confusions." *Scientific American* 223:30–36.

271

hear the word *radio.* Michael F. Dorman, James E. Cutting and Lawrence J. Raphael, 1975. "Perception of Temporal Order in Vowel Sequences with and without Formant Transitions." *Journal of Experimental Psychology: Human Perception and Performance* 104(2):121–129. Referred to in Mari Riess Jones, 1976. "Time, Our Lost Dimension: Toward a New Theory of Perception, Attention and Memory." *Psychological Review* 83(5):323–355.

272

streaming reflects a special feature of the human brain. Mari Riess Jones, 1981. "Only Time Can Tell: On the Topology of Mental Space and Time." *Critical Inquiry,* Spring, pp. 557–576.   Mari Riess Jones, 1976. "Time, Our Lost Dimension: Toward a New Theory of Perception, Attention and

Memory." *Psychological Review* 83(5): 323–351. Mari Riess Jones, 1978. "Auditory Patterns: The Perceiving Organism. In E. C. Carterette and M. P. Friedman, eds. *Handbook of Perception*. Vol. 8. New York, Academic Press, pp. 255–287.
278
"this development is significant." Wolfgang Stefani, 1981. 'The Psycho-Physiological Effects of Volume, Pitch, Harmony and Rhythm in the Development of Western Art Music. Implications for a Philosophy of Music History. Unpublished MA thesis, Andrews University School of Graduate Studies, p. 149.
278
The "expressive" element. Constant Lambert, 1948. *Music Ho!* London, Penguin Books.
280
not a sound wave but a "time wave." Emil Zuckerkandl, 1973. *Sound and Symbol. Music and the External World*. Translated by Willard A. Task. Princeton, N.J., Princeton University Press.
282
"the many different 'psychological' times." Elliott Carter, 1983. "Music and the Time Screen." In John W. Grubbs, ed. *Current Thought in Musicology*. Austin, University of Texas Press.
282
"whirling out of control or running down." David Schiff, 1983. *The Music of Elliott Carter*. New York, Eulenberg Books.
282
"perhaps makes it very hard for the listener." Elliott Carter. Interview by Leighton Kerner. *Saturday Review*, December 1980, p. 41.
282
"heard over a distant hill." Robert P. Morgan, 1975. "Spatial Form in Ives." In H. Wiley Hitchcock and Vivian Perlis, eds. *An Ives Celebration*. Urbana, University of Illinois Press.
283
"By denying the listener a tonal frame of reference." Henry Pleasants, 1967. *The Agony of Modern Music*. New York, Simon and Schuster, p. 135.

SIXTEEN: TEMPORAL INTELLIGENCE

286
mother acts as an external pacemaker. Beatrice Beebe, Daniel Stern, Joseph Jaffe, 1979. "The Kinesic Rhythm of Mother-Infant Interactions." In Aaron W. Siegman and Stanley Feldstein, eds. *Of Speech and Time. Temporal Speech Patterns in Interpersonal Contexts*. Hillsdale, New Jersey, Lawrence Erlbaum Associates, pp. 23–24.
286
more complex, more idiosyncratic and remote from physical dimensions. Mari Riess Jones, 1976. "Time, Our Lost Dimension. Toward a New Theory of Perception, Attention and Memory." *Psychological Review* 83(5):323–355.

287

"when it lodges in the queer element of the human spirit." Virginia Woolf, 1928. *Orlando: a Biography*. New York, Harcourt Brace and Co., p. 98. "Time, unfortunately, though it makes animals and vegetables bloom and fade with amazing punctuality, has no such simple effect upon the mind of man. The mind of man, moreover, works with equal strangeness upon the body of time. An hour, once it lodges in the queer element of the human spirit, may be stretched to fifty or a hundred times its clock length; on the other hand, an hour may be accurately represented on the timepiece of the mind by one second."

288

Absolute time and music. Daniel Stern, 1977. *The First Relationship*. Cambridge, Massachusetts, Harvard University Press. p. 89.

288

a "magic range." Beatrice Beebe, 1982. "Micro-Timing in Mother-Infant Communication." In Mary Ritchie Key, ed. *Nonverbal Communication Today*. New York, Mouton.

288

people go far beyond mere intuition. William J. Friedman, interview with the author. And see William J. Friedman, ed., 1982. *The Developmental Psychology of Time*. New York, Academic Press.

289

"I think one of the most interesting results." William J. Friedman, interview with the author.

290

Time "is a construction." Paul Valéry, 1973. *Cahiers*. Paris, La Gallimard, p. 1303.

290

a "family" of competencies. Howard Gardner, 1983. *Frames of Mind. The Theory of Multiple Intelligences*. New York, Basic Books.

291

a reality "no less tremendous." Howard Gardner. *Ibid*. p. 187.

291

Literary inspiration of Joan Didion. Roger Shepard, 1978. "The Mental Image." *American Psychologist* 33:125–137.

291

Joyce Carol Oates's daydreams. Don O'Briant 1986. "A little wilder Joyce Carol Oates," *The Atlanta Constitution*, April 23, p 1C.

292

Kindergarten teacher and class. Lorraine Harner, 1981. "Children's Understanding of Time." *Topics in Language Disorders*, December 1981, pp. 51–65.

293

children and the dual nature of the present. Lorraine Harner, 1980. "Comprehension of Past and Future Reference Revisited." *Journal of Experimental Child Psychology* 29:170–182. Lorraine Harner, 1975. "Yesterday and Tomorrow. Development of Early Understanding of the Terms." *Developmental Psychology* 11:864–865. Lorraine Harner, 1981. "Children Talk About the Time and Aspect of Actions." *Child Develop-*

ment 52:498–506. Lorraine Harner, 1982. "Talking About the Past and Future." In W. J. Friedman, ed. *The Developmental Psychology of Time.* New York, Academic Press. Lorraine Harner, 1982. "Immediacy and Certainty: Factors in Understanding Future Reference." *Journal of Child Language* 9:113–124.

295
Young children's maps poorly integrated. Nancy L. Hazen, Jeffrey J. Lockman and Herbert Pick, Jr., 1978. "The Development of Children's Representations of Large-Scale Environments." *Child Development* 49:623–635.

297
Time sense in chimps. David Premack, 1983. "The Codes of Man and Beasts." *The Behavioural and Brain Sciences* 6:125–167. David Premack, 1976. *Intelligence in Ape and Man.* Hillsdale, New Jersey, Lawrence Erlbaum Associates.

297
"If we had been on our toes." David Premack, interview with the author.

298
"We cannot say that it is just a matter." David Premack. *Ibid.*

SEVENTEEN: DARWIN AND THE RETURN OF MARTIN GUERRE

301
Question mark memory. Endel Tulving, 1985. "How Many Memory Systems Are There?" *American Psychologist* 40(4):385–398. "We could speculate that fragment completion is basically a procedural memory task, or basically a semantic memory task, but such conclusions are purely conjectural. At the present stage of our knowledge, it is no less plausible to entertain the hypothesis that fragment completion reflects the operation of some other, as yet unknown, memory system, perhaps a precursor to episodic memory. We could refer to this unknown system as the QM system (QM for Question Mark) and keep our eyes and minds open for evidence for and against its hypothetical existence."

305
"wander around freely in the structure." Endel Tulving, 1983. *Elements of Episodic Memory.* Oxford Psychology Series No. 2. Oxford, England, Oxford University Press, p. 38.

306
List of differences between semantic and episodic memory. Endel Tulving, 1983. *Elements of Episodic Memory. Ibid.* p. 35.

307
"If you go backwards in time." Endel Tulving, interview with the author.

307
"Nevertheless, we know that the computer." Endel Tulving, 1983. *Elements of Episodic Memory. Ibid.* p. 53.

308
A *gedanken* experiment. Endel Tulving, interview with the author.

308
Adaptive value of episodic memory. Endel Tulving, 1985. "Memory and Consciousness." *Canadian Psychology* 26(1):1–11.

309
"the human tape recorder." Ulric Neisser, 1981. "John Dean's Memory: A Case Study." *Cognition* 9:1–22.

311
"The semantic memory system handles temporal concepts." Endel Tulving, 1983. *Elements of Episodic Memory. Ibid.* p. 42.

312
The revival of narrative. Karen Winkler, 1984. " 'Disillusioned' with Numbers and Counting, Historians Are Telling Stories Again." *Chronicle of Higher Education* 28(16):5–6.

312
a manifesto that is now famous. Lawrence Stone, 1979 "The Revival of Narrative: Reflections on a New Old History." *Past and Present.* 85:3–23.

312–313
"It is just those projects." Lawrence Stone. *Ibid.* p. 12.

313
"A belated recognition of the importance of power." Lawrence Stone. *Ibid.* p. 10.

314
Natalie Davis's narrative style. Natalie Zemon Davis, 1983. *The Return of Martin Guerre.* Cambridge, Massachusetts, Harvard University Press.

314
"my intent was to make the story flow back." Natalie Davis, interview with the author.

315
"For me it is always extremely interesting." Natalie Davis, interview with the author.

316
the search for a science of history. Sir Isaiah Berlin, 1981. *Concepts and Categories.* Henry Hardy, ed. New York, Penguin Books pp. 103–141.

EIGHTEEN: TAKING TIME APART

322
a tall, thin, undernourished man." Richard Restak, 1979. *The Brain, The Last Frontier.* New York, Warner Books, p. 273.

323
Studies of Korsakoff patients by Robert Hicks and his colleagues: Robert E. Hicks, 1985. "Chronognosia: Dissociation of Timing and Temporal Integration Amnesia." Unpublished manuscript. Robert E. Hicks, C. Thomas Gaultieri, James P. Mayo and Randall Evans, 1984. "Deranged Memory and Temporal Information Processing." Unpublished manuscript. Robert E. Hicks, Mario Perez-Reyes, James P. Mayo and C. Thomas Gaultieri, 1984. "Cannabis, Atropine, and Temporal Information Processing." Unpublished manuscript.

323
The account of Dr. Hicks's discoveries and his own comments are based on interviews with the author.

325
the "specious present." William James, 1890. *The Principles of Psychology.* Vol 1. New York, Dover Publications, p. 609.

326
Adam at the primal moment. *Ibid.* p. 641.

327
"Self-knowing" consciousness and the frontal lobes. Endel Tulving, 1985. "Memory and Consciousness." *Canadian Psychology* 26:5.

330
"It's like swimming in the middle of a lake." Endel Tulving, "Memory and Consciousness." *Ibid.* p. 4.

337
A "perspective projection." William James, 1890. *The Principles of Psychology* Vol. 1. *Ibid.* p. 630.

338
"Time is puzzling." John Lucas, 1973. *A Treatise on Time and Space.* London, Methuen, p. 5.

NINETEEN: THE CODES OF TIME

340
Lost on the Apollo moonwalk. R. M. Downs and David Stea, 1973. "Cognitive Maps and Spatial Behaviour: Process and Products." In R. M. Downs and David Stea, eds. *Image And Environment.* London, Arnold, p. 8.

340
"see time spacious before him." Thomas Mann, 1924. *The Magic Mountain.* Translated by H. T. Lowe-Porter. New York, Vintage Books, p. 103.

340
"Great spaces of time." Thomas Mann. *Ibid.* p. 104.

340–341
"Experience is what I agree to attend to." William James, 1890. *The Principles of Psychology.* Vol. 1. New York, Dover Publications, p. 402.

342
"Time and space seem less well organized in the dream." Ernest Hartmann, 1973. *The Functions of Sleep.* New Haven and London, Yale University Press, p. 137.

344
Chess masters often a little paranoid. Ernest Hartmann. *Ibid.* p. 157.

344
Experiments by Robert Ornstein. Robert Ornstein, 1969. *On the Experience of Time.* Harmondsworth, England, and Baltimore, Maryland, Penguin Books.

346
"Emma Woodhouse, handsome, clever and rich." Jane Austen, 1816. *Emma.* New York, New American Library, p. 1.

348

A random experience has less redundancy. Robert Ornstein, 1969. *On the Experience of Time. Ibid.*, pp. 64–67.

349

a successful day seems shorter. J. J. Harton, 1939. "An Investigation of the Influence of Success and Failure on the Estimation of Time." *Journal of General Psychology* 21:51–62.

350

"An early bird who caught a very strong worm." Robert M. Mulligan and H. R. Schiffman, 1979. "Temporal Experience as a Function of Organisation in Memory." *Bulletin of the Psychonomic Society* 14(6):417–420.

350

Parts of brain involved in redundant codes. Karl Pribram, 1966. "Limbic Lesions and the Temporal Structure of Redundancy." *Journal of Comparative and Physiological Psychology* 61(3):368–373.

352

The flow of time "appears to be an emergent property." Sir Alan Cottrell, 1977. In Ronald Duncan and Miranda Weston-Smith, eds. *The Encyclopaedia of Ignorance*. New York, Pocket Books, pp. 134–135.

353

"so closely bound up" Thomas Mann, 1924. *The Magic Mountain. Ibid.* p. 104.

TWENTY: PUTTING TIME TOGETHER

355

Homer's *Iliad* and subjective consciousness. Julian Jaynes, 1976. *The Origin of Consciousness in the Breakdown of the Bicameral Mind*. Boston, Houghton Mifflin Company.

355

"It is impossible for us." Julian Jaynes. *Ibid.* p. 72.

357

"If I ask you to think of the last hundred years." *Ibid.* p. 60.

358

"So we arrive at the position." *Ibid.* p. 41.

359

Melges's temporal integration inventory. Frederick Melges, Jared Tinklenberg, Leo Hollister and Hamp Gillespie, 1970. "Temporal Disintegration and Depersonalisation During Marihuana Intoxication." *Archives of General Psychiatry* 23:204–210.

362

"I do not even wish to live with eternal things." Bertrand Russell, letter to Goldsworth Lowes Dickinson August 26, 1902. In Bertrand Russell, 1951. *The Autobiography of Bertrand Russell. The Early Years 1872–World War One*. New York, Bantam Books. It should be noted that Russell also wrote, on April 3, 1902, in a letter to Gilbert Murray: "Unless the contemplation of eternal things is preserved, mankind will become no better than well-fed pigs. But I do not believe that such contemplation on the whole tends to happiness. It give moments of delight, but these are outweighed by years of effort and depression."

362

"the straitjacket of existence." Søren Kierkegaard, 1941. *Concluding Unscientific Postscript*. Translated by D.F. Swenson and W. Lowrie. Princeton, New Jersey, Princeton University Press.

364

Sartre on Faulkner. Jean-Paul Sartre, 1962. *Literary and Philosophical Essays*. Translated from the French by Annette Michelson. New York, Collier Books, p. 85.

365

Heidegger's three dimensions. Martin Heidegger, 1962. *Kant and the Problem of Metaphysics*. Bloomington and London, Indiana University Press.

365

"timeishness." Magda King uses this term to express the sense of "time-originating" in her book *Heidegger's Philosophy, A Guide to His Basic Thought*. New York, Dell Publishing Co., 1964, p. 171.

367

"This dwelling in the past." John Mcquarrie, 1973. *Existentialism*. New York, Penguin Books, p. 202.

367

Bernard Aaronson's experiments. Bernard S. Aaronson, 1968. "Hypnotic Alterations of Space and Time." *International Journal of Parapsychology* 10:5–36.

368

Kidnapped in a baby carriage. Jean Piaget, 1973. *The Child and Reality*. Translated by Arnold Rosin. New York, Penguin Books, pp. 43–44.

369

The object of psychoanalysis. Jerome Bruher, 1986. *Actual Minds, Possible Worlds*. Cambridge, Massachusetts, and London, England, Harvard University Press, p. 9.

369

Time in *JR*. Susan Strehle, 1982. "Disclosing Time: William Gaddis's *JR*." *Journal of Narrative Technique* 12(1):1–14.

TWENTY-ONE: "THERE'S AN IDEA MISSING"

372

In a famous lecture written in 1946. Jean-Paul Sartre, 1946. *Existentialism Is a Humanism*. Translated by Philip Mairet, 1948, in *Existentialism and Humanism*. London, Methuen and Co., Ltd.

372

"In the nothingness." Mary Warnock, 1970. *Existentialism*. New York, Oxford University Press, p. 98.

376

Jerry Fodor's theory. Jerry Fodor, 1983. *Modularity of Mind: An Essay on Faculty Psychology*. Cambridge, Massachusetts, MIT Press.

377

"Martini-bikini" illusion. Roger Shepard, 1978. "The Mental Image." *American Psychologist*, February 1978, pp. 125–137.

378

some games to play. Jerry Fodor, 1983. *Modularity of Mind. Ibid.* p. 55.

379

"No doubt there are *some* limits." *Ibid.* p. 55.

381

Metaphors "are the threads by which the mind holds on." Hannah Arendt, 1971. *The Life of the Mind.* New York, Harcourt Brace Jovanovich, p. 109.

381

Einstein's imaginary journey. In Ronald W. Clark, 1972. *Einstein. The Life and Times.* New York, Avon Books, p. 113.

383

Three distinct cognitive systems underlie knowledge of the months. William J. Friedman, 1983. *Journal of Experimental Psychology Learning Memory and Cognition* 9:650–666.

383

each day of the week as a different shape. John Buchan, 1932. *The Gap in the Curtain.* London, Hodder and Stoughton, p. 63.

384

"You cannot, absolutely cannot think of time." Julian Jaynes, 1976. *The Origin of Consciousness in the Breakdown of the Bicameral Mind.* Boston, Houghton Mifflin Company, p. 60.

384

"What, then, is time?" St. Augustine. *Confessions.* Translated with an introduction and notes by John K. Ryan, 1960. Garden City, New York, Image Books, p. 287.

385

"You could have predicted several years ago." Jerry Fodor, interview with the author.

385

"The main problem in psychology." Jerry Fodor. *Ibid.*

386

"Speaking crudely." Jerry Fodor, 1983. *Modularity of Mind: An Essay on Faculty Psychology.* Cambridge, Massachusetts, MIT Press, p. 105.

387

Toscanini's tempo. Manfred Clynes. "The Pure Pulse of Musical Genius." *Psychology Today*, July 1974, pp. 51–55.

387

Children use mental lists. William J. Friedman, 1983. "Image and Verbal Processes in Reasoning About the Months of the Year." *Journal of Experimental Psychology. Learning Memory and Cognition.* 9:650–666.

387

a feeling for narrative may emerge. Carl G. Gustavson, 1955. *A Preface to History.* New York, McGraw-Hill Book Company, Inc.

Index

Aaronson, Bernard, 367
absolute time, 287–88
  Newtonian concept of, 33–35, 61
Acetabularia mediterranea
  (mermaid's wineglass), 75–76,
  122
acetylcholine, 175, 180
Acrasiales fungi, chemical clocks
  in, 220–21
action, vs. reaction, 230–31
adaptation, 45–46
Adler, Alfred, 204
adrenal glands, 102–5
adrenaline, 222, 343
age, temporal identity and, 86
algae, weekly rhythms in, 75–76,
  132
Ali, Muhammad, 229–30
allergies, 91
Amis, Kingsley, 173
amnesia, Korsakoff's psychosis,
  321–31, 376
amphetamines, 333, 335, 344
amphibians, 257
amplitude, 117–18, 122
  of expectancy wave, 332–33
  of melatonin, 144
analogies, 380–81, 384
animals:
  built-in guidance of, 157–62
  "cold-blooded," 43, 52–54
  dormant periods of, 48
  light sensitivity in, 141
  see also specific animals
Aplysia, cyclic AMP in, 227
ara-C, 90, 91
Arendt, Hannah, 381
Aristotle, 28–30, 63–64

Armstrong, Edwin Howard
  ("Feedback"), 225–26
Aserinsky, Eugene, 187
asthma attacks, 91–92
attention, 339–45, 352, 373
  dreams and, 341–44
  nerve cells and, 180–81
  selective, 340–41
  sleep and, 177, 178, 182–83
Augustine, Saint, 384
Austen, Jane, 346–48
Autobiographical Notes (Einstein),
  24

babies:
  REM cycles of, 187, 188
  time sense of, 286–87
Bach, J. S., 270, 273, 279
Bailyn, Bernard, 313
Barrow, Isaac, 33–34
basic rest-activity cycle (BRAC),
  187–88, 196–98
bats, hearing of, 255–56
Becker, Ernest, 55
Beebe, Beatrice, 286, 288
bees:
  humans compared to, 168
  learning of, 158–61, 268
  memory of, 301
behaviorism, 369, 374
Belousov, B. P., 219, 220
Belousov-Zhabotinsky reaction,
  219
benzene ring, discovery of, 73
bereavement, biological clocks
  and, 110–11
Bergson, Henri, 45, 46, 384
Berlin, Sir Isaiah, 316

Berlioz, Hector, 44
Bernard, Claude, 44–49, 66–69,
  144, 389
biogenic amines, 175, 178
biological clocks (biological time),
  11–19, 25, 66–79, 94–114
  body temperature and, 66, 71–
    72, 85
  brain and, 203–15
  circadian, see circadian clocks
  computer investigation of, 15, 16,
    76–77, 97, 119–20
  as constraints, 13–15
  coordination of, 113
  cyclic AMP as mechanism of,
    227
  drugs and, 87–91, 109–10
  existentialism and, 361
  of fetus, 156–57
  homeostasis compared to, 61–62,
    66–68, 74–75, 79, 155–56
  independence of, 69–70, 74–75
  as internalization of external
    environment, 68–69, 72
  light and, 135–54
  logic of, 67–68
  master, 39–40, 94–114
  meal schedules and, 108–10
  as mermaid's tail, 18
  as modular, 389
  as multiplicity, 168
  psychological time contrasted
    with, 17–19
  range of entrainment and, 140–
    142
  as silent orchestra, 15–16, 106–
    7
  social contact and, 110–11
  survival strategies and, 13, 14
  weekly rhythms and, 75–79
  "wiring" between, 112
biological rainbow, 115–34
  components of, 120–21
biological reality:
  blind spots in, 266, 269, 271
  relevance and, 268
  space vs. time in, 248–49
biological structures, discovery of,
  63–65
biorhythms, mistaken "discovery"
  of, 124

birds:
  learning of, 161
  migration of, 161–62
black holes, 38
Blake, William, 291
blind people, time cues and, 140
blood, 51–52
blood pressure, changes in, 66, 67,
  76
bone marrow cancer, 88, 89–90
BRAC (basic rest-activity cycle),
  187–88, 196–98
brain, 203–15, 248, 389–90
  of bees, 158, 159–60
  biological rhythms and, 210–11
  cells of, 54, 334
  changes in alertness of, 66, 67
  as complex society of systems,
    184
  cortex of, see cortex
  efficiency of, 204–6
  evolution of, 209–10, 214, 327
  frequency spectrum of, 118–19
  hearing and, 253–55, 260–61
  as hologram, 96–97
  in homeostasis, 49, 52, 54
  hypothalamus, see hypothalamus
  Korsakoff's psychosis and, 326–
    328
  linguistic views of, 214–15
  master clock in, 94–102
  memory and, 59
  neocortex, 209
  norepinephrine pathways in, 343
  pineal gland and, 145, 146
  psychological sameness and, 55–
    56
  in REM sleep, 177, 194
  reptile "eye," 257
  sleep-dream clock in, 175–76
  temporal fission and, 270–74
  thalamus, 262, 326–27, 328
  time models constructed by, 18,
    40
  vertebrate, 256, 260–61
  "visceral," 209
brainstem, 196, 209, 343
Bray, William, 218–19
Bridgman, P. W., 56–57
Brothers Karamazov, The
  (Dostoevsky), 45

Bruner, Jerome, 369
Bruno, Giordano, 30–31
Buchan, John, 383

caffeine, 152, 208
Cage, John, 253
calendar week, origins of, 77–79,
  83
Calvin, M., 227
Campbell, Scott, 171, 172
cancer, 88, 89–90, 109–10, 132
Cannon, Walter, 49, 53, 66, 69
Canterbury Tales (Chaucer), 70
Čapek, Milič, 31
carbachol, 152
Carter, Elliott, 281–82
Caruso, Enrico, 225–26
Carver, Robert, 255
catecholamines, 343, 358
cats, REM cycle of, 199
Caughey, John, 192
cells:
  brain, 54, 334
  discovery of, 64
  division of, 88–90, 103
  early, 50
  specialization of, 51, 89
cell theory, 64
cerium, 219
Chaucer, Geoffrey, 70
chemical clocks, 216–28
  in communications, 219–28
  discovery of, 218–19
  in slime molds, 220–21
chemistry:
  darkness and, 148–49
  frequencies in, 127
  light and, 142–48
  spectrum in, 118
  time and, 142–49
chemotherapy, 88–91
children:
  mental maps of, 295–96
  temporal knowledge of, 292–94,
    333, 384, 386–87
  see also babies
chimpanzees, time sense research
  and, 297–98
Chomsky, Noam, 214–15
chronobiology, 67, 75, 88, 121
Churchill, Sir Winston, 210

circadian clocks, 70–71, 106–11,
  119–22, 283–84
  autonomy of, 231
  depression and, 151
  drugs and, 151–54
  lateral geniculate nucleus and,
    148
  light and, 136–43, 148
  as metaphors, 388, 389
  niches and, 258–59
  sleep and, 184, 198–99
  subjective day and night in, 138–
    140
circasemiseptans, 132
circaseptans, 130–33
circular movement, 28–32
Cisplatin, 88
civilization, homeostasis and, 53,
  54
cloakroom ticket theory, 23–24, 32,
  35
CNV (contingent negative
  variation), 332–35, 337
cockroaches:
  biological clocks of, 112–13
  entrainment to time cues in, 107–
    108
coffee, 152, 212, 213
cognition, 373–88
  time, 386–87
Cole, K. C., 131
Coltrane, John, 270
communications:
  chemical clocks in, 219–28
  cyclic AMP and, 221–23, 227–28
  mother-child, 286, 288
  time and, 216–18
  see also language; music; speech
computers, 65–66, 374, 385
  biological clocks investigated by,
    15, 16, 76–77, 97, 119–20
  clocks of, 216–18
  extraterrestrial signals
    distinguished by, 116
  memory of, 307–8
  rhythm of, 218
  seven-day rhythms detected by,
    76–77
concentration, 50–51
  glucose, 50, 52
confabulating, 323, 325

Confessions (Saint Augustine), 384
consciousness, 162–64, 188
  en-soi, 362–63
  memory and, 304, 307
  pour-soi, 362–64
  self, 354–60
  space in, 356–57
  stream of, 60, 181–83, 223, 236, 330
constancy, principle of, 60–61, 66
context, time sense and, 339, 344–345
contingent negative variation (CNV), 332–35, 337
conversation, 229–46
  dinner-table, 239–41
  eavesdropping on, 269–70
  hesitation in, 243, 244–45
  listening in, 237
  meaningless sounds in, 243
  plays compared to, 235, 244
  repeated words in, 243
  rhythm of, 235–46
  sameness vs. change in, 245
  stream of, 236
  tempo of, 241–45
  wrong timing in, 238–39
Copernicus, Nicolaus, 30, 94
cortex, 196, 209, 233, 256, 262
  frontal lobes of, 327–28, 330, 334, 335–36, 358
cortisol, 86, 99, 102–4, 109, 111
  rhythm of, 115
Cottrell, Sir Alan, 24–25, 352
Cullers, Kent, 116–17, 125
cybernetics, 211
cyclic AMP, 221–23, 227–28

dance music, 275, 276–77
Darwin, Charles, 44, 64
Davies, Peter Maxwell, 126
Davis, Natalie, 313–15
daydreams, 190–92, 198
Dean, John, 309–10
decentering, 293–94
de Forest, Lee, 225–26
Delcomyn, Fred, 233–34
depression, 149–54
  seasonal nature of, 149–50
  treatment of, 151–54

Descartes, René, 144
Dialogue Concerning the Two Chief World Systems (Galileo), 32
diapause, 72
diatom, 72
Dickinson, Emily, 149–50
Didion, Joan, 291
dinosaurs, 257–58, 264
  disappearance of, 53–54, 262
distance senses, 256, 264
  see also hearing; smell; vision
DNA, 64, 65
  synthesis of, 89–90, 103
Doctor Faustus (Mann), 251
dopamine, 334–35, 336
Dostoevsky, Fëdor, 45
Drake, Stillman, 267
dreams, 73, 99, 167–85
  attention and, 341–44
  daydreams, 190–92, 198
  lack of context and, 344
  melatonin and, 146
  non-REM, 189–90
  psychological theory of, 203–4
  REM, *see* REM dreams
drugs:
  anticancer, 88–91, 109–10
  biological clocks and, 87–91, 109–110
  biological time manipulated with, 151–54
  jet lag treated with, 152–53, 211
  light as, 149, 151
  marijuana, 330–33, 359–60
  psychological time and, 330–37
duration, 344–52
  Ornstein's investigation of, 344–350

ears, hearing and, 253–54, 260
Ebbinghaus, Hermann, 205
echoes, 254–55
Edison, Thomas, 142, 172
Ehrenkranz, Joel, 147
Ehret, Charles, 212
Einstein, Albert, 19, 23–25, 34–39, 122, 200, 363, 381
  Barrow compared to, 33
  Newton admired by, 28, 35
  Plato compared to, 27

Einstein, Albert (*cont.*)
  relativity theory of, 19, 25, 35–39, 135–36
  Spinoza's influence on, 38–39
electroencephalograph (EEG), 119, 174, 184, 191, 198, 199
electron microscopes, 64–65
Emma (Austen), 346–48
endotoxin, 131
English language, stressed syllables in, 238
en-soi, 362–63
entropy, principle of, 219
enzymes, 221–22, 227
eosinophils, 103
epinephrine, 85
episodic memory, 301, 303–17, 354, 368
  certainty and, 308–9
  fitness and, 309, 317
  in history, 311–16
  semantic memory compared to, 303–8, 310–11
Erickson, Frederick, 237–41, 245
"even tenour" hypothesis, 34
evoked potentials, 332
evolution, 247–65
  biological clocks and, 12–18
  of brain, 209–10, 214, 327
  change as essence of, 14
  homeostasis as strategy of, 44–56
  of intelligence, 256–65
  of mammals, 257–63
  of memory, 306–7
  of pineal gland, 145
  range of entrainment and, 141–142
  of siesta gate, 194
  sleep and, 177, 194, 209–10
  time-important strategy in, 42–43, 63–79
  time-unimportant strategy in, 42–62
existentialism, 46, 360–66, 371–73
expectancy wave, 332–36
extroverts, 85, 105, 106, 208
eye, optical vs. thermal, 136

Faulkner, William, 364
feedback, 74–75, 223–26
  in adrenal system, 104
  as control mechanism, 223–24
  depression and , 149
  motor skills and, 232–34
  regeneration vs. oscillation in, 225–26
  as source of oscillation, 224–26
  as unnecessary to rhythmic behavior, 233–34
"feed-sidewards," 104
fetus, time sense of, 156–57
Fliess, Wilhelm, 124
fluid matrix, 49–52
Fodor, Jerry, 376, 378, 379, 384–86
food consumption, timing of, 92
Foulkes, David, 189–90
Fourier, Joseph, 116–17
Fourier analysis, 116–19, 132
freedom, 155–66, 364, 390–91
  dreams and, 204, 358
  existential, 372–73
  homeostasis and, 48, 49, 53–55, 155–56
  of humans, 162–66, 372–73
  nature of, 165
  thinking and, 378–79
free life, 48
frequency, 117–20, 125–30
  of melatonin, 144
  spectrum, 118–19, 125–30
Freudian theory, 370, 379
Friedman, William J., 288–89, 383
frogs, senses of, 248
fruit flies, effects of light on, 137
Fuller, Buckminster, 118
future time, 293–94, 364–70

GABA (brain messenger), 152, 180
Gaddis, William, 369–70
Galileo Galilei, 31–32, 97–98, 267
ganglion, 256
Gap in the Curtain, The (Buchan), 383
Gardner, Howard, 290, 291
geometrical time, 31–34
geometry, projective, 58–59
Gertz, John, 199–200
glucagon, 109
glucose, 50, 52, 222
glucostats, 52
glycogen, 222
Gold, Thomas, 24

Goossens, Eugene, 283
Gould, James, 159–61, 164, 269
grammar, universal, 214–15
gravity, time affected by, 37–38
Greeks, ancient, 77
    pineal gland as viewed by, 144
    time as viewed by, 26–30
Green Man, The (Amis), 173
growth hormone, 86, 99, 177, 179

Halberg, Franz, 70, 77, 79, 92, 97–
    98, 102, 104, 134
    on biological rhythms, 120–23
    on harmonic relationships, 128,
    130
    on spectral compromise, 132
Haldol, 334–35
Hamlet (Shakespeare), 163–64,
    378
harmonics, 126–28, 130
Harner, Lorraine, 293
Hartmann, Ernest, 178, 342–44
Hawking, Stephen, 33
hawks, 259
hearing, 248–56, 259–61, 264
    psychological components of,
    253
    relativity theory of, 272–73
    space and, 252–56
    see also music; speech
Heidegger, Martin, 361, 365–67
Hicks, Robert, 323–24, 328, 329,
    330, 332, 334, 335, 336
Higgins, Joseph, 220
Hirsh, Ira, 252
history, historians, 387–88
    narrative, 311–16
    new, 311–12
Hobson, Allan, 175–76, 179, 182–
    183, 192
    on coffee breaks, 213
    on jet-lag prevention, 211–12
Hofer, Myron, 110
homeostasis, 44–56, 61, 62
    biological clocks compared to, 61–
    62, 66–68, 74–75, 79, 155–56
    Cannon's views on, 49, 53, 66
    derivation of word, 44, 49
    drug administration and, 87–88
    feedback in, 74, 223
    fluid matrix and, 49–52

freedom and, 48, 49, 53–55, 155–
    156
    hypothalamus and, 95, 136
    internal sameness theory and, 44–
    49
    sleep and, 169, 179
Homer, 355–56
Hooke, Robert, 64, 66
Horace, 170
hormones, 52, 66, 86–87, 143
    cortisol, 86, 99, 102–4, 109, 111,
    115
    epinephrine, 85
    growth, 86, 99, 177, 179
    meal schedules and, 109
    melatonin, 143–47, 150, 259
    pituitary, 104
    temporally active, 111
hospitals, subjective time in, 138–
    139
Hrushevsky, William, 128–29
Hubble, Edwin, 39
humanism, universal, 361–62
human nature, 372–73
    time and, 156, 164–65
humans:
    cell specialization in, 51
    freedom of, 162–66, 372–73
    homeostatic processes in, 49–52
    learning of, 164
    melatonin in, 146
    threshold to entrainment by light
    in, 141–42
    variability of body rhythms in, 86
    weekly rhythms in, 76–77
    see also specific topics
Hume, David, 92
Huxley, Aldous, 291
hypothalamus, 95–98, 104
    homeostasis and, 95, 136
    neuroendocrine transducers in,
    143
    suprachiasmatic nuclei (SCN) in,
    95–99, 105–9, 136, 143, 148
    temperature "eye" of, 136
    ventromedial nuclei (VMH) of,
    108–9

identity, temporal, 83–93, 128, 214
    age and, 86
    defined, 84

identity, temporal (*cont.*)
  personality and, 85
  relativity principle and, 86–87
  Weinberg's views on, 84
Iliad (Homer), 355–56
imagination, 18–19, 73–74, 163
immune system, 67, 76, 89–90, 91,
  131
impulsive people, 208
infants, *see* babies
inference, statistics of, 123–24, 132
information:
  brain and, 203–15
  computer clocks and, 216–18
  conversations and, 229–46
  limitations of senses and, 247–48
  in messages, 345–48
  psychological time and, 345–48
  time and, 201–317
infradian rhythm, 121, 122
insects, diapause in, 72
insulin, 52, 66, 109
intelligence:
  evolution of, 256–65
  visual imagery and, 290–91
intelligence, temporal, 285–98, 361
  behaviorist view of, 369
  as family of competencies, 291–
  292
  space and, 294–96
internal sameness:
  Bernard's theory of, 44–49
  see also homeostasis
introverts, 85, 105, 208
Ives, Charles, 253, 282–83

Jacob, François, 209, 247–48, 266
James, William, 60–61, 181, 236,
  263, 325–26, 337, 338
  on experience, 340–41
Jaynes, Julian, 355–58, 384
Jerison, Harry, 259–62, 264, 294
jet lag, 106, 110–11, 129, 138, 140–
  141
  Hobson's views on, 211–12
  manual dexterity and, 206
  treatment of, 152–53, 211–12
Jewish week, 77, 78
Johns Hopkins Peabody Institute,
  255
Jones, Mari Riess, 272–73, 286

JR (Gaddis), 369–70
Julesz, Bela, 251–52

Kekulé, Friedrich, 73
kidneys, 76, 198
  anticancer drugs and, 88
  changes in, 66, 67
Kierkegaard, Søren, 362
Kleitman, Nathaniel, 173, 187–88,
  190, 196, 205
Koestler, Arthur, 32
Korsakoff's psychosis, 321–31, 376
Koyré, Alexander, 32
Kripke, Daniel, 190–91

Lambert, Constant, 249–50, 278
language:
  learning theory and, 137–38, 214–
  215
  meaning in, 267
  perception of, 377–78
  syntax of, 291
  see also speech
"larks," 85, 105, 208
Lashley, Karl, 231–33
lateral geniculate nucleus (LGN),
  148
Lavie, Peretz, 192–93, 195, 199,
  211
Lawrence, Gary, 322–23
learning, 157–62, 223
  of bees, 158–61, 268
  critical periods and, 158–62, 164
  language, 137–38, 214–15
  nerve cells and, 180–81
  sleep and, 177, 178
leukemia, 89–90
Lewis, C. S., 163
life:
  Bernard's categories of, 47–48
  free, 48
  oscillating, 47–48, 69
  spatial structure of, 63–65
  temporal structure of, 65–79
life time, science of, 25, 39–41
light, 135–54
  artificial, 141–42, 144, 150
  circadian system and, 136–43,
  148
  as drug, 149, 151
  indoor, 144, 150

melatonin and, 143–47, 150
in relativity theory, 135–36
spectrum of, 120
speed of, 35–37, 136
white, 120
Linnaeus, Carolus, 160
listening, 237
logic, 57–58, 309, 311
long-term memory, 177–78, 206–8,
  223, 299, 301–17, 382
  episodic, see episodic memory
  Korsakoff's psychosis and, 322
  procedural, 301–4, 306–7, 357–
  358
  question mark (QM), 001
  semantic, see semantic memory
Lucas, John 338
luteinizing hormone, 198
lymph, 51–52

Macbeth (Shakespeare), 164
McCarley, Robert, 175–76
Mcquarrie, John, 367
Magendie, François, 144
Magic Mountain, The (Mann), 340
mammals:
  auditory system developed by,
  260–61
  dinosaurs compared to, 53–54
  evolution of, 257–63
  nocturnal life style of, 257–62
  reptiles compared to, 257, 260–
  261
  unborn, 156–57
  see also humans
Mann, Thomas, 251, 340, 353
manual dexterity, 206
marijuana, time distorted by, 330–
  333, 359–60
Marx, Groucho, 241–42
Marx, Karl, 312
Masonic Cancer Center, 88
master clocks, 39–40, 94–114
  blows to concept of, 96–104
  of computers, 216–18
  criteria of, 95
  history of ideas about, 26–40
  of music, 275–79
  suprachiasmatic nuclei (SCN) as,
  95–99, 105–9
  X and Y model of, 99–102, 110

mathematical time, 13, 31–39
  geometrical, 31–34
mathematics:
  Fourier analysis, 116–19, 132
  statistics, 122–25
mating behavior, 55
May, Rollo, 165
meals:
  rhythm of, 235, 239–41
  as time cue, 108–10
meaning, 361
  time distortions and, 364, 367
mechanical periodicity, 78
melatonin, 143–47, 259
  depression and, 150
  molecular structure of, 143
  seasonal changes and, 145–46
  as sedative, 143–44
Melges, Frederick, 359–60
memory, 97, 101–2, 205–8, 223,
  299–317, 339, 344–45, 352
  of bees, 158–60
  computer, 307–8
  consciousness and, 304, 307
  evolution of, 306–7
  Korsakoff's psychosis and, 321–
  331, 376
  long-term, see long-term memory
  marijuana and, 330
  predictability vs.
  unpredictability in, 345–48
  short-term, 178, 179, 206–7, 223,
  299–301, 310, 322, 373, 382
  sleep and, 177–79, 182–83, 206
  types of, 59
mental faculties, as family of
  competencies, 290–92
metabolism, 108, 133, 156, 179
metacognition, 154
metaphors, 380–83, 387–89
microscopes, 64–65
Mildenberger, Karl, 229–30
Miller, George, 373
Mitchell, Edgar, 340
modules, 376–80, 382
Monday morning blahs, 71
Montaigne, Michel de, 4
Moore-Ede, Martin, 98, 109, 138,
  152, 213
mother-child communication, 286,
  288

motion:
  Aristotelian view of, 29
  circular, 28–32
  internalization of laws of, 72–73
  local, 31–32
  in relativity theory, 36–37
  straight-line, 29, 31–32, 34
motor skills, Lashley's study of,
    231–33
Muggeridge, Malcolm, 162–63
Mumford, Lewis, 78
muscles, REM sleep and, 173, 177
music, 266–84
  absolute timing in, 288
  beat in, 275–79
  biological clocks compared to, 15–
    16
  body rhythms compared to, 125–
    130
  conversation compared to, 235,
    242–44
  dance, 275, 276–77
  establishment of large forms of,
    278–79
  family of competencies and, 291
  meter in, 275, 277–82
  modern, 274, 281–83
  motor skills and, 232–33
  painting compared to, 249–50
  pitch, loudness, and time link in,
    271–72
  polyphonic, 275–76
  redundancy in, 348
  rhythm in, 273–74, 277, 279, 280–
    281, 283
  small-scale structure of, 267
  temporal fission (streaming) in,
    270–74
  time structure of, 16, 17, 40, 249–
    251

NASA Biomedical Research
    Division, 207
Neisser, Ulric, 309–10
neocortex, 209
nerve cells (neurons), 143, 152–53,
    233
  required for attention and
    learning, 180–81
nervous system, 248
  arousal rhythm of, 207

  as communication medium, 143
  in homeostasis, 49, 52, 54
  light and, 135
  REM sleep and, 173, 198
  see also brain
neuroendocrine transducers, 143
  pineal gland, 104, 143–48, 259
neuropeptide Y, 148
neurotransmitters, 67, 130, 143,
    150, 334
  acetylcholine, 175, 180
  dopamine, 334–35, 336
  GABA, 152, 180
  neuropeptide Y, 148
  norepinephrine, 175–76, 181,
    343
  in REM sleep, 180
  serotonin, 143, 175–76, 181
Newton, Sir Isaac, 28, 33–37, 120
niches, ecological, 258–59, 262
"90-minute enigma, the," 199
Nixon, Richard, 309, 310
norepinephrine, 175–76, 181,
    343

Oates, Joyce Carol, 291
objective time, 287, 365–66
  mathematical time compared to,
    13
  mental measures of, 18–19
  in relativity theory, 37
Obusek, Charles, 268–69
Odyssey, (Homer), 356
optics, Newtonian revolution in,
    120
Origin of Species (Darwin), 44
Ornstein, Robert, 344–45, 348–50
oscillating life, 47–48, 69
oscillation, 217–19, 224–28
  biochemical, 227–28
  in feedback, 224–26
  rhythmic movements and, 233–
    234
owls, 259
  hearing of, 255
"owls," 85, 105, 106, 208

painting:
  music compared to, 249–50
  Renaissance, 58–59
past time, 293–94, 364–70

Pauly, John, 102
personality:
    information processing and, 208
    temporal identity and, 85
phases of waves, 117, 144
phonemes, 268–70
phosphodiestrases, 227
physics:
    invariance in, 58–59
    relativity theory and, 19, 25, 35–39, 135–36
    resonance in, 128
    time of, 19, 23–25, 35–39, 61
Piaget, John, 368–69
pilots, work schedules of, 213
pineal gland, 104, 143–48, 259
    see also melatonin
pituitary gland, 104
planetary week, 77–78
planets, frequencies of, 126
Plato, 26–30, 33, 35, 39, 94, 98
plays, conversation compared to, 235, 244
Pleasants, Henry, 283
polyphony, 275–76
potassium, excretion of, 86, 88, 99
pour-soi, 362–64
Powell, Ernest, 96–97
Premack, David, 297–98, 304
present time, 293–94, 365, 367, 368
Pribram, Karl, 96–97, 350–51, 359
primates, 194
    reproductive behavior in, 55
primitive propositions, 57–58
Principles of Psychology, The (James), 60–61
privileged time, see master clocks
procedural memory, 301–4, 306–7, 357–58
proteins, 67, 89, 199
Proust, Marcel, 317
psychological dreams, 203–4
psychological sameness, 55–61
psychological time, 17–19, 25, 72–74, 287, 289–98, 330–38, 344–353
    biological time contrasted with, 17–19
    complexity of, 19
    in conditions of emergency, anticipation, and danger, 335

contingent negative variation (CNV) and, 332–35, 337
    as dividing line, 296–97
    duration and, 344–52
    existentialism and, 360–66
    marijuana's exaggeration of, 331–333
    master clocks and, 40, 41, 289
    music and, 281–82
    self-consciousness linked to, 354–360
    theory of mind and, 40

question mark (QM) memory, 301

radios, chemical, 220
rapid eye movement (REM), see REM cycles; REM dreams; REM sleep
rats, in sleep experiments, 170
reaction, vs. action, 230–31
reaction times, 230
redundancy, 347–50
relativity theory, 19, 25, 35–39, 135–136
    of hearing, 272–73
REM cycles, in waking, 188, 189
REM dreams, 173, 180, 182–83, 189–92, 198, 341–44
    daydreams compared to, 190–92, 198
    loss of self in, 358
    as ultradian rhythm, 192
REM sleep, 173, 174, 176–80, 182–184
    brain and, 173, 198
    discovery of, 187
    initiation of, 197
    memory and, 206
    neurotransmitters in, 180
    as wakeability gate, 194–95
Renaissance:
    painting in, 58–59
    time as viewed in, 30–31
reproduction, 156–57
    pineal gland and, 145, 146
    seasonal rhythms of, 145, 146, 147, 156
reptiles, 257–58
    extinction of, 53–54, 261–62

reptiles (*cont.*)
  mammals compared to, 257, 260–261
  vision of, 257, 260, 262
resonance, 127–29
Restak, Richard, 322
Return of Martin Guerre, The (Davis), 313–15
"Revival of Narrative, The" (Stone), 313
rhythm:
  biological rainbow and, 115–34
  biological timetables and, 155–166
  brain and, 210–11
  circadian, *see* circadian clocks
  circasemiseptan, 132
  circaseptan, 130–33
  of conversation, 235–46
  infradian, 121, 122
  light and, 135–54
  master clocks and, 94–114
  mixture of sameness and change in, 115
  in music, 273–74, 277, 279, 280–281, 283
  randomness in, 122–23
  search for extraterrestrial intelligence compared to, 115–117
  sleep and, 167–200
  statistics of inference and, 123–124, 132
  tempo compared to, 242
  temporal identity and, 83–93
  ultradian, 121, 122, 192–93, 196–200
  wakefulness and, 186–200
Richardson, Bruce, 188
Romer, Alfred, 261
Rordorf, Willy, 78
Russell, Bertrand, 57–58, 362
Russia:
  body rhythm research in, 86
  signal detection in, 116

sameness:
  biological, *see* homeostasis
  in logic, 57–58
  in physics, 58–59
  psychological, 55–61

Sartre, Jean-Paul, 18, 163, 264, 390–391
  existential philosophy of, 361–365, 371–72
scalar unit timing, 288
Scheving, Lawrence, 96–98, 130
schizophrenia, 335
Schleiden, Matthias, 64
Schwann, Theodor, 64
Science, 187, 233
SCN (suprachiasmatic nuclei), 95–99, 105–9, 136, 143, 148
Scollon, Ron, 236, 241–43
Search for Extraterrestrial Intelligence (SETI), 116
self:
  consciousness of, 354–60
  depersonalization of, 359–60
  semantic memory, 301–2, 354–55, 357–58
  episodic memory compared to, 303–8, 310–11
  in history, 311–12, 314, 316
senses, 247–65
  limitations of, 247–48
  see also hearing; smell; vision
serotonin, 143, 175–76, 181
seven-day (circaseptan) rhythms, 130–33
sexual behavior, 55, 103
Shakespeare, William, 163–64, 378
Shepard, Alan, 340
Shepard, Roger, 72–73, 215, 291, 377
Sherrington, Sir Charles, 256
shift work:
  adjustment to changes in, 106, 107, 110–11, 152, 154, 213
  research on, 188
siesta time, 193–94
Siffre, Michel, 100–102, 110, 274
Simpson, Hugh, 86–87
singularity, 38
Six Bagatelles (Webern), 279
skin cells, 89
sleep, 167–200
  body temperature and, 85, 171
  cell division during, 89
  as complement of wakefulness, 188
  cycles of, 186–87

depression and, 150, 151
fragmentation of, 171
functions of, 169, 177
as homeostatic mechanism, 169,
    179
melatonin and, 143, 146
REM, see REM sleep
slow-wave, 174, 177
stage one, 173
stage two, 173
stage three, 173–74
stage four, 174, 179
states of, 167–68
ultradian rhythms and, 192–93,
    196–200
see also dreams
sleepability, 193–94, 209
sleep-deprivation studies, 184, 188
sleep-dream clock, 175–76, 178,
    183
slime molds, chemical clocks in,
    220–21
Slonimsky, Nicolas, 283
slow-wave sleep, 174, 177
smell, 256, 259, 261, 264
snails:
    biological clocks in, 112
    cyclic AMP in, 227
    social contact, body clocks and,
        110–11
Sonnenschein, David, 190–91
sound, see hearing; music; speech
space:
    in consciousness, 356–57
    hearing and, 252–56
    large-scale structures of, 267
    mastery of, 290
    in Platonic universe, 27–28
    in relativity theory, 35, 37–38,
        135–36
    temporal intelligence and, 294–
        296
    temporalization of, 252–65
    vision and, 248
space-time, 37–38
spatial structures, discovery of, 63–
    65
spectrum:
    of frequencies, 118–19, 125–30
    of light, 120
    rainbow, 120–21

speech, 266–71
    phonemes in, 268–70
    redundancy in, 348
    temporal fission and, 271
    time patterns of, 16–17
    see also language
Spinoza, Benedict, 38–39
spontaneity, tempo and, 244
statistics, 122–25, 132
Stefani, Wolfgang, 278
stereo record players, limitations
    of, 252–53, 255
Stern, Daniel, 229–31
Stevenson, Jim, 116, 117
Stockhausen, Karlheinz, 274–75
Stone, Lawrence, 312–13
straight-line motion, 29, 31–32, 34
Stravinsky, Igor, 283
streaming (temporal fission), 270–
    274
stress, 208, 222, 343
subjective time, see psychological
    time
suicide, seasonal patterns of, 150
Sunday (Rordorf), 78
suprachiasmatic nuclei (SCN), 95–
    99, 105–9, 136, 143, 148
Sutherland, Earl, 222
system dominance, 211

tau, 105–6
    light and, 136, 140, 151
    of nocturnal animals, 258–59
tautologies, 57–58
temperature:
    homeostasis and, 51–54, 136
    of water, 51
temperature, body, 49–50, 66, 71–
    72, 96, 109, 136
    brain efficiency and, 204, 205
    sleep and, 85, 171
    temporal fission (streaming), 270–
        274
tempos, conversational, 241–45
terrestrial dynamics (local motion),
    31–32
testing, timing in, 238–39
thalamus, 262, 326–27, 328
THC (tetrahydrocannabinol), 360
thinking:
    continuous, 60

thinking (*cont.*)
  freedom and, 378–79
  logical, 57–58, 309, 311
  non-REM dreams compared to, 189–90
  sameness and, 56–57, 60
  unconscious, 358
Thomas, Lewis, 131
Thorington, Luke, 144
Three Mile Island, nuclear accident at, 213
tides:
  body clocks and, 72
  harmonics and, 126
Timaeus (Plato), 26
time:
  "biologizing" of, 142, 203
  dethronement of, 25–41
  flow of, 24, 33–35
  history of ideas about, 25–41
  importance of, 14–15, 42–43, 63–79
  inborn sense of, 286–87
  as multiplicity, 289, 296
  spatializing of, 32, 356–57, 383–384
  synthesis of information about, 11
  unimportance of, 14–15, 42–62
  see also specific topics
time-binding, 260
timeishness, 365–66
Torricelli (Galileo's pupil), 33
Toscanini, Arturo, 54
Tovey, Donald, 280
Toynbee, Arnold, 54
transplant surgery, 76, 91
Tulving, Endel, 304–8, 311, 327, 329–30

ultradian rhythm, 121, 122
  sleep and, 192–93, 196–200
universal grammar, 214–15
universal now, Einstein's rejection of, 35
unnatural signals, 117

Valéry, Paul, 290
vertebrates, 256–58, 260–61
  pineal gland in, 145
  time sense of, 297–98

vision, 248–51, 256, 257, 264, 376–377
  of mammals, 262
  night, 259
  reptile, 257, 260, 262
  time and, 249–50
Voltaire, 372

Wainwright, S. D., 145
wakeability gates, 194–95
wakefulness, 186–200
  basic rest-activity cycle and, 187–188, 196–98
  sleep as complementary to, 188
Warnock, Mary, 372
Warren, Richard, 267–69
water, 49–52
  internalization of, 68
  temperature of, 51
Watergate hearings, memory and, 309–10
Webb, Wilse, 169–73, 209
Weber, Eugen, 316
Webern, Anton, 279
weekly rhythms, 75–79, 83
  circaseptans, 130–33
Wehr, Thomas, 149, 151, 153–54
Weinberg, Gerald, 84
Weisskopf, Victor, 127
Weitzman, Elliot, 183–84, 198
Wheeler, John, 38–39, 40
Wilson, A. T., 227
Woolf, Virginia, 287
workers:
  performance cycles of, 211
  see also shift work

X and Y model, 99–102, 110, 150–151
  manual dexterity and, 206
  sleep and, 171, 196, 197

zeitgebers (time-givers), 70, 105–7, 210, 212
Zeitmasze (Stockhausen), 274–75
Zerubavel, Eviatar, 78
Zhabotinsky, A. M., 219, 220
Zucker, Irving, 75
Zuckerkandl, Emil, 280